Charged Aerosol Detection for Liquid Chromatography and Related Separation Techniques

Charged Aerosol Detection for Liquid Chromatography and Related Separation Techniques

Second Edition

Edited by

Ian N. Acworth
Consultant
Melrose, MA, USA

Paul H. Gamache
Consultant
Nashua, NH, USA

WILEY

Published by John Wiley & Sons, Inc., Hoboken, New Jersey.

Library of Congress Cataloging-in-Publication Data

Names: Acworth, I. N. (Ian N.) 1958–editor | Gamache, Paul H., 1957–editor | John Wiley & Sons publisher
Title: Charged aerosol detection for liquid chromatography and related separation techniques / edited by Ian Niel Acworth, consultant, Melrose, MA, USA [and] Paul Henri Gamache, consultant, Nashua, NH, USA.
Description: Second edition. | Hoboken, New Jersey : Wiley, [2027] | Includes index.
Identifiers: LCCN 2026030916 (print) | LCCN 2026030917 (ebook) | ISBN 9781394300600 cloth | ISBN 9781394300624 adobe pdf | ISBN 9781394300617 epub
Subjects: LCSH: Liquid chromatography | Atmospheric aerosols | Aerosols | Electrostatic precipitation | Separation (Technology)
Classification: LCC QD79.C454 C43 2026 (print) | LCC QD79.C454 (ebook)
LC record available at https://lccn.loc.gov/2026030916
LC ebook record available at https://lccn.loc.gov/2026030917

Cover Design: Wiley
Cover Image: Pobytov/Getty Images

Set in 9.5/12.5pt STIXTwoText by Straive, Chennai, India

Contents

List of Contributors

Ian N. Acworth
Consultant
Melrose, MA
USA

Wesley W. Barnhart
One Amgen Center Drive
AMGEN Inc.
Thousand Oaks, CA
USA

David Bauer
Amgen Research
Cambridge, MA
USA

Juan Bian
Synthetic Molecule Analytical Chemistry
Genentech Inc.
South San Francisco, CA
USA

Andrei Blasko
Clearview Pharma Solutions, LLC
Dallas, TX
USA

Ton Brooijmans
DSM Coating Resins
Waalwijk
The Netherlands

Charles R. Bupp
One Amgen Center Drive
AMGEN Inc.
Thousand Oaks, CA
USA

Paul Cools
DSM Coating Resins
Waalwijk
The Netherlands

and

The Dow Chemical Company
Freeport, TX
USA

Benjamin Eggart
Thermo Fisher Scientific
Research and Development
Germering
Germany

Paul H. Gamache
Consultant
Nashua, NH
USA

Imad A. Haidar Ahmad
One Amgen Center Drive
AMGEN Inc.
Thousand Oaks, CA
USA

Troy T. Handlovic
One Amgen Center Drive
AMGEN Inc.
Thousand Oaks, CA
USA

Ulrike Holzgrabe
Institute of Pharmacy
University of Würzburg
Wuerzburg
Germany

Frank Hrovat
Synthetic Molecule Analytical Chemistry
Genentech Inc.
South San Francisco, CA
USA

Ping Hu
School of Chemistry and Molecular Engineering
East China University of Science and Technology
Shanghai
China

Yan Jin
Thermo Fisher Scientific (China) Co., Ltd.
Chromatography Mass Spectrometry Division
Shanghai
China

Brian A. Lanman
One Amgen Center Drive
AMGEN Inc.
Thousand Oaks, CA
USA

Gabriel M. Leme
One Amgen Center Drive
AMGEN Inc.
Thousand Oaks, CA
USA

Haizhen Liang
Beijing Institute of Radiation Medicine
Beijing
China

Jessica Lin
Synthetic Molecule Analytical Chemistry
Genentech Inc.
South San Francisco, CA
USA

Katherine S. Lovejoy
Thermo Fisher Scientific
Product Applications
Germering
Germany

and

Dionex Softron GmbH, a Part of
Thermo Fisher Scientific
Product Applications
Germering
Germany

Baiping Ma
Beijing Institute of Radiation Medicine
Beijing
China

Claudia Seidl
One Amgen Center Drive
AMGEN Inc.
Thousand Oaks, CA
USA

Guobin Shen
Thermo Fisher Scientific (China) Co., Ltd.
Chromatography Mass Spectrometry Division
Shanghai
China

Frank Steiner
Thermo Fisher Scientific
Product Applications
Germering
Germany

and

Dionex Softron GmbH, a Part of
Thermo Fisher Scientific
Product Applications
Germering
Germany

Kelly Zhang
Synthetic Molecule Analytical Chemistry
Genentech Inc.
South San Francisco, CA
USA

Preface

Charged aerosol detection (CAD), introduced commercially in 2005, is now widely described as the "gold standard" universal detector for use with liquid-phase separations. In addition to its dynamic range, ease of use, reproducibility, and sensitivity, this technique is very often chosen for its broad detection scope (i.e., universality) and ability to provide analyte-independent sensitivity (i.e., uniformity). In that regard, this book's cover illustration, which depicts the "universe of chemistry," aligns very well with its content. Since its introduction, CAD has been widely adopted as evidenced by over 1500 peer-reviewed articles that encompass several fields of application. These descriptions and the pioneering work of Dr. Stan Kaufman, Dr. Roy Dixon, and many others have led to an increased understanding of the theory and practice of CAD. The capabilities of CAD technology have also continued to evolve along with those of relevant separation technologies. The use of CAD combined with UV detection and mass spectrometry (MS) has, in particular, experienced substantial growth.

As with the first edition, this book is intended to serve as a primary resource for analytical chemists across a wide range of disciplines. It was developed through collaboration with numerous scientists who have expertise in the research and application of separation science, detection technologies, and aerosol measurement. CAD can be coupled with a range of separation techniques in numerous configurations where many variables can influence its performance. The practical value of CAD also depends on its intended use and the alternative and complementary tools available for a given application. The primary objectives of this book are to explain how the fundamental principles and operating variables of CAD influence its performance under a variety of analytical conditions, while providing expert insights and diverse perspectives on applying CAD to solve specific analytical challenges.

This book is organized into three main sections, each comprising several chapters. The chapters are authored by experts in their respective fields, and each begins with a summary and table of contents to provide readers with a concise overview of the material presented. Section 1 includes a theoretical description (Chapter 1), a review of current literature (Chapter 2), guidance for practical use (Chapter 3), and practical strategies toward achieving uniform response and accurate quantification in the absence of individual calibration standards (Chapter 4). These chapters are highly complementary, providing discussions of the advantages and limitations of CAD, recent advances in its applications and technology, and best practices for achieving optimal performance. Until the 2017 publication of this book's first edition, the relationship between detector response and analyte concentration was not fully understood and often described as proportional to aerosol particle surface area. This was found to be a significant oversimplification as first detailed in Chapter 1 of the book's first edition. This updated theory is re-emphasized here and throughout this book as it underlies the practical use of the instrument's power function setting. Importantly, this parameter can significantly impact detector performance, including linearity, peak shape, and signal to noise ratio. Chapter 2 discusses the evolution of the technology, compares CAD with other universal detectors, and provides a broad overview of key applications reported in the literature since mid-2016. Here the focus is on scientific papers that describe method development, performance evaluation, comparison to other techniques, troubleshooting, and validation. Many of these applications are then further detailed in later sections of the book. Chapter 3 provides practical recommendations including those related to instrument settings, mobile phase and column selection, and data analysis. This information should be highly valuable to any user and includes techniques to develop methods, to monitor performance during routine use, and to troubleshoot. The power function is a key focus of this chapter, and important new tools are described to quickly optimize this setting. Chapter 4 examines variables that may influence the observed response uniformity of CAD-based methods including those related to the chromatographic method, detector settings, and pre-analytical considerations. The main focus is on practical strategies toward achieving accurate quantification in the absence of individual standards, a very important topic exemplified throughout the book.

Section 2 consists of separate chapters, each focusing on specific analyte classes: lipids (Chapter 5), ions (Chapter 6), carbohydrates (Chapter 7), pharmaceutical excipients (Chapter 8), traditional herbal medicines (Chapter 9), and small molecule pharmaceuticals (Chapter 10). Taken together, these chapters include descriptions of the use of CAD with many different separation techniques including normal-phase, size exclusion, hydrophilic interaction, mixed-mode, and supercritical fluid chromatography and its combination with UV and MS. For each analyte class, an informative review is provided on experimental approaches to both separate and quantitate species present in different sample matrices and in different areas of application (e.g., biosciences, chemical, food and beverage, natural products, and pharmaceutical) that encompass fundamental research to routine regulated methods. These chapters contain mostly new information as they are mainly based on scientific articles published since mid-2016. Numerous examples are presented as case studies in which the practical and theoretical concepts introduced in Section 1 are applied to explain experimental results, provide guidance for improving analytical performance, and illustrate the advantages and limitations of CAD relative to other detection techniques.

Section 3 provides an in-depth exploration of several major application areas of CAD. With the exception of Chapter 15, the content in this section is entirely new and complements the material presented elsewhere in the book. Cross-references between chapters reinforce fundamental concepts and promote a more comprehensive understanding of the technology. Chapter 11 reviews CAD's role in the important field of lipid-based drug delivery, which encompasses simpler formulations (e.g., microsomes) to more complex systems (e.g., mRNA-based therapeutics). This includes discussion of analytical challenges encountered across diverse formulation types and key considerations of sample preparation and separation strategies for use of CAD for quantitative analysis in applications such as in-process control, stability assessment, and formulation development. Chapter 12 discusses the analysis of polysorbate excipients in biopharmaceutical products. The heterogeneous composition and susceptibility of polysorbates to degradation present risks to product quality and poses substantial analytical challenges. The advantages and limitations of CAD are examined, along with emerging trends such as advancements in CAD technology, high-throughput applications, and the integration of multi-detector systems. The integration of CAD in multi-detector systems is further explored in Chapter 13, where practical considerations for comprehensive analysis of complex samples are described in detail with a focus on safety assessment of pharmaceuticals. Chapter 14 further describes the integration of CAD into chromatography-based workflows in drug discovery and the advantages it offers for streamlining compound quantitation and analysis. Its utility in accelerating drug development, extending detection capabilities, and supporting high-throughput purification platforms such as the microgram high-throughput purification platform are discussed. Chapter 15 describes the analysis of industrial polymers with CAD, often in combination with MS. This field involves the use of solvents and separation techniques that are rather unique to this field, thereby demonstrating the broadness of applicability of CAD and the potential for its use in other fields of study.

It is our sincere hope that this book will serve as a valuable resource for all those working in the field of liquid-phase separations and will help stimulate further research in all aspects of universal detection.

Nashua, August 2026

Ian N. Acworth
Consultant, Melrose, MA, USA
Paul H. Gamache
Consultant, Nashua, NH, USA

Acknowledgment

This book would not have been possible without the significant efforts and expertise of all contributors. We are also very grateful to Stan Kaufman whose insight to CAD theory as described in the first edition of the book still serves as a central foundation for this second edition. Finally, we would like to thank Paul Cools and Ton Brooijmans whose chapter on industrial polymers from the first edition is included here in its entirety. This chapter is based on their work conducted at DSM Coating Resins, Waalwijk, The Netherlands.

Section 1

Fundamentals of Charged Aerosol Detection

1

Principles of Charged Aerosol Detection

Paul H. Gamache[1] *and Benjamin Eggart*[2]

[1] *Consultant, Nashua, NH, USA*
[2] *Thermo Fisher Scientific, Research and Development, Germering, Germany*

CHAPTER MENU

1.1 Summary

This chapter provides a brief history and detailed overview of charged aerosol detection (CAD) and a semiempirical model describing its response and expected performance under various analytical conditions. CAD and other evaporative aerosol detectors (EADs) involve the same successive steps: primary spray droplet formation from an eluent stream, conditioning by inertial impaction to remove droplets too large to evaporate during passage through the instrument, and evaporation of remaining droplets to form residue particles, each comprised of nonvolatile background impurities and any nonvolatile analyte present. Detection of the residue particles produces the detector signal. In CAD, the aerosol is given a charge dependent on the particle size, and the total charge carried by the aerosol is measured as a current; in evaporative light scattering detection (ELSD), the aerosol is detected by its light scattering properties. Both detection methods produce a response that is approximately mass-flow dependent. The analyte dry bulk density affects the residue particle size for a given eluting mass, which has a minor effect on the mass sensitivity of both detector types. Other analyte properties, in particular optical properties (e.g., particle refractive index [RI]), for the ELSD likewise affect the sensitivity. Detection selectivity for EADs is based on differences in vaporization of components within an eluent. Accordingly, these techniques are expected to have very similar detection scope, eluent requirements, and solvent dependency of response. The unique characteristics of CAD are due to the aerosol measurement technique, which involves diffusion charging of residue particles and detection of the current due to deposition of particles with their charge in an aerosol–electrometer filter. Aerosol charging by diffusion mechanisms is well known to have only a minor dependence on particle material (i.e., analyte properties), which is the basis for uniform response capabilities of CAD. Like ELSD, CAD response (e.g., peak area versus mass injected [m_{inj}]) can be described by a power law function with a variable exponent b. Linear response, never perfectly achieved by either method, would correspond to $b = 1$. For both techniques, the exponent b is at its maximum at the lowest m_{inj} and decreases with increasing m_{inj}. This is attributed to smaller residue particles that have a higher power law exponent (β_1) of response and are more prevalent with low m_{inj} and the low concentration that occurs near the edges of any peak. For ELSD, this corresponds to Rayleigh light scattering for particle diameters (d) typically <50 nm, where $\beta_1 = 6$, while for CAD this corresponds to aerosol charging of $d < \sim 9$ nm, where $\beta_1 \sim 2.25$. For CAD, the lower d transition and smaller β_1 (closer to 1) underlies the widely observed lower detection limits, wider dynamic range, and less complex response curve than ELSD. Newer CAD designs produce an even smaller relative proportion of residue particles of $d < \sim 9$ nm, further simplifying the response curve, enabling lower sensitivity limits and a wider quasi-linear response range.

1.2 History and Introduction to CAD

CAD was first described in 2001 by Kaufman at TSI Inc. in a provisional patent application that ultimately led to US patent 6,568,24 [1]. This device was termed an evaporative electrical detector (EED) and was based on coupling liquid chromatography (LC) and other separation techniques with TSI's well-established electrical aerosol measurement (EAM) technology [2]. Around the same time, Dixon and Peterson at California State University were pursuing a similar avenue of innovation with a laboratory-built device that coupled LC with an earlier generation of TSI's EAM instruments. Dixon and Peterson described their device, termed aerosol charge detector (ACD), in the *Journal of Analytical Chemistry* in 2002 [3]. In both instances, the primary objective was to exploit the advantages, well described in aerosol science literature [4], of EAM over direct light scattering for measuring the very small (i.e., low nm diameter range) particles typically produced by LC detectors. The approach was therefore mainly geared toward addressing some of the limitations of ELSD, which at the time had been used for LC detection for about 20 years. Subsequent collaboration between TSI and ESA Biosciences, Inc. led to the introduction of the first commercial instrument, the Corona® CAD®, in 2005 [5]. While there are some differences among these early EAM-based LC devices and with newer commercial instruments, the basic detection process remains the same. Therefore, Kaufman's patent disclosure and Dixon's article are acknowledged as the primary theoretical descriptions of CAD.

Since its commercial introduction in 2005, CAD has been widely adopted for a broad range of chromatographic applications (see Chapter 2). CAD and other EAD techniques, including ELSD and condensation nucleation light scattering detection (CNLSD) [6], are described as "universal" since response depends primarily on aerosol particle size and number concentration (e.g., particles/cm^3 of gas) rather than individual analyte properties. These "common property" measurement characteristics provide significant advantages over other devices whose *detection scope* (*viz.*, range of chemicals for which a useful response can be obtained) and *sensitivity* (*viz.*, signal output per unit mass or concentration) are highly dependent on analyte nature, such as optical properties (e.g., ultraviolet (UV) absorption, fluorescence (Fl)) or propensity to form gas-phase ions (e.g., electrospray ionization with mass spectrometry [ESI-MS]). While UV detection remains a primary technique for many LC analyses, its detection scope is limited to compounds with a sufficient UV chromophore, and its sensitivity varies widely among analytes. Likewise, the detection scope and sensitivity of MS strongly depend on ion source and operational conditions, analyte (e.g., basicity, surface activity), and eluent conditions (e.g., pH).

A significant need exists for LC detectors that can be used with a range of chromatographic conditions and can provide *both* broad detection scope *and* sensitivity that is independent of analyte properties (defined here as *uniform response*). Uniform response provides the possibility to quantify multiple analytes with a single calibration standard or calibration model [7–9]. This is desired for many analyses where individual standards are unavailable. Considering the entire chemical scope of potential analytes, the wide diversity of solvents and additives used with LC, and the common presence of impurities, it is easy to recognize that an LC detector whose sensitivity is completely independent of chemical and physical properties will not be very useful, since signal will always remain constant. That is, some level of *selectivity* or *specificity* is required to obtain a differential response between analytes and eluent components. For EADs, selectivity is based on differential vaporization.

In this book's first edition, Chapter 1 provided an updated description of CAD theory [10], which included new insights into the nature of CAD's response. This addressed a discrepancy noted in Kaufman's original patent [1], where response was found to vary more closely in proportion to the square root of concentration rather than the expected cube root. This chapter will reiterate and further emphasize the principles that explain CAD's response curve and those that explain its response uniformity. The objective is to provide a theoretical foundation from which to describe best practices for use of CAD (Chapter 3). Unless otherwise stated, this discussion will be primarily based on the Corona™ Veo™ and Vanquish™ series CAD (Thermo Fisher Scientific Inc.) design, here referred to collectively as "VCAD P."

1.2.1 Aerosol Basics

To facilitate a discussion of CAD theory, basic information about aerosols is described below. For readers interested in additional background to aerosol technology, references [4] and [11] are recommended.

An aerosol is a suspension of solid particles and/or liquid droplets in a gas and is often described as a two-phase system (gas and condensed). For LC detection, the gas includes the flow that is supplied to the detector and any component from the liquid eluent (e.g., the bulk solvent) that is vaporized. Evaporation involves a transition from liquid droplets to, typically, solid particles, but, in some cases, the "stable particles" that remain after evaporation are still in liquid form (e.g., oils). Pneumatic nebulizers used with CAD and other EADs produce "polydisperse" aerosols with broad (>1 order of magnitude) and often nonsymmetrical size distributions. A "snapshot" of the aerosol at any location within the detection flow path will consistently exhibit a broad size distribution, truncated by an impactor. Consequently, the time-dependent signal will result from contributions across a spectrum of particle sizes.

Table 1.1 provides some perspective on the number of molecules comprising a dried particle of a given diameter, calculated by Equations 1.1 and 1.2 and assuming that all particles are spherical:

$$V = \left(\frac{1}{6}\right)\pi d^3 \tag{1.1}$$

$$\text{Number of molecules} = \frac{V\rho N_A}{M} \tag{1.2}$$

where V = volume (cm^3); d = particle diameter (cm); ρ = density (g/cm^3); M = molar mass (g/mol); and N_A = Avogadro's constant (6.022×10^{23} mol^{-1}). For a high-molar-mass protein such as albumin (66.5 g/mol), a d = 10 nm particle may consist of only a few molecules, while the same-size particle may consist of >2500 glucose or >10,000 NaCl molecules.

The condensed phase of an aerosol typically spans a diameter range of approximately 1 nm to 100 μm, determined by the droplet size distribution generated during the pneumatic nebulization process. In most of this range ($d >$ ~20 nm), many of the properties and behaviors of a condensed-phase component are more

Table 1.1 Estimated number of molecules for a given particle diameter.

	Molar mass (g/mol)	ρ Density (g/cm^3)		Number of molecules			
			d (nm)	2	5	10	50
Albumin	66,500	1.37		0	1	6	812
Insulin	5808	1.37		1	9	74	9.30×10^3
Polysorbate 80	1310	1.07		2	32	257	3.22×10^4
Azithromycin	749	1.18		4	62	497	6.21×10^4
Cholesterol	387	1.05		7	107	856	1.07×10^5
Glucose	180	1.54		22	337	2.7×10^3	3.37×10^5
NaCl	58	2.17		94	1.46×10^3	1.17×10^4	1.46×10^6

determined by physical size than chemical composition. This includes motion in gravitational and electrical fields, and, for CAD and other EAM techniques, interaction with gas-phase ions through Brownian diffusion. This relationship is fundamental to the process of diffusion charging, which is central to the response uniformity of CAD and will be discussed in more detail within this chapter.

1.3 Charged Aerosol Detection Process

1.3.1 Spray Formation and Drying

The overall scheme of spray drying (nebulization, removal of large droplets, evaporation) and EAM (aerosol charging, ion removal and current measurement) with CAD is depicted in Figure 1.1. It should be noted that all EADs use the same basic principle of spray drying a liquid eluent to form an aerosol residue and thus share many of the upstream components and characteristics with CAD. Since detection selectivity is, in all cases, based on relative vaporization of components within an eluent (e.g., analyte versus solvent), the observed detection scope and eluent requirements are, as expected, practically the same for these detectors [12, 13].

1.3.1.1 Nebulization

With pneumatic nebulization, a high-velocity gas flow provides the energy to break up a slower liquid stream into droplets. With the widely used concentric nebulizer, the eluent flows through a central capillary that is surrounded by a conduit with a much faster gas flow. Parallel flows converge at the nebulizer tip, where high gas velocities generate the shear load and force necessary to overcome the liquid's viscosity and surface tension, thereby breaking it up into fine droplets of various sizes. The spray is directed toward an impactor surface, which imposes a size cut on the generated droplet size distribution.

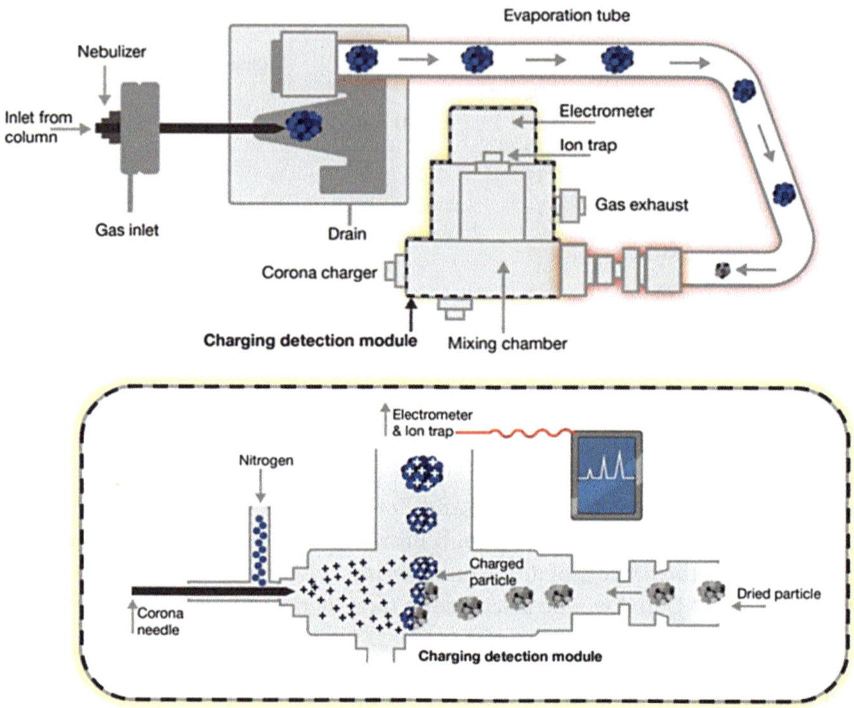

Figure 1.1 CAD involves pneumatic nebulization of eluent, removal of large droplets within a spray chamber, droplet evaporation, diffusion charging of resultant aerosol residue within a mixing chamber using an opposing ion jet formed via corona discharge, removal of excess ions and high mobility charged particles in an ion trap, and measurement of aggregate charge of aerosol particles with a filter/electrometer. The figure depicts a single multicomponent droplet transitioning to a single residue particle for sake of clarity. In reality, $>10^6$ droplets and residue particles are typically produced per second. *Source:* Reproduced with permission of Thermo Fisher Scientific, Inc.

Pneumatic nebulization is widely characterized as a physical process (i.e., independent of solute concentration) that produces polydisperse aerosols exhibiting a nearly lognormal size distribution. While average diameter is usually sufficient to describe a monodisperse aerosol, geometric mean diameter (GMD) and geometric standard deviation (GSD) are more appropriate for lognormal distributions. GMD (i.e., nth root of the product of n diameters) typically provides a better estimate of central tendency than arithmetic mean. A true lognormal distribution is symmetrical on a logarithmic scale, and therefore count median diameter (CMD) and GMD are equivalent. GSD (i.e., inverse log of the standard deviation of the log-transformed values) is likewise often preferred to estimate width or shape of the distribution.

The characteristics (e.g., GMD, GSD, etc.) and stability of the *primary aerosol* (*viz.*, formed at the nebulizer tip) are critical to the performance of CAD and other aerosol-based detectors. Several semiempirical models have been developed to predict aerosol characteristics generated by different nebulizer designs and as a function of solvent and conditions [14–16]. The widely described Nukiyama–Tanasawa (N–T) model is shown in Equation 1.3. As with other models, there are reported limitations to its predictive accuracy for a given nebulizer design. Even so, it is useful to qualitatively describe the main control parameters of the primary aerosol, which can be classified according to LC conditions (i.e., eluent volumetric flow rate, surface tension, and viscosity), nebulizer dimensions (i.e., axial velocity of gas and liquid in the nebulizer), and nebulizer conditions (i.e., gas volumetric flow rate). Table 1.2 lists the parameters and property values of common LC solvents:

$$\text{SMD} = \frac{585\sigma^{0.5}}{(v_g - v_l)\rho^{0.5}} + 597\left[\frac{\eta}{(\sigma\rho)^{0.5}}\right]^{0.45} \times \left(1000 \times \frac{Q_l}{Q_g}\right)^{1.5} \tag{1.3}$$

The N–T model predicts the Sauter mean diameter (SMD), which is the diameter of a spherical droplet whose ratio of volume to surface area is equal to that of the complete spray sample. For a liquid flow rate of 0.6 mL/min and nominal dimensions of the VCAD P concentric nebulizer, SMD (μm) predictions are 9.5 for water and 7.0 for acetonitrile (CH_3CN). This illustrates that, for a given condition, solvents with lower surface tension and viscosity tend to produce smaller droplets. To model CAD response (Section 1.4), we have found that N–T, predicted SMDs, along with empirical data from a scanning mobility particle sizer (SMPS) spectrometer (TSI Model 3938), are suitable to predict the primary aerosol characteristics for both water and CH_3CN. SMDs for water and CH_3CN were converted to GMDs (μm) of 3.6 and 2.7, respectively, using equations described in Hinds book [4] and assuming a GSD for both solvents of 1.85.

Table 1.2 Parameters and property values for N–T model.

Parameter	Water	Methanol	Acetonitrile
σ = liquid surface tension (dyn/cm)	72.8	22.6	29
ρ = liquid density (g/cm^3)	1.00	0.79	0.78
η = dynamic viscosity (P, poise)	0.010	0.006	0.003
Q_l = volumetric flow rate of liquid (L/min)	LC flow rate		
Q_g = volumetric flow rate of gas (L/min)	Typically 2.75		
V_l = liquid axial velocity (m/s)	Defined by Q_l and nebulizer capillary i.d.		
V_g = gas axial velocity (m/s)	Defined by Q_g and nebulizer annular area		

1.3.1.2 Removal of Large Droplets

Aerosol transport is defined here as the fraction of the aerosol that is transported to the evaporation tube and thus remains to form an aerosol residue and be measured. Importantly, changes in aerosol transport during gradient elution chromatographic methods can lead to baseline drift and significant changes in analyte sensitivity (e.g., peak area/ng).

Water is the most common LC solvent and has relatively low volatility. From the saturation vapor pressure (P_s) at reference conditions (e.g., 23.8 Torr at 25 °C), it can be estimated that a ~ 25 μL/min flow of water should saturate a 1.0 L/min flow of dry gas. The flow path of an aerosol detector, operating in a typical HPLC system, with flow rates ranging from a few μL/min up to 2 mL/min, can thus be easily saturated (100% relative humidity [%RH]), which would inhibit evaporation, leading to deficient results. Especially with aqueous eluents, the fraction of solvent entering the evaporation tube must therefore be reduced. Hinds [4] showed that droplet lifetime is proportional to D^2 (i.e., surface area) and strongly depends on %RH. For example, at a given %RH, the lifetime of a 1.0 μm droplet is ~100-fold less than that of a 10 μm droplet. Also, the lifetime of a given droplet size is ~1000-fold lower at 50% RH than in saturated conditions. The graph (Figure 1.2) illustrates the evaporation period of a 20 μm droplet in a 30 °C environment at constant

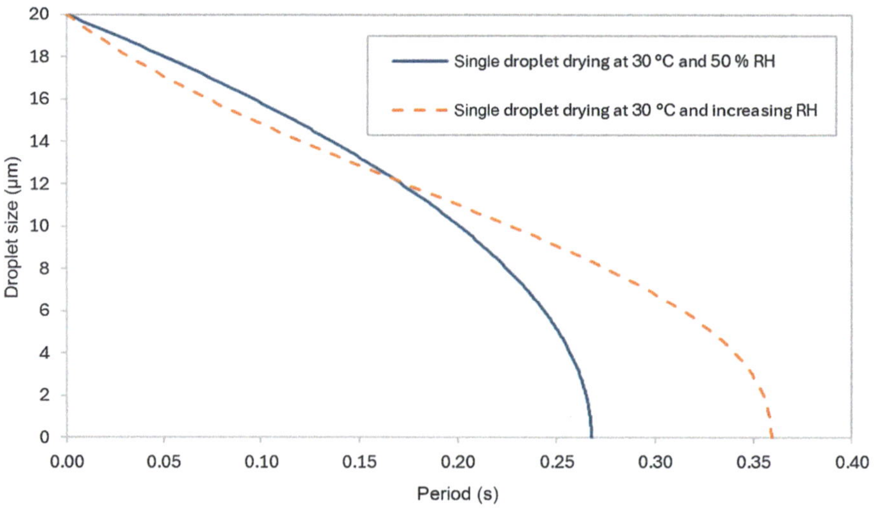

Figure 1.2 Drying of a 20 μm large droplet in an environment of 30 °C. The solid line shows the drying period at a constant relative humidity of 50%, and the dotted line shows the drying time in a slowly saturating environment due to the evaporation of droplets from an eluent flow of 60 μL/min.

relative humidity of 50%, compared to the evaporation period under increasing relative humidity due to the evaporation of the sprayed eluent at a rate of 60 μL/min.

The VCAD P spray chamber is designed so that the spray is directed toward a surface, which is arranged to guide the flow in an upward and reversed direction. This serves as an inertial impactor, where larger droplets with higher inertia are unable to follow the sharply curved streamlines of the flow and therefore collide with the chamber wall, coalesce, and are directed to waste. SMPS measurements indicate that, for a 0.6 mL/min flow of water, a maximum droplet diameter (D) of ~5 μm is transported to the evaporation tube. Under these conditions, ~90% of the liquid volume is typically directed to waste.

Evaporation can start at the point of nebulization and is considered here in the context of aerosol transport. A reduction in D before impaction would lead to higher transport than predicted by impactor cut size (D_{cut}). A main factor influencing evaporation rate (R_e) is the ratio of the partial pressure of the vapor to P_s. Measurements with an SMPS indicate that, for more volatile solvents, D_{cut}—and therefore transport—is effectively increased due to pre-impactor evaporation. For example, D_{cut} determined for CH_3CN was ~11 μm, compared with ~5 μm for water. For a flow rate of 0.6 mL/min, ~30% of pure CH_3CN is directed to waste (70% transport), compared with 10% transported for water. Not only does pre-evaporation have an impact, but also the fact that most organic solvents have lower surface tension and viscosity. The shear and thrust forces during pneumatic nebulization generally produce smaller droplets. This means that the log normal droplet size distribution leads to smaller GMD. The influence of solvent properties on aerosol transport is therefore a function of both nebulization and pre-impactor evaporation. In general, aerosol transport should be highest for solvents with low viscosity and surface tension, since these properties are typically associated with relatively high P_s and, by Equation 1.3, with production of smaller droplets. These relationships underlie the differences in sensitivity observed with aqueous versus organic mobile phases and the changes that occur in gradient elution methods. Figure 1.3 shows dried particle size distributions, showing an approximately ninefold higher number concentration and a slightly higher GMD for "pure" CH_3CN as compared with pure water. While this appears to be mainly attributed to aerosol transport (sevenfold difference as previously noted), it suggests that, in this experiment, CH_3CN also had a higher concentration of nonvolatile impurities (see Section 1.5.2).

1.3.1.3 Evaporation

Understanding the specific process of aerosol evaporation and factors influencing relative solute partitioning between gas and condensed phases can be useful in method development. This section will discuss the process of aerosol evaporation

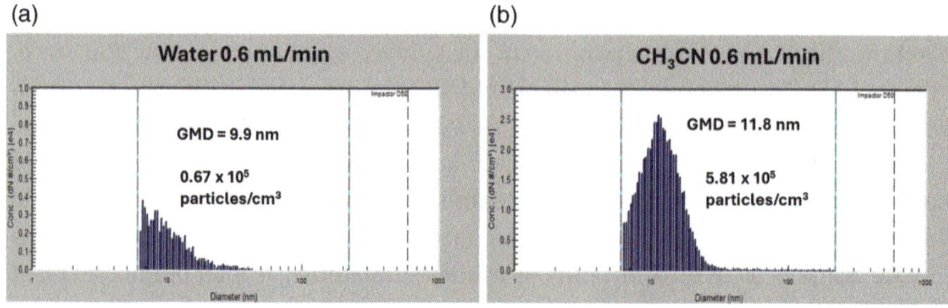

Figure 1.3 Particle size distributions measured using an SMPS from the outlet of the evaporation tube of the Corona Veo for a constant 0.6 mL/min flow of water (a) and CH_3CN (b).

with a focus on the solute properties and analytical conditions that may influence the size and stability of aerosol particles.

An aerosol droplet produced by nebulizing an LC eluent is often a binary (e.g., aqueous–organic) solvent mixture containing a variety of solutes (i.e., analyte, additives, and impurities). Evaporation decreases droplet size, which concentrates the less volatile components and dynamically changes the solvent composition. The process enhances solute–solute interactions, precipitation, and other effects, producing a condensed phase.

1.3.1.4 Volatility and Detector Response

Several studies have examined detection scope and sensitivity for a diverse range of chemicals in relation to common indicators of volatility. This includes M, P_s, boiling point (B_p), and enthalpy of vaporization (ΔH_v). Most studies have shown an approximate value above or below which solutes are reliably detected, and these are typically described as *nonvolatile* (e.g., $B_p > {\sim}400\,°C$, [7, 17] or $M > 350$ g/mol *and* $\Delta H_v > 65$ kJ/mol [9]). Near these limits (e.g., B_p from 300 to 400 °C, [7]), weaker correlation (e.g., lower sensitivity or no response) has been observed. Solutes within this range are typically considered *semivolatile,* and those beyond are considered *volatile*. The lines between these categories, however, are certainly blurred, and there are unexpected outliers. Additional insight is provided from these and other studies [9, 18–20] that have examined the influence of variables related to solvents, additives (e.g., pH modifiers, ion-pairing agents), conditions (e.g., liquid flow rate, evaporation T), and solute properties (e.g., log D, pK_a, charge state) on LC-aerosol detection scope, sensitivity, and uniformity of response. The important effect of mobile phase additives on the behavior of ionizable species will be discussed in more detail later in this chapter and in Chapter 3.

Since volatility may be significantly influenced by experimental conditions, it is useful to describe a solute in terms of its expected behavior for a range of conditions or its observed behavior within a given method. For example, an analyte may behave as a nonvolatile with an evaporation temperature of 35 °C but as a semi volatile at higher temperatures. Solute volatility may also be influenced by instrument design. For example, the response curves (Figure 1.4) for caffeine and theophylline are more similar with the VCAD P design than with the first-generation (UCAD) design. The lower sensitivity and different curve shape observed for caffeine with the UCAD design is highly characteristic of an analyte that behaves as a semivolatile. As described in the first edition of this book [10], the VCAD P design allows transport of larger droplets, which produce

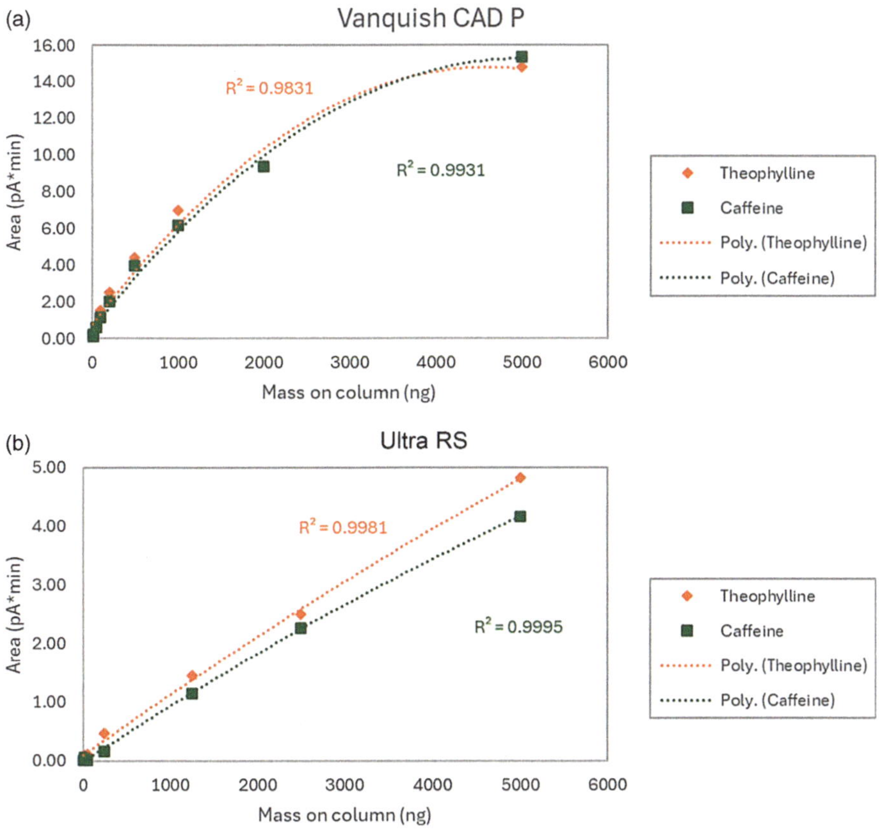

Figure 1.4 Response curves obtained with different CAD instrument designs. Under the same chromatographic conditions (isocratic 20% CH_3OH, 1.0 mL/min), the Corona ultra RS response for caffeine (more volatile) is lower and exhibits a higher power law exponent (Sections 1.4 and 1.5) than the less volatile theophylline. Latest VCAD P response curves are similar for both analytes even at higher T_e (35 °C) than for the early ultra RS detector (UCAD) (presumed $< \sim 20$ °C due to cooling from nebulization). This is attributed to the larger overall particles measured with the improved VCAD P design.

larger particles than the first-generation (UCAD) instrument. Since evaporation rate is inversely proportional to surface area, it follows that a greater proportion of caffeine should remain in the condensed phase with the VCAD P design. This leads to caffeine behaving more like a nonvolatile with VCAD P despite the higher evaporation temperature used in this example.

With the latest VCAD P design, a temperature coupling mode (patented) was introduced. This mode allows the evaporation temperature and the subsequent process temperature to be controlled separately. For example, droplets can be evaporated into particles within the drying tube by setting the appropriate evaporation temperature and then protected from particle evaporation by a lower temperature after exiting the evaporation tube. This is particularly beneficial for semi-volatile analytes. On the other hand, an increased temperature after the evaporation tube can prevent particles from recondensing in nearly saturated atmospheres. This advantage might be particularly relevant for methods requiring higher evaporation temperatures, for example, when higher molar mass additives are required for chromatographic separation. In discussing factors that determine the size, composition, and stability of dried aerosol particles we assume that evaporation of the bulk solvent is essentially complete after the evaporation tube. Equation 1.4 is widely used to describe the relationship among dried particle diameter (d), primary aerosol droplet diameter D, nonvolatile solute mass concentration C_s, and solute "bulk" density (ρ_s). This indicates that the volume of each dried particle is proportional to the starting volume and the (v/v) concentration of nonvolatile solutes in its corresponding primary droplet, corrected for bulk density. The particle's diameter is then proportional to the cube root of its volume. It is assumed that all droplets include some level of nonvolatile impurity, and the subscript (s) refers to solute, which is taken to mean "nonvolatile impurity + analyte, if present":

$$d = D\left(\frac{C_s}{\rho_s}\right)^{1/3} \tag{1.4}$$

Key aspects of Equation 1.4 include:

1) The exponent (1/3) indicates a nonlinear relationship between d and C_s. Table 1.3 provides an example of this relationship for a given D and range of solute concentrations. This is a main factor that determines both the shape of a response curve and chromatographic peak shape, as will be discussed in Section 1.4.
2) Particle bulk density (ρ_s) influences dried particle diameter (d) weakly because of the cube root (e.g., Table 1.3). Since ρ_s differs among analytes and response depends on d, it then follows that ρ_s is a factor that may contribute to differences in sensitivity (e.g., peak area/ng) among analytes. A study by Matsuyama et al. [21] clearly showed the dependence of CAD response on analyte density. This topic will be further discussed in Section 1.5.5.

Table 1.3 Dried particle size dependency on solute concentration and density.

	Analyte A	Analyte B
$D = 5.0\,\mu m$	$\rho = 1.0$	$\rho = 2.0$
(µg/mL)	d (nm)	d (nm)
0.1	23	19
1.0	50	40
10	108	86
100	233	185
1000	501	398

1.3.2 Aerosol Charging

Aerosol charging can occur by several mechanisms. Most relevant to CAD are static electrification, field charging, and diffusion charging. ESI is an example of static electrification of liquids, which is based on an electrostatic charge at the surface that is carried away as aerosol droplets separate from the bulk liquid. The mechanisms of field and diffusion charging, by contrast, involve *collisions* between ions and particles within an aerosol. Field charging, which takes place in the presence of a strong electric field (i.e., potential gradient), can produce a high charge density on aerosol particles, but the level and distribution of charges significantly depend on the particle material (i.e., dielectric constant). Diffusion charging involves collision of ions and aerosol particles when there is no appreciable electric field and is also the dominant mechanism for $d < 100$ nm even in the presence of an electric field [4]. An important advantage of diffusion charging is that the process has minimal dependence on the particle material [22], which provides the basis for uniform response capabilities of CAD. Also, while field charging (i.e., mean charge per particle) is largely proportional to particle surface area (d^2), unipolar diffusion charging was found empirically to be proportional to $d^{1.133}$ over a wide range of particle sizes. This relationship is a main factor that underlies the shape of CAD response curves.

1.3.2.1 Unipolar Diffusion Charging

CAD is based on *unipolar* diffusion charging in which the aerosol is exposed to ions of only one polarity. CAD uses positive charges. Positive corona discharges generate a higher concentration of ions and often have better ion penetration in the charging region compared to negative corona discharges. A higher ion concentration results in more efficient and faster particle charging. Additionally, the use of positive charge carriers compared to negative charge carriers demonstrates increased stability and controllability. The magnitude of charge depends mainly on the ion concentration produced by the ion source, time of aerosol exposure to ions, and particle size (i.e., proportional to $d^{1.133}$). The theory describes a process of charging that results from random Brownian motion of ions in proximity to aerosol particles. It is assumed that (i) each ion–particle collision leads to charge transfer to the particle, and that (ii) charge is not transferred back to adjacent gas molecules. Atmospheric pressure chemical ionization (APCI) used with liquid chromatography–mass spectrometry (LC–MS) also uses a corona discharge, where charging is based on collisions with gas-phase ions. A main difference between these techniques is the relative size of the colliding "reagent" ion and analyte species. Premnath et al. [23] described an insightful study investigating the intersection between the particle diffusion charging that predominates in CAD and the chemical ionization mechanisms dominant in APCI. With APCI, ions collide with vapor molecules, and the transfer of charge is highly dependent on the activation energy for charge transfer and steric effects during collision. Also, an equilibrium can be achieved because charge can rapidly transfer back to vapor molecules. Their studies showed that back reaction rates were highest for the smallest cluster ion size and decreased to negligible for clusters of $d_p > \sim 0.5$ nm.

The design used in all UCAD and VCAD P instruments is based on that described by Medved et al. [2]. A positive corona discharge is produced by creating a nonuniform electrostatic field between a sharp needle tip and an orifice plate. The needle is positive with respect to the orifice plate. At the needle tip, a high enough field strength is established to break down the normally insulating surrounding gas such that it becomes conductive. In the region near the tip, electrons flow with high enough velocity to knock electrons from gas molecules and create positive ions and additional free electrons. These electrons are, in turn, accelerated toward the positive-polarity needle tip, creating enough energy to knock electrons off other gas molecules. This leads to a self-sustaining chain reaction called the Townsend avalanche that creates high concentrations of electron-positive ion pairs. Some electrons recombine with positive ions, releasing energy in the form of blue and UV photons, which serve to ionize other molecules and help sustain and stabilize the electron avalanche. The CAD ion source is isolated from the aerosol particles and supplied with particle-free gas, either N_2 or air. Several studies have examined the nature of ions produced by corona discharge with these surrounding gases. The most consistently reported primary ions are $N_2^{+\bullet}$ and $N_4^{+\bullet}$, and, with even trace levels of water vapor (e.g., 3 ppm) [24], the most abundant secondary ions are hydrated protons $H_3O^+(H_2O)_n$ [25, 26].

Most CAD instruments use a corona jet design, where positive ions produced as described above are repelled from the needle tip, attracted toward the negative-polarity orifice plate, and are additionally entrained by a flow of gas through a subsonic orifice into a separate mixing chamber with conductive walls. In this manner, a steady turbulent flow of positive ions is established within a region with little to no electric field, thus facilitating aerosol charging by diffusion mechanisms only. The aerosol produced by spray drying LC eluent is introduced to the mixing chamber and opposes (collides with) the corona ion jet. In this process, convective mixing of the turbulent jets allows diffusion charging to proceed unhindered. This technique differs from the diffusion charger used in Dixon's description of ACD [3], which lacked the described convective mixing component. A recent paper described a "direct corona charging" design, where the corona discharge is located directly within the analyte aerosol flow path, i.e., at the outlet of the evaporation tube [27]. While performance was mainly evaluated with relatively simple eluents, the authors report high sensitivity and stability. With this device, the corona discharge needle maintains direct physical interface with aerosol particulates. Accumulation of contaminant species on the electrode surface induces temporal and spatial variations in discharge characteristics, resulting in nonuniform charging fields and compromised measurement reproducibility. These aerosol–electrode interactions facilitate progressive degradation of the corona discharge apparatus. Particulate deposition and subsequent electrochemical processes at the electrode surface contribute to dimensional alterations, material erosion, and modification of work function parameters. Such degradation necessitates increased maintenance frequency and potential system recalibration to maintain operational stability.

Medved et al. [2] studied transmission efficiency (fraction of particles leaving versus entering the charger) and charging efficiency (fraction of particles charged) of the corona jet device as a function of d over the range of 3–50 nm. They showed high transmission efficiency and, by inference, minimal particle loss. Several studies with this device [2, 28–31] have shown that, in accordance with theory, mean charge per particle is found to be proportional to $d^{1.133}$ over the range from 10 nm to 1.0 μm and for $d < \sim 9$ nm proportional to surface area (d^2) or a higher exponent. This well-established relationship is a main factor that determines both the shape of a response curve and the chromatographic peak shape obtained with CAD.

According to diffusion charging theory, particle dielectric constant may influence the charging of particles in the low nm diameter range. Shin et al. [32] used the corona jet charger to study compact, roughly spherical aerosol particles of sucrose, NaCl, and Ag, which have dielectric constants (ε) of 3.3, 6.1, and infinity (metallic), respectively. Over a range of 10–200 nm, they observed a difference in slope of mean charge per particle versus d of 10.5% between Ag and sucrose and a difference of 5.2% between Ag and NaCl. Li et al. [30] reported similar results of ~3% difference between NaCl and Ag particles over a mean d range of 20–200 nm. In both studies, material dependency was less than predicted from theory. This may be attributed to the high efficiency of the specific corona jet charger for which it is theorized that once particles acquire a first charge, the resultant material-independent Coulombic force becomes the dominant factor of the interaction potential. As previously described, aerosol particles encountered with LC are rarely expected to consist of a single pure substance, a factor that may serve to dampen the influence of analyte dielectric constant on response uniformity. Considering this effect and the previously described dependency, it seems reasonable to assume that analyte dielectric constant may be a relatively minor factor influencing the response uniformity of CAD.

1.3.2.2 Summary of Aerosol Charging

It is well established that for diffusion charging with CAD, mean charge per particle is proportional to $d^{1.133}$ over the range from 10 nm to 1.0 μm, while for $d < {\sim}9$ nm is proportional to a higher power of d. The contribution of these smaller particles to the shape of CAD's response curve was first described in 2017 in Chapter 1 of the first edition of this book. This addressed a discrepancy, noted in Kaufman's original patent [1], where response was found to vary more closely in proportion to the square root of concentration rather than the expected cube root. This has significant implications for the shape of a CAD response curve. Studies of particle material dependence on diffusion charging suggest that differences in analyte dielectric constant may have a minor influence on response uniformity (discussed in Section 1.5). The final steps of the CAD process, after aerosol charging, are removal of excess ions within an ion trap and measurement of aggregate charge with a filter/electrometer. These steps will be described in more detail in Section 1.4.

1.3.3 Summary of CAD Process

Here, we briefly summarize the CAD process by tracing discrete volumes of eluent to a column outlet and then by describing subsequent aerosol characteristics at different stages within the detection flow path. Figure 1.5 depicts a column with three solute bands at different positions along a column and a chromatogram that represents complete separation. Isocratic conditions are assumed, where the concentration of a nonvolatile solute within a given volume is the sum of its time-dependent analyte concentration and a constant impurity concentration. The time-dependent analyte concentration results from the mass injected and the dilution and dispersion of the separation process (described in Section 1.4).

The key assumptions described previously are:

1) The size distribution of a primary aerosol is solute independent. Accordingly, all droplets, *at a given instant*, have the same solute mass concentration and differ only in size.

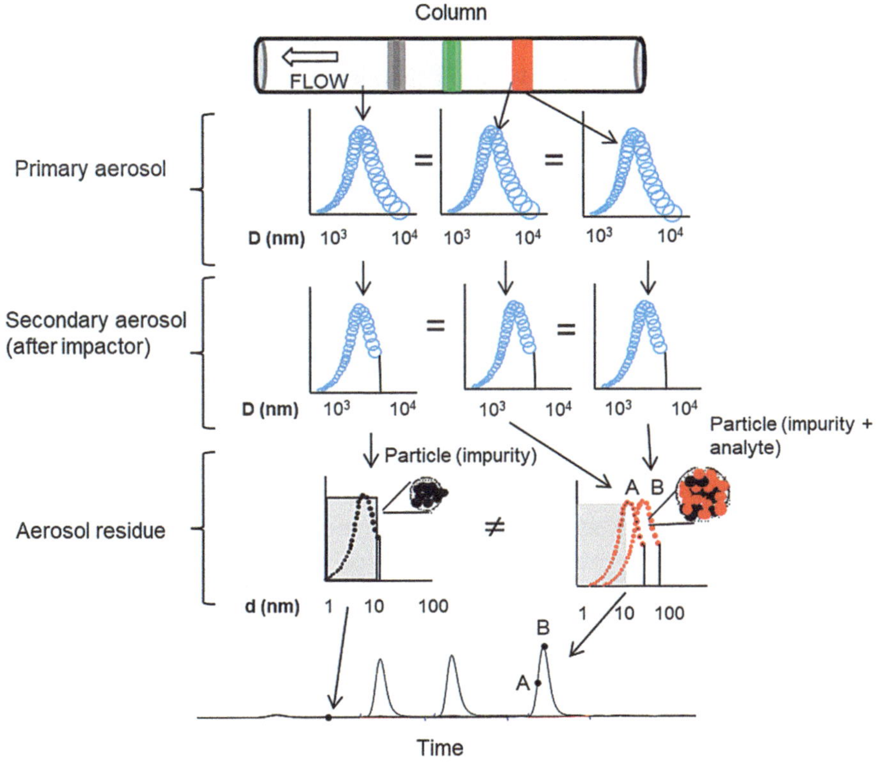

Figure 1.5 CAD is depicted for an isocratic separation where three discrete eluent volumes (baseline and two points within a single analyte solute band) are projected to the outlet of a column and traced through the main steps of the process. Particle cartoons are magnified to show impurity or analyte + impurity components. Shaded areas of each aerosol residue distribution reflect the relative proportion of $d < \sim 9$ nm, which have a higher aerosol charging power law exponent, thereby influencing the overall power law exponent of CAD response.

2) Primary aerosol droplet distributions remain constant throughout isocratic elution but change during solvent gradients due to the changing solvent properties.
3) Removal of primary aerosol droplets by inertial impaction above a certain cut size to form a secondary aerosol is solute independent but depends on eluent properties that influence both nebulization (Equation 1.3) and pre-impaction evaporation.
4) The effective diameter of each dried aerosol residue particle is equal to the diameter of its corresponding primary droplet times the cube root of v/v concentration of its nonvolatile constituents divided by the cube root of its particle bulk density.

Note: All of the previous assumptions are also generally valid for any EAD.

5) Signal (current) is based on diffusion charging of dried aerosol particles and measurement of aggregate charge reaching the electrometer filter, where mean charge per particle is proportional to $d^{1.133}$ over the range from 10 nm to 1.0 μm, while for $d < \sim 9$ nm is proportional to a higher exponent. Loss of particles in the ion trap also plays a role (further described in Section 1.4). Analyte dielectric constant is a relatively minor factor influencing the response uniformity of CAD.

Figure 1.5 shows that, for isocratic conditions, all three of the illustrative primary aerosols have the same droplet size distribution (Assumption 2) and all droplets within a given primary aerosol have the same solute concentration (Assumption 1). Since the "effective" impactor cut size remains constant in isocratic conditions (Assumption 3), all secondary aerosols also have identical size distributions. Number concentrations are the same for all dried aerosols and identical to that of the secondary aerosols. The size of each residue particle reflects the v/v concentration of nonvolatile solute in its corresponding primary droplet (Assumption 4, for $\rho_s = 1.0$). Representative residue particles are magnified to depict a small particle consisting of many background impurity species and a larger particle with the same number of background species whose size is increased by the presence of many analyte species. The residue distributions for the two volumes within a single solute band reflect a size shift proportional to a 1/3 power law exponent of the instantaneous nonvolatile solute v/v concentrations at peak maximum and at ~50% of peak maximum. The difference in response is mainly due to the difference in the charge accepted and carried by the particles. Differences between these distributions in the relative proportion of particles below $d \sim 9$ nm underlie changes in the overall power law exponent of CAD response over a range of analyte amounts, as further described in Section 1.4.

1.4 CAD Response Model

To construct a model of CAD, we will consider the successive steps of primary spray droplet formation from an eluent stream, impaction to remove droplets larger than a specified cut size, evaporation of the remaining droplets to form residue particles of the analytes, charging of the residue particles, removal of excess ions, and detection of the current due to deposition of the particles with their charge in an aerosol–electrometer filter. Each step of this process depends on the droplet or residue particle size, and each of these dependencies is evaluated based on a theoretical model or on empirical data.

The model predicts primarily the time-dependent response of the detector to the concentration of analyte in the eluent. This response function is then used to predict the time-dependent current during passage of an eluting peak, and the current is integrated over the peak to predict the net charge delivered as a function of the injected analyte mass. For simplicity, we assume only isocratic conditions.

1.4.1 Primary Droplet Size Distribution

Droplets of the eluent are formed by a pneumatic nebulizer as discussed earlier. We assume that the analyte concentration is sufficiently low so as not to influence properties of the mobile phase such as surface tension, viscosity, and so on. This ensures that the behavior of the nebulizer, and thus the size distribution of these "primary droplets," is not affected by the presence or absence of an analyte solute band.

We assume further that this primary droplet size (D) distribution is lognormal. The fundamental parameters of this distribution are, besides overall amplitude, the CMD, and the GSD. The normalized form is:

$$\text{Lognormal}(D, \text{CMD}, \text{GSD}) = \frac{e^{-1/2 \cdot \left[\frac{\ln(D/\text{CMD})}{\ln(\text{GSD})}\right]^2}}{\sqrt{2 \cdot \pi} \cdot D \cdot \ln(\text{GSD})} \tag{1.5}$$

and the normalization property is:

$$\int_0^\infty \text{Lognormal}(D, \text{CMD}, \text{GSD}) dD = 1 \tag{1.6}$$

1.4.2 Impactor

An impactor is integrated within the spray chamber to remove droplets that are too large to be fully evaporated in the downstream evaporation tube, thus forming a "secondary aerosol." In the inertial impactor, the primary droplets and carrier gas are made to flow around an obstacle, whereby the inertia of droplets above a specified diameter D_{cut} causes them to collide with the obstacle. The liquid from these droplets is led to a waste drain. We assume an ideal impactor with a sharp cutoff so that the secondary aerosol droplet size distribution is described by the same lognormal function below the cutoff and is zero above the cutoff. In reality, however, impactors do not exhibit such sharp cutoff characteristics but rather demonstrate a more gradual collection efficiency curve across the transition region. The actual value of the cutoff is dependent on details of the nebulizer geometry and on the mobile phase density, viscosity (see Section 1.3.1), and surface tension and is determined empirically.

The mean droplet volume in the primary distribution may be written as:

$$V_{\text{dp}} = \int_0^\infty (x \cdot \text{CMD})^3 \cdot \text{Lognormal}(x \cdot \text{CMD}, \text{CMD}, \text{GSD}) \cdot \text{CMD}\, dx \tag{1.7}$$

and the mean volume of the droplets remaining after impaction is:

$$V_{dc} = \int_0^{\frac{D_{cut}}{CMD}} (x \cdot \text{CMD})^3 \cdot \text{Lognormal}(x \cdot \text{CMD}, \text{CMD}, \text{GSD}) \cdot \text{CMD}\, dx \quad (1.8)$$

The fraction of eluent volume that is converted to aerosol is V_{dc}/V_{dp}, and the remainder goes to the waste drain.

1.4.3 Drying and Residue Formation

The secondary droplets from the nebulizer and impactor are led through a heated drying tube, where each droplet evaporates to leave a residue particle consisting of the nonvolatile solutes in the droplet. The simplest intuitive picture of this process requires that the volume of a droplet's residue particle is equal to the droplet volume times the volume/volume (v/v) concentration C of its nonvolatile constituents. Under the approximation that the residue particle is spherical, its diameter will be $d_{res}(D, C) = D \cdot C^{\alpha}$ with $\alpha = 1/3$. Empirical studies of the shift in the measured residue size distribution versus (v/v) concentration persistently show a power law with an exponent $\alpha \sim 0.2$. There seems to be no known systematic bias in these measurements, but also no rationale for a value such as 1/5 is apparent. In view of this, we will retain α as a parameter in the model formalism but will continue to assume $\alpha = 1/3$ for present purposes.

1.4.3.1 Residue Particle Parameters

The residue particles will have a size distribution identical to that of the secondary droplets, scaled according to $d_{res}(D, C) = D \cdot C^{\alpha}$. Thus, the parameters of the distribution become:

$$d_{cut} = D_{cut} \cdot C^{\alpha} \quad \text{cmd} = \text{CMD} \cdot C^{\alpha} \quad (1.9)$$

but the dimensionless GSD is identical to that of the droplets. Thus, the residue size distribution is lognormal(d, cmd, GSD) for $d < d_{cut}$ and zero otherwise. It should be noted that the terms "count mean diameter" and "geometric standard deviation," as used here, apply only to the mathematical form of the distribution without the cutoff. Figure 1.6 shows the residue particle size distribution according to this model for isocratic elution with a fixed v/v concentration of nonvolatiles.

1.4.4 Charging of Residue Particles

The dried aerosol is injected as a jet into a mixing chamber along with an opposing jet of gas containing positive ions formed in a corona discharge in a separate chamber. The electric field in the mixing chamber is negligible, thus the ion–particle encounters are governed by diffusion. The mean charge per particle is dependent on particle diameter, and the dependence has been found empirically for the charger design used in all CAD models [2, 28–31] to follow a power law with an

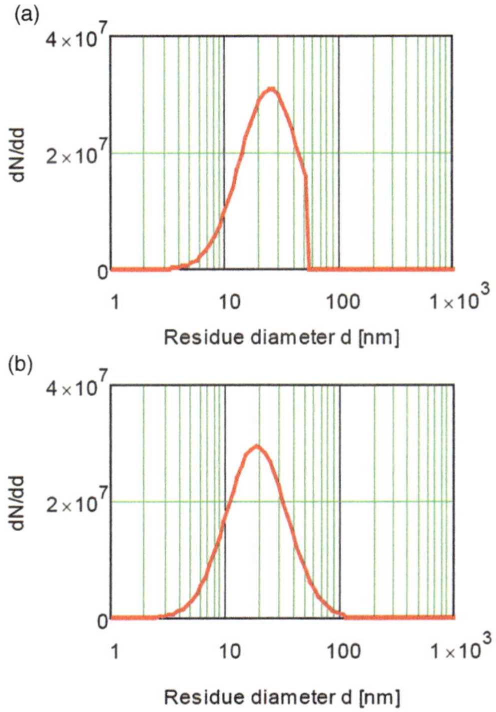

Figure 1.6 Distribution of dry residue particle diameters for v/v concentration = 10^{-6}. (a) Mobile phase = water; CMD = 3.6 μm; GSD = 1.85; D_{cut} = 5.0 μm, (b) mobile phase = CH_3CN; CMD = 2.7 μm; GSD = 1.85; D_{cut} = 11 μm.

exponent $\beta_2 = 1.133$ for particles larger than about $d_0 = 9$ nm. For smaller particles, the exponent is larger, the limited data being consistent with $\beta_1 = 2.25$. For the model, it will be useful to have an empirical function that represents this behavior. A reference point is that for particles of diameter $d_{ref} = 100$ nm, the mean charge per particle is $n_{ref} = 3.9$ electronic charge units. We use dimensionless units for the diameter, defining:

$$x = \frac{d}{d_0} \quad x_{ref} = \frac{d_{ref}}{d_0} \tag{1.10}$$

Then the mean number of charge units per particle is represented by:

$$n(x) = n_{ref} \cdot \frac{x^{\beta_1}}{x_{ref}^{\beta_2}} \quad \text{if } x < 1, \quad \text{and} \quad n(x) = n_{ref} \cdot \left(\frac{x}{x_{ref}}\right)^{\beta_2} \text{otherwise} \tag{1.11}$$

Figure 1.7 shows how $n(x)$ changes with residue particle diameter.

1.4.5 Ion Removal

Ions formed in the corona discharge of the unipolar charger are injected into the mixing chamber by a gas flow. These ions quickly accumulate small clusters of solvent vapor molecules, and it is these clusters that diffuse to the residue particles

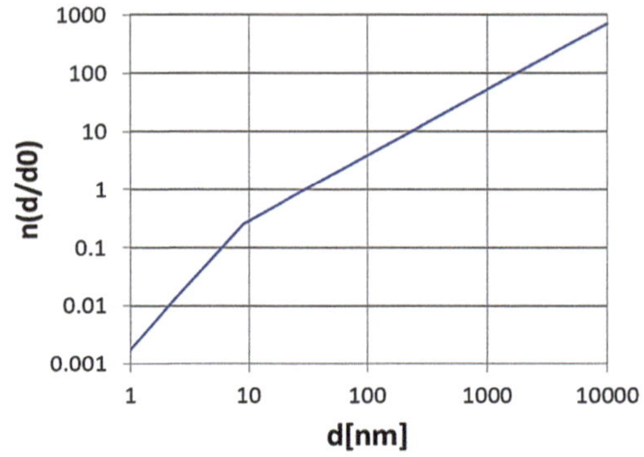

Figure 1.7 Mean number of electronic charge units per particle versus particle diameter in nanometers.

Table 1.4 Ion trap parameters.

Inner electrode radius (mm)	r_1	0.75
Outer electrode radius (mm)	r_2	2.7
Length (mm)	L	37
Applied voltage (V)	V	20
Total gas flow (L/min)	F_{gas}	3.75

and impart their charge. The electrical mobilities of these species are known to be greater than about $0.5\,\text{cm}^2/\text{V}\cdot\text{s}$. Excess ions can travel downstream with the charged aerosol particles, but it is important to keep them from reaching the electrometer detector, where they could contribute a large and fluctuating background current. To this end, a simple coaxial electrostatic precipitator is included as an "ion trap" immediately downstream from the mixing chamber, where the charging takes place. The trap has the parameters in Table 1.4.

Besides its function in removing ions, the ion trap removes some of the residue particles. Its transmission for charged particles and ions is calculated in this section.

For simplicity, we assume fully developed laminar flow in the annular space of the ion trap. The gas velocity is parallel to the axis denoted by z and is given by:

$$v_z(r) = \frac{2 \cdot F}{\pi \cdot \left(r_2^2 - r_1^2\right)} \cdot \frac{r_2^2 - r^2 - 2 \cdot r_{\text{maxv}}^2 \cdot \ln(r_2/r)}{r_2^2 + r_1^2 - 2 \cdot r_{\text{maxv}}^2} \tag{1.12}$$

where the radius of maximum v_z is:

$$r_{\text{maxv}} = \sqrt{\frac{r_2^2 - r_1^2}{2 \cdot \ln(r_2/r_1)}} \tag{1.13}$$

The electric field in the annular space is radial, outward directed, and given by:

$$E_r(r) = \frac{V}{r \cdot \ln(r_2/r_1)} \tag{1.14}$$

A particle or ion of electrical mobility Z moves downstream at the gas flow velocity, parallel to the z axis, while the electric field drives its migration across the flow at radial velocity $v_r(r, Z) = E_r(r) \cdot Z$. If the particle enters the precipitator at $z = 0$ and at a starting radial coordinate r_0, it follows a trajectory given by

$$z(r, r_0, Z) = \int_{r_0}^{r} \frac{V_z(r')}{V_r(r', Z)} dr' \quad r_1 < r_0 < r_2 \tag{1.15}$$

All ions or particles having electrical mobility greater than a critical value Z_{crit} will be precipitated onto the outer electrode of the ion trap. The value of Z_{crit} can be found by solving $z(r_2, r_1, Z) = L$ (L = trap length) for Z. With the parameters given by Table 1.4, $Z_{crit} = 0.172\,\text{cm}^2/\text{V}\cdot\text{s}$. Thus, the trap removes all of the ions, as desired. It was verified by observing the electrometer current while varying the trap voltage V that ion current observed at $V = 0$ vanished when $V = 10\,\text{V}$ was reached. The value 20 V was chosen to assure removal of all ions.

1.4.5.1 Attenuation of Particle Signal by Ion Trap

For particles with mobility $Z < Z_{crit}$, those entering outside a critical radial coordinate $r_{crit}(Z)$ will also be collected on the precipitator wall and lost; those entering at $r_0 < r_{crit}(Z)$ will pass through and can carry their charge into the electrometer to produce a signal. To obtain the value of r_{crit} given a particle diameter, we use the Millikan relationship giving the electrical mobility of a spherical particle of diameter d:

$$Z(d) = \frac{e \cdot C_c(\lambda, d)}{3 \cdot \pi \cdot \eta \cdot d} \text{ where } C_c(\lambda, d) = 1 + \frac{2\lambda}{d} \cdot \left[1.165 + 0.483 \cdot e^{-0.997 \cdot \left(\frac{d}{2\lambda}\right)}\right] \tag{1.16}$$

C_c is the Cunningham slip correction, λ is the gas mean free path, η is the gas viscosity, and e is the electronic charge. Thus, to find $r_{crit}(d)$, we solve $z(r_2, r, Z(d)) = L$ for r.

Having found $r_{crit}(d)$, we note that the transmission for particles of diameter d is just the ratio of the cross-sectional area within $r_{crit}(d)$ to the total cross-sectional area of the trap's annular passage:

$$T(d) = \frac{r_{crit}(d)^2 - r_1^2}{r_2^2 - r_1^2} \tag{1.17}$$

Table 1.5 Parameters and results for the rate of residue particle formation for water and CH_3CN.

Mobile phase		Water	CH_3CN
Flow rate (mL/min)	F_{liq}	0.6	0.6
Mean droplet volume		Femtoliter	
Before impactor	V_{dp}	134.1	56.6
After impactor	V_{dc}	12.7	37.9
Fraction of droplet counts after impactor			
	N_{cut}	0.703	0.989
Fraction of droplet volume after impactor			
	ϕ	0.095	0.669
Primary droplet rate (/s)	r_d	7.46E+07	1.77E+08
Residue particle rate (/s)	r_p	5.24E+07	1.75E+08

1.4.6 Signal Current

Eluent flows into the nebulizer at a rate F_{liq}. For our model calculations we take $F_{liq} = 0.6\,\text{mL/min}$. Table 1.5 shows the mean droplet volume, the fraction of droplets remaining after impact, and the resulting rate of residue particle formation for example cases of water and CH_3CN.

The particles so formed are charged, and their charge is carried to the electrometer's collection filter, where the arriving charge is measured as a current. The mean number of charge units carried to the electrometer by the entire distribution of residue particles produced at a given concentration C, including loss in the precipitator, is:

$$n_{mean}(C) = \int_{\frac{d_{crit}}{d_0}}^{\frac{C^{\alpha}\cdot D_{cut}}{d_0}} n(x) \cdot T(d_0 \cdot x) \cdot \text{Lognormal}(d_0 \cdot x, \text{cmd}, \text{GSD}) \cdot d_0\, dx \tag{1.18}$$

where the limits of integration are set by d_{crit}, where $T(d) = 0$, and d_{cut}, where the size distribution is reduced to zero because of the impactor. The detector current (Figure 1.8) is then given by:

$$i(C) = e \cdot n_{mean}(C) \cdot r_p \quad \text{where} \quad r_p = \left(\frac{F_{liq}}{V_{dp}}\right) \cdot N_{cut} \tag{1.19}$$

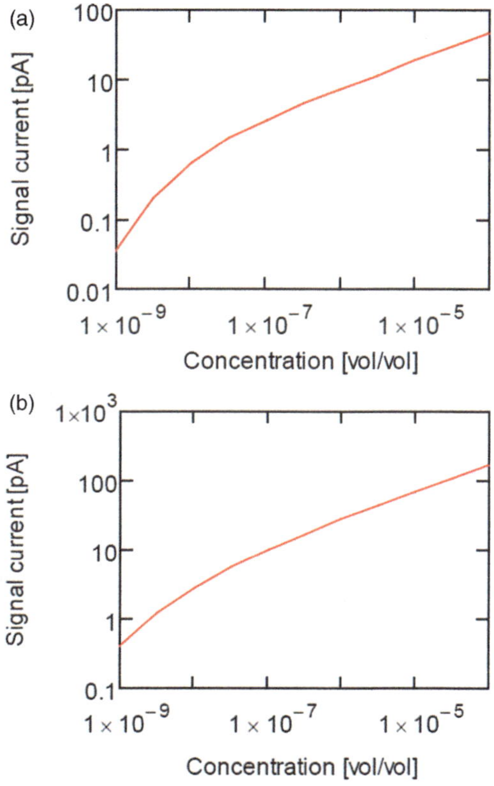

Figure 1.8 CAD detector signal as a function of concentration. Mobile phase: (a) Water; (b) CH_3CN.

1.4.7 Signal from an Eluting Peak: Peak Shape

The eluent consists of a mobile phase, dissolved mobile phase impurities (which we ignore for this model), and the analytes, which have been separated by the column. Following Equation 1.20, the instantaneous v/v concentration resulting from injected mass m_{inj} of an analyte whose bulk density is ρ may be represented by a Gaussian peak in concentration:

$$C(t, m_{\text{inj}}) = \left(\frac{m_{\text{inj}}}{\rho \cdot F_{\text{liq}} \cdot \sqrt{2 \cdot \pi} \cdot \sigma_{\text{t}}} \right) \cdot \exp\left[-\frac{1}{2} \left(\frac{t - t_{\text{r}}}{\sigma_{\text{t}}} \right)^2 \right] \tag{1.20}$$

where t_{r} is the column retention time and σ_{t} is the peak's standard deviation in time. The time-dependent CAD signal current pulse is then given by $I(t) = i[C(t, m_{\text{inj}})]$. The nonlinear response $i(C)$ results in a broadening of the current-time profile compared with the concentration-time profile, and due to the change in

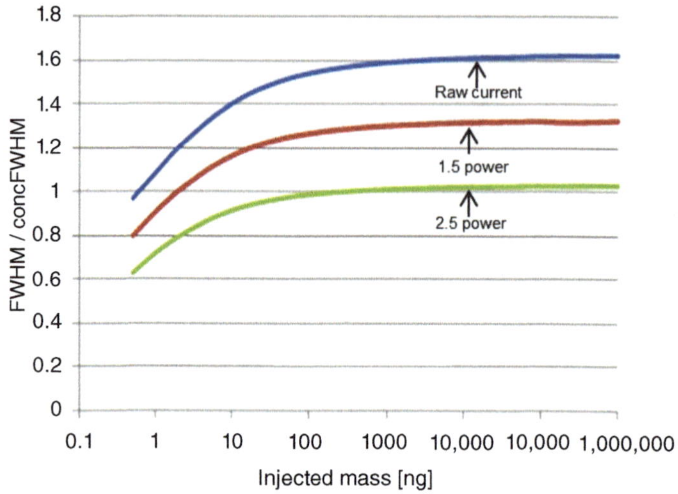

Figure 1.9 Ratio of peak full width at half maximum (FWHM) for "raw" current predicted by the CAD response model versus a "true" Gaussian concentration distribution over a range of m_{inj} and for different power function values.

slope of $i(C)$ as the peak concentration increases, the broadening becomes more significant with increasing injected mass. Figure 1.9 plots the ratio of peak widths for "raw" current predicted for CAD versus a "true" Gaussian concentration distribution at different power function values. This shows the significant broadening from an inherent (sublinear) response and the effects of current$^{1.5}$ and current$^{2.5}$ power function values, which will be further discussed in later sections.

1.4.8 Peak Area Versus Injected Mass

The total charge delivered to the electrometer during the elution of a peak is the area of the current versus time function $I(t)$. In the model, this can be integrated for a range of injected mass values, leading to the following expression for peak area or total charge:

$$Q(m_{inj}) = \int i[C(t, m_{inj})]dt \tag{1.21}$$

The result of this calculation is shown in Figure 1.10 for CH_3CN with $\rho = 1.4\,g/cm^3$.

1.4.9 Summary

We have presented a model of the salient characteristics of the CAD detector, tracing its behavior from the formation of spray droplets of eluent through impaction to remove the largest droplets, drying to form residue particles of

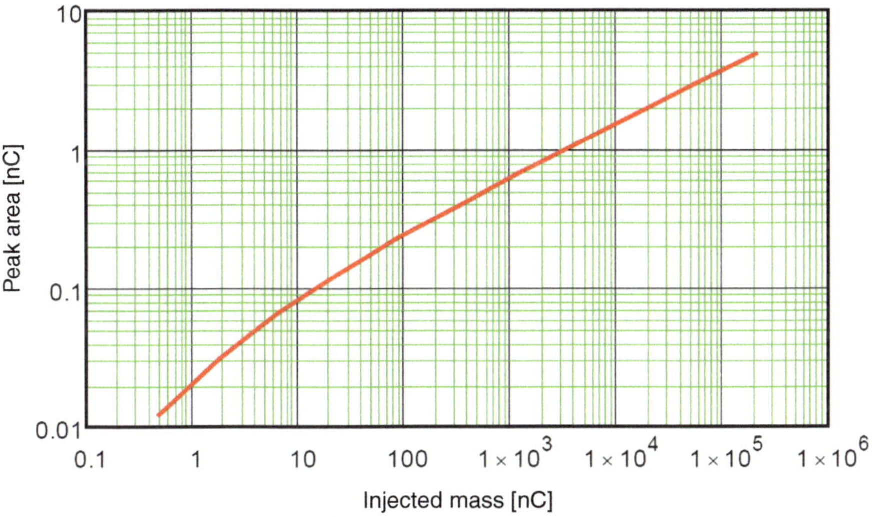

Figure 1.10 Peak area is shown in nanocoulombs for injected mass in the range of 0.5 ng to 200 µg. Mobile phase: CH_3CN; analyte bulk density: 1.4 g/cm^3.

the analyte, charging the resulting aerosol in a unipolar charger, and removing the ions with attendant loss of some of the smallest residue particles. The resulting current shows a slight decrease in slope over the very wide dynamic range of the detector, which gives rise to a small degree of broadening of the time-dependent eluting peaks compared with the actual peak width of the analyte concentration. To make use of this model, some empirical knowledge of the behavior of the nebulizer and impactor is required, as well as the charging characteristics of the unipolar charger.

1.5 Expected Performance

The expected performance of CAD, considering analyte properties, detector design, and experimental variables, is discussed here along with some comparisons to other EADs. Since all EADs use the same basic principle to produce an aerosol residue, a key focus will be on characteristics of the downstream aerosol measurement technique that explain differences in performance among these devices.

1.5.1 Response Curve: Shape and Dynamic Range

Figure 1.11 plots peak area response (A) versus analyte mass injected (m_{inj}), observed with VCAD P and ELSD under identical conditions. The response can be described over segments of the range of each detector by a power law function:

$$A = a(m_{\mathrm{inj}})^b \tag{1.22}$$

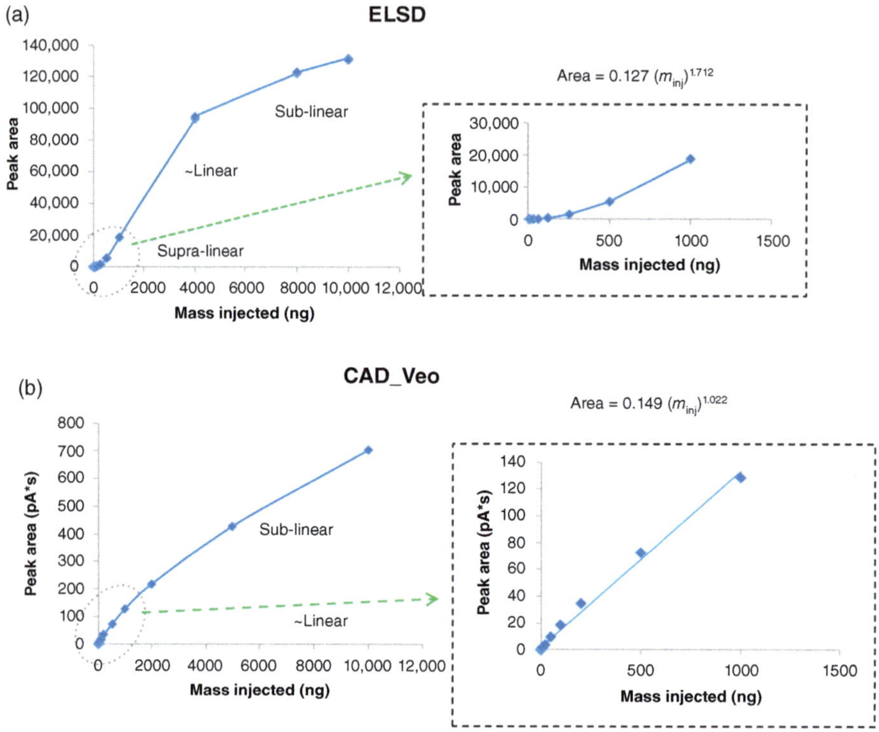

Figure 1.11 (a) ELSD peak area versus m_{inj} theophylline—0.008–10 μg (left); 0.008–1.0 μg (right). Mobile phase 20% CH_3OH; 1.0 mL/min; 20 × 4.0 mm, 3 μm C18 column. Signal units depend on attenuation setting and thus not specified. (b) VCAD peak area versus m_{inj} theophylline—0.002–10 μg (left); 0.002–1.0 μg (right). Conditions as in (a). Note: VCAD P signal output is $pA^{1.5}$.

The coefficient a indicates overall sensitivity and the variable exponent b describes curve shape. The response is linear when $b = 1.0$ and a is then the slope of A/m_{inj}. Nonlinear response is a deviation from a straight line either curving upward (*supralinear*, $b > 1$) or downward (*sublinear*, $b < 1$). The more b differs from 1.0, the greater the degree of curvature in either direction. ELSD response exhibited a characteristic sigmoidal shape. At the lower range (m_{inj} 0.008–1.0 μg), response was supralinear with significant curvature ($b = 1.712$) and transitioned to sublinear ($b = 0.31$) toward higher m_{inj} (8.0–10 μg). VCAD P response, with signal output $pA^{1.5}$, was also sublinear at the higher range but was approximately linear over nearly three orders of magnitude, from m_{inj} 0.002 to 1.0 μg (i.e., $b = 1.022$, coefficient of determination [$r^2 = 0.993$] from linear regression).

The variable exponent b for CAD response is derived from:

- Spray drying exponent ($\alpha = 1/3$), which is the same for CAD and ELSD

- Aerosol charging exponents (β), which depend on particle diameter (d), where $\beta_2 = 1.133$ for $d > {\sim}9$ nm and $\beta_1 \sim 2.25$ for $d < {\sim}9$ nm

Likewise, the variable exponent b for ELSD is derived from the same spray drying exponent ($\alpha = 1/3$) and β exponents from different light scattering mechanisms that depend on the relationship between d and incident light wavelength (λ) [33, 34]:

- Refraction and reflection: $\beta_3 = 2.0$ for $d/\lambda > 2.0$
- Mie: $\beta_2 = 4.0$ for ${\sim}2.0 > d/\lambda > 0.1$
- Rayleigh: $\beta_1 = 6.0$ for $d/\lambda < 0.1$

CAD and ELSD continuously measure polydisperse aerosols whose distributions shift to a larger size with increasing m_{inj} and toward a peak's apex. The signal results from an entire distribution of residue particles at each point, and peak area is the signal integrated over a distribution of analyte concentrations across the width of a peak. ELSD response thus typically involves a variable mixture of light scattering mechanisms, while CAD likewise involves measurement of a variable proportion of charged particles both above and below $d = 9$ nm. Theoretical limits for the overall exponents b can be approximated from the individual exponents α and β, where:

- VCAD P "intrinsic" response: $0.38 < b < 0.75^{*}$
- VCAD P with signal output $\text{pA}^{1.5}$ $0.57 < b < 1.13$
- ELSD: $0.67 < b < 2.0$

For both EADs, the smallest b is observed for the highest m_{inj}, where residue distributions contain predominantly larger particles (lower β). This explains the downward curving, or sublinear response curves, at the high end of the m_{inj} range of both detectors. The response of both techniques then transitions toward higher b toward the lowest m_{inj}, where the residue distributions now contain a greater proportion of small particles (higher β). The approximate particle sizes, d, where each technique transitions toward highest β, are:

- CAD: $d < {\sim}9$ nm
- ELSD: $d < {\sim}50$ nm (assuming $\lambda = 500$ nm, the approximate center of visible light)

The particle diameter transition point for ELSD is approximately fivefold higher (125-fold higher in volume) than CAD. This, together with the much higher β ($\beta_1 = 6$), underlies the commonly observed signal "drop," which is steeper (greater curvature) and occurs at higher m_{inj} levels than with CAD. This explains the better limits of detection (LODs) most often observed with CAD versus ELSD: here <1 ng for CAD versus ~7.8 ng for ELSD.

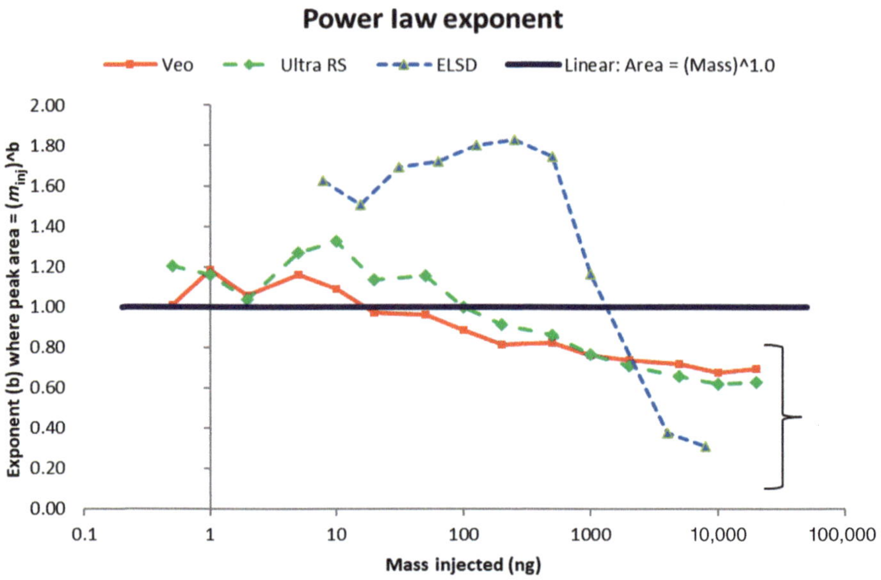

Figure 1.12 Power law exponent b versus m_{inj} obtained under identical conditions (see Figure 1.11) with UCAD, VCAD P, and an ELSD. Note: VCAD P signal output is $pA^{1.5}$ (see below).

A plot of b versus m_{inj} in Figure 1.12 more clearly shows the greater complexity of ELSD response with an approximately six-fold change in b (0.31–1.80) over a dynamic range of only ~10^3. By comparison, VCAD P shows a smaller ~1.75-fold change (0.68–1.18) and more gradual change in b over a much wider dynamic range of 10^5.

1.5.1.1 Signal Output—Power Function

The newer VCAD P design allows transport of larger droplets, which produce dried aerosol distributions with lower relative proportions of smaller particles ($d <$ ~9 nm) than first-generation UCAD models. This VCAD P design has the effect of reducing the intrinsic b range, simplifying the response curve, and shifting any signal drop to even lower mass levels. Since the intrinsic response of the VCAD P design remains sublinear throughout its dynamic range (see theoretical limits above), the instrument's digital signal processing algorithm applies a fixed exponent or internal *power function* (PF) of $pA^{1.5}$ to the instantaneous signal. Figure 1.12 plots b versus m_{inj} obtained under common test conditions with ELSD, UCAD, and VCAD P. The range for UCAD was 0.62–1.33 and for VCAD P with signal output of $pA^{1.5}$ was narrower from 0.68 to 1.18. Supralinear response was observed for all three detectors with approximate b_{max} of 1.8 for ELSD, 1.33

for UCAD, and 1.18 for VCAD P. This indicates a more gradual signal drop with VCAD P and a supralinear response that is evident at lower mass ranges than the other detectors.

The fixed exponent PF of 1.5 was chosen as an approximation to provide linear response for typical methods and quantitation ranges and to facilitate method transfer between first- and second-generation designs. The VCAD P design also provides a user-defined setting called the "Power Function Value" (PFV), which is a key parameter for optimizing and extending the linear response range for a given method. The currently released VCAD P simplified the settable power function and the internal power function. The VCAD P outputs pA^b, where b can be set from 1.50 to 2.40. Furthermore, the Chromeleon™ Chromatography Data System allows recalculation of data recorded with a power function $b1$ to another power function $b2$. An optimal setting can simplify calibration, improve peak shapes, and improve the accuracy of both quantitation and determination of detection limits. Recommendations for use of the PFV are detailed in Chapter 3.

1.5.1.2 Semivolatile Analytes

Figure 1.13 shows that caffeine has lower sensitivity (A/m_{inj}) and, in Figure 1.4, a higher b than theophylline. This is highly characteristic of an analyte that behaves as a semivolatile, where its evaporation influences response. Since evaporation rate R_e is inversely proportional to particle surface area, this effectively increases b, especially toward the lower end of the m_{inj} range. As a result, analytes that behave as semivolatiles usually produce sigmoidal response curves with lower and more variable sensitivity than analytes that behave as nonvolatiles. Figure 1.13 also shows that better detection limits were obtained with VCAD P for both analytes. This is attributed to the larger residue particles produced by the VCAD P design, which, as previously described, shifts any signal drop to lower mass levels for both analytes and slows the evaporation rate for the more volatile analyte, CAF.

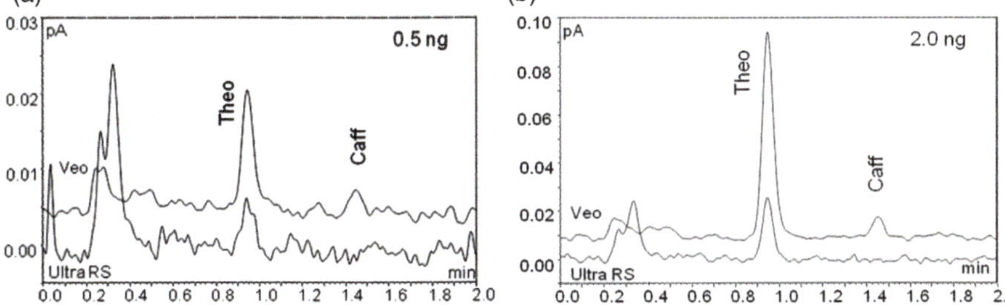

Figure 1.13 Comparison of Corona ultra RS and Corona Veo near the limit of detection for theophylline (Theo) and caffeine (Caff); m_{inj} = 0.5 ng each (a); m_{inj} = 2.0 ng (b).

1.5.2 Sensitivity Limits

Sensitivity limits (i.e., LOD and LOQ) can be determined by various procedures, including visual inspection, SNR, and relationship between the slope of a calibration curve and standard deviation of the response. Since CAD response may become supralinear toward these limits, accurate estimation of LOD and LOQ usually requires actual analysis of appropriately low analyte levels rather than estimation from higher analyte levels. The latter can lead to misleading results, especially when assuming linear response.

With any EAD, an important consideration is the concentration of nonvolatile and semivolatile impurities within an eluent. Figure 1.14 plots CAD baseline current and noise as a function of the v/v concentration of theophylline, which was intentionally added to the mobile phase. Here, the noise, measured in pA peak to peak, is observed to be a consistent fraction (~6%) of the baseline signal over the measured concentration range. This clearly indicates that noise can be reduced by minimizing the concentration of nonvolatile impurities in the eluent. Evaporation temperature (T_e) is an instrumental setting that can be used to optimize signal versus noise. For a given method, increasing T_e can often reduce the baseline current and noise, as long as the analytes of interest are significantly less volatile than the impurities in the mobile phase. In practice, the nature of impurities is usually unknown, and therefore experimentation is required to optimize performance.

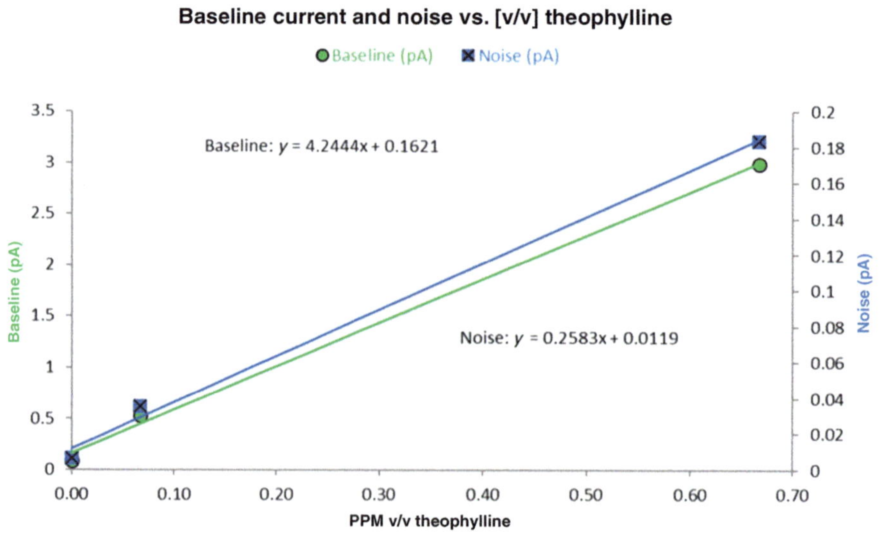

Figure 1.14 Baseline current and noise as a function of the v/v concentration of theophylline (THEO), intentionally added to the mobile phase. Conditions as in Figure 1.11.

The topics of optimization and recommended practices to minimize the concentration of impurities are further discussed in Chapter 3.

1.5.2.1 Eluent Impurities

The possible influence of impurities on detector response is often difficult to assess since the exact nature and concentration are usually unknown. Impurities commonly observed with LC–MS may be relevant also to EADs. This includes Na, Cl, K, Fe, phthalates, polyethylene glycol, polysiloxanes, and fluoropolymers arising from various sources, including solvents, additives, system components, and general labware [35]. Recommended practices to achieve a low and consistent background level are specifically discussed in Chapter 3. With good laboratory practices, impurities can usually be kept at concentrations low enough to produce particles, in the absence of analyte, in the low nm size range.

1.5.3 Peak Shape

As described, sublinear response artificially distorts the chromatographic peak by broadening of the instantaneous signal versus time profile compared with the "true" mass concentration versus time profile. Likewise, the supralinear response that is inherent with ELSD leads to what is commonly called "peak shaving" [36]. While this type of distortion is more pronounced with ELSD due to its higher exponential response, the broadening observed with CAD can adversely affect resolution and is another reason to optimize the PFV. Figure 1.9 shows the significant broadening predicted from an intrinsic (sublinear) response and the effects of $pA^{1.5}$ and $pA^{2.5}$ signal outputs. Figure 1.15 compares the peak widths obtained with serial UV and V CAD P (with $pA^{1.5}$) detectors. This shows a slight peak broadening with VCAD P that gradually increases over the 100 to ~5000 ng mass range. The significant broadening at the highest mass range is attributed to column and UV flow cell overload, as it is more pronounced with UV.

1.5.4 Mass Versus Concentration Sensitivity

Detectors can be classified as either *mass-flow sensitive* (response proportional to analyte mass per unit time) or *concentration sensitive* (response proportional to analyte concentration within the eluent at a particular time). For true mass-flow sensitivity, peak area is independent of flow rate, and an increase should result in temporal peak sharpening with increased height; for true concentration sensitivity, peak height is independent of flow rate, and thus an increase should result in lower peak area. The dependency of CAD response on flow rate is of particular interest given the relatively common use of post-column flow splitting (e.g., to MS) and post-column solvent addition. Gorecki et al. [12] first described "inverse gradient" compensation with CAD, which will be discussed in more detail in Chapter 4.

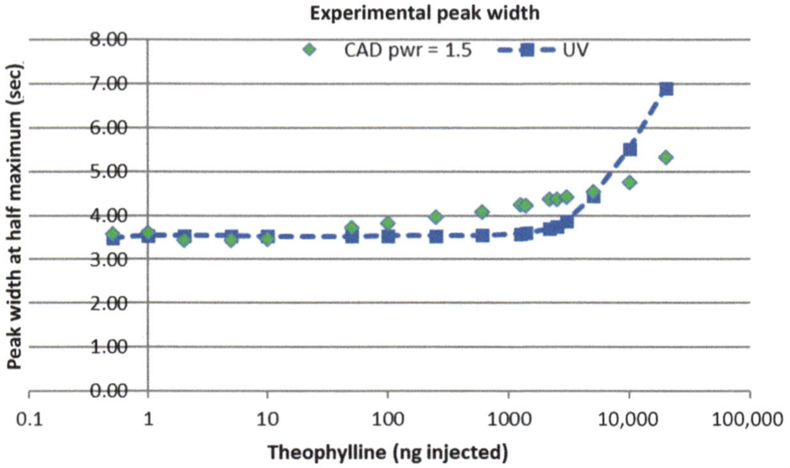

Figure 1.15 Peak width (FWHM) obtained with a UV detector in series with CAD (power function of $pA^{1.5}$) over a m_{inj} range of 0.5–10,000 ng theophylline, showing gradual broadening of CAD peak area due to nonlinear response. Broadening of both signals at highest m_{inj} is attributed to column overload. Conditions as in Figure 1.11.

They showed that CAD behaved approximately as a mass-flow-sensitive device, since peak area response obtained with 50% CH_3CN without compensation was very similar to that with compensation at twice the liquid flow. However, deviation from true mass-flow-sensitive behavior is expected, since only a fraction of the primary aerosol is typically transported to the electrometer. Most studies indicate that CAD and other EADs do not produce a true mass-flow-dependent response and, as with many other techniques, the extent of deviation from this behavior depends somewhat on analytical conditions [37].

The extent to which CAD behaves as a mass-flow-dependent device is also of interest when scaling chromatographic methods to larger or smaller internal diameter (i.d.) columns. A common approach is to approximately scale both flow rate and injection volume in proportion to column volume. Figure 1.16 shows an example of this, where the sample concentration was also adjusted so that m_{inj} was the same. Since, with a smaller i.d. column, analytes elute in a smaller eluent volume, their peak concentration is higher. The similar response obtained for equivalent m_{inj} of both analytes clearly shows that CAD behaves primarily as a mass-flow-dependent device. The slightly higher response for the lower flow condition, however, indicates some deviation from true mass-flow response. Since nebulizer gas flow was the same, this may be due to production of a finer primary aerosol (e.g., lower SMD by Equation 1.3) and higher mass transport in the lower flow condition. It is significant to note that lower baseline current and noise, and therefore higher SNR, was achieved with the lower flow condition. This may be attributed to a lower mass flow of nonvolatile impurity.

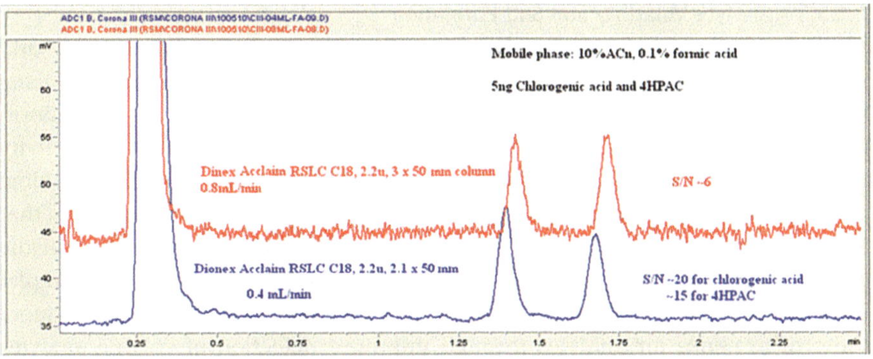

Figure 1.16 Chromatograms for the same m_{inj} (5 ng) of chlorogenic acid (CA) and 4-hydroxyphenylacetic acid (4HPAC), showing approximate mass-flow versus time-dependent response of CAD with lower noise and thus higher signal-to-noise ratio at the lower flow rate. Top trace: flow rate—0.8 mL/min, 3.0 mm i.d. × 50 mm L, 2.2 μm C18 column; bottom trace: flow rate—0.4 mL/min, 2.1 mm i.d. × 50 mm L, 2.2 μm C18 column.

1.5.5 Response Uniformity

Uniform response is defined here as sensitivity (e.g., peak area/m_{inj}) that is independent of analyte properties. The ability to obtain analyte-independent response provides the possibility to quantify multiple analytes with a single calibrator or calibration model, which is highly desirable for many analyses (e.g., drug impurities, natural products, lipids, polymers). The following variables are expected to significantly affect EAD response uniformity: (i) solvent gradients, (ii) analyte volatility and salt formation, (iii) analyte density, and (iv) dependency of aerosol detector response on particle material. Since the first three of these factors should similarly influence all EADs, this discussion will mainly focus on characteristics of the downstream aerosol measurement technique that may explain differences in response uniformity between these devices.

1.5.5.1 Solvent Gradient Effects

Gradient effects are due to changes in aerosol formation and transport, which are solvent dependent and analyte independent. All EADs share a necessity to reduce solvent load to ensure complete downstream solvent evaporation. While there are variations in design, the shared constraints suggest that the overall magnitude of solvent effects on response uniformity should be quite similar. Several studies have shown up to a 10-fold, and often nonlinear, change in analyte sensitivity as a function of aqueous–organic composition [8, 38]. Inverse gradient post-column addition, as described in Chapter 4, is often used to compensate for these changes.

1.5.5.2 Analyte Volatility and Salt Formation

The relationships between analyte volatility and both detection scope and response uniformity are expected to be similar for all EADs. In all cases, reducing the evaporation temperature (T_e) is generally expected to produce more uniform response and a broader detection scope since, under these conditions, selectivity is effectively reduced. A consequence of this, however, may be higher baseline current and noise due to more sensitive detection of semivolatile impurities that are likely to be present in most experiments. Also, as discussed below, acidic and/or basic mobile phase additives can promote salt formation during droplet evaporation. For all EADs, this can lead to a broader detection scope and more uniform response for analytes through improved detection of volatile analytes that are ionized in solution. On the other hand, salt formation can also lead to higher sensitivity for ionized nonvolatile analytes compared with those that are neutral. The effect of this on response uniformity can generally be minimized by using lower-MW mobile phase additives, since they should contribute less to overall particle mass [19, 20, 39].

1.5.5.3 Ionizable Solutes

Aerosol droplet evaporation involves a rapid increase in the concentration of less volatile components, thereby favoring solute–solute interactions. Several studies have shown that volatile ionic mobile phase additives can, during spray drying, combine with other ionic solutes (e.g., analytes, impurities, sample matrix components, and other oppositely charged additives) to form less-volatile salts. Volatile additives can be used to increase detection scope and influence sensitivity for ionizable analytes [18–20, 40, 41]. Salt formation can also be disadvantageous if leading to higher background signal and noise (e.g., volatile basic impurity + volatile acidic modifier), baseline drift, ghost peaks, and decreased response uniformity. While the choice of mobile phase additive is usually driven by requirements of the separation technique (i.e., pH modification, ion pairing), possible influences on detection should also be considered. Most studies have shown that the extent to which additives contribute to CAD signal largely depends on their molar mass and the extent of ionization of the interacting species [18, 19]. Practical implications of these interactions will be discussed in Chapters 3 and 4.

1.5.5.4 Analyte Density

Equation 1.4 and Table 1.3 indicate that solute density (ρ_s) influences dried particle diameter (d_p) and thus should have a similar, albeit minor, influence on the sensitivity of all EADs. Matsuyama et al. [21] showed an inverse relationship between analyte density and CAD response, which follows from Equation 1.4, since analytes with higher density produce smaller residue particles and thus lower response. A simple way to compensate for differences in analyte density is therefore to multiply the obtained response (peak area or peak height) for each analyte by the cube root of its density. For ionizable analytes, it may be preferable to use density of the predicted salt form (e.g., basic analyte + acidic mobile phase additive) rather than that of the free analyte to compensate for differences among analytes.

1.5.5.5 Dependence of Aerosol Measurement Technique on Residue Particle Material

The influence of solvent gradients, analyte volatility, salt formation, and analyte density on response uniformity is expected to be similar among all EADs. When comparing the response uniformity of these devices, a main differentiating factor may therefore be the relative dependency of the downstream aerosol detector's response on residue particle material (i.e., primarily determined by analyte properties). Aerosol charging by diffusion mechanisms is well known to have only a minor dependence on particle material. Accordingly, several studies have shown that, after accounting for other variables (e.g., solvent dependency, volatility), CAD response exhibits a minor dependence on analyte properties [5, 7, 12, 21, 38] and consistently provides more uniform response than ELSD [13, 17, 42–47]. ELSD makes use of photometers to measure the combined angular light scattering of all aerosol particles within the detection flow path. Angular light scattering intensity is a very complex function of the light source intensity, scattering angle of measurement, aerosol particle concentration, RI of the particle material, particle shape, and ratio of particle diameter to the wavelength of incident light (d_p/λ) [34, 48]. The dependence of ELSD response on analyte RI has been previously described [34, 49]. Oppenheimer et al. [48] estimated that ELSD response may vary by a factor of 2 or more for analytes encompassing an RI range of 1.4–1.6. The RI of light-absorbing analytes also depends on their absorption coefficient [4]. Analyte-specific absorption and Fl have been reported to influence ELSD response, where variations are more pronounced when using monochromatic (e.g., laser diode) compared with polychromatic light sources [8, 50].

The technique of CNLSD is similar to ELSD in which the condensation of liquid vapor onto the dried aerosol particles is used to amplify particle size into a region of higher scattering efficiency [51]. The response of CNLSD, assuming use of serial diffusion screens to suppress background and to modulate the dynamic range, is a product of particle size distribution, particle number concentration, diffusion screen transmission efficiency, and condensation nucleation detection efficiency curve [52]. CNLSD commonly provides better sensitivity than ELSD, but studies have shown that the signal significantly depends on analyte nature [13, 53]. Figure 1.17 compares CAD and CNLSD response for 24 chemically diverse nonvolatile species analyzed by flow injection analysis with a constant solvent. In the absence of a chromatographic separation, peak area

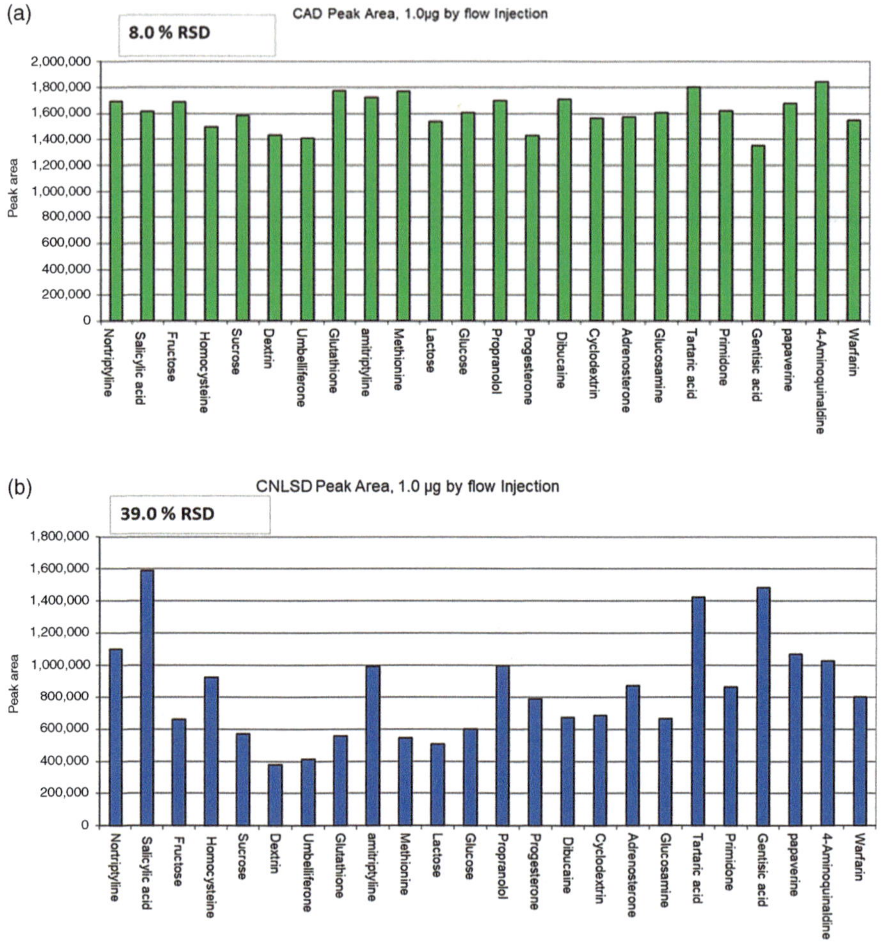

Figure 1.17 Peak area response obtained with CAD (a) and CNLSD (Quant, NQAD; b) from flow injection analysis of 1.0 μg of each analyte dissolved in mobile phase (50% aqueous CH_3OH, v/v). Flow rate = 1.0 mL/min. Evaporation temperature: CAD ambient; NQAD 35 °C.

measurements from both detectors were corrected to account for the counterion content of a few analytes that were obtained as salts (amitriptyline, dibucaine, nortriptyline, and propranolol). While no correction was made for differences in analyte density, this is expected to be a minor factor that applies similarly to both detection techniques. Variation in response (%RSD) among analytes was found to be ~8.0% for CAD and ~39.0% for CNLSD. The much larger variation observed with CNLSD is assumed to be mainly attributed to the downstream aerosol measurement technique. While the response of CNLSD should be less dependent on analyte RI, response variation may be attributed to differences in the wettability and solubility of the particle material for a given condensing fluid (e.g., water, *n*-butanol). This is supported by several studies of atmospheric aerosols where, for example, with water as the condensing fluid, significant differences in particle counting efficiency were found among organic aerosols, between water-soluble inorganic salts and water-insoluble aerosols, and between aerosols with and without trace impurities [54–56].

In summary, the more uniform response for nonvolatile analytes that is commonly observed with CAD compared with other detectors, including ELSD and CNLSD, is mainly attributed to the widely described minor dependence of diffusion charging on aerosol particle material. Primary factors that influence the response uniformity of all EAD techniques include solvent dependency, analyte density, and volatility. Also, to obtain accurate quantification of multiple analytes with a single calibrator or calibration model (see Chapter 4), additional factors should be considered, including detector linearity and precision over the mass range of interest; pre-analytical variables such as analyte purity, stability, and solubility; and analytical variables not specifically related to the detection technique, such as column recovery, chromatographic resolution, and peak shape.

References

1 Kaufman, S.L. and Inventor (2003). Evaporative electrical detector. Patent No: US 6,568,245.

2 Medved, A., Dorman, F., Kaufman, S., and Pöcher, A. (2000). A new corona-based charger for aerosol particles. *Journal of Aerosol Science* 31: 616–617.

3 Dixon, R.W. and Peterson, D.S. (2002). Development and testing of a detection method for liquid chromatography based on aerosol charging. *Analytical Chemistry* 74 (13): 2930–2937.

4 Hinds, W.C. (2012). *Aerosol Technology: Properties, Behavior, and Measurement of Airborne Particles*. John Wiley & Sons.

5 Gamache, P.H., McCarthy, R.S., Freeto, S.M. et al. (2005). HPLC analysis of nonvolatile analytes using charged aerosol detection. *LC GC North America* 23 (2): 150–161.

6 Allen, L.B. and Koropchak, J.A. (1993). Condensation nucleation light scattering: a new approach to development of high-sensitivity, universal detectors for separations. *Analytical Chemistry* 65 (6): 841–844.

7 Hutchinson, J.P., Li, J., Farrell, W. et al. (2010). Universal response model for a corona charged aerosol detector. *Journal of Chromatography A* 1217 (47): 7418–7427.

8 Mathews, B., Higginson, P., Lyons, R. et al. (2004). Improving quantitative measurements for the evaporative light scattering detector. *Chromatographia* 60 (11–12): 625–633.

9 Squibb, A.W., Taylor, M.R., Parnas, B.L. et al. (2008). Application of parallel gradient high performance liquid chromatography with ultra-violet, evaporative light scattering and electrospray mass spectrometric detection for the quantitative quality control of the compound file to support pharmaceutical discovery. *Journal of Chromatography A* 1189 (1): 101–108.

10 Gamache, P.H. and Kaufman, S.L. (2017). Principles of charged aerosol detection. In: *Charged Aerosol Detection for Liquid Chromatography and Related Separation Techniques* (ed. P.H. Gamache), 1–65. John Wiley and Sons.

11 Kulkarni, P., Baron, P.A., and Willeke, K. (2011). *Aerosol Measurement: Principles, Techniques, and Applications*. John Wiley & Sons.

12 Gorecki, T., Lynen, F., Szucs, R., and Sandra, P. (2006). Universal response in liquid chromatography using charged aerosol detection. *Analytical Chemistry* 78 (9): 3186–3192.

13 Hutchinson, J.P., Li, J., Farrell, W. et al. (2011). Comparison of the response of four aerosol detectors used with ultra high pressure liquid chromatography. *Journal of Chromatography. A* 1218 (12): 1646–1655.

14 Kahen, K., Acon, B.W., and Montaser, A. (2005). Modified Nukiyama–Tanasawa and Rizk–Lefebvre models to predict droplet size for microconcentric nebulizers with aqueous and organic solvents. *Journal of Analytical Atomic Spectrometry* 20 (7): 631–637.

15 Mugele, R. and Evans, H. (1951). Droplet size distribution in sprays. *Industrial and Engineering Chemistry* 43 (6): 1317–1324.

16 Rizk, N. and Lefebvre, A. (1984). Spray characteristics of plain-jet airblast atomizers. *Journal of Engineering for Gas Turbines and Power* 106 (3): 634–638.

17 Cohen, R.D. and Liu, Y. (2014). Advances in aerosol-based detectors. *Advances in Chromatography* 52: 1.

18 Cohen, R.D., Liu, Y., and Gong, X. (2012). Analysis of volatile bases by high performance liquid chromatography with aerosol-based detection. *Journal of Chromatography. A* 1229: 172–179.

19 Russell, J.J., Heaton, J.C., Underwood, T. et al. (2015). Performance of charged aerosol detection with hydrophilic interaction chromatography. *Journal of Chromatography. A* 1405: 72–84.

20 Sinclair, I. and Gallagher, R. (2008). Charged aerosol detection: factors for consideration in its use as a generic quantitative detector. *Chromatography Today* 1: 5–9.

21 Matsuyama, S., Orihara, Y., Kinugasa, S., and Ohtani, H. (2015). Effects of densities of brominated flame retardants on the detection response for HPLC analysis with a corona-charged aerosol detector. *Analytical Sciences* 31 (2): 61–65.

22 Fissan, H., Neumann, S., Trampe, A. et al. (2007). Rationale and principle of an instrument measuring lung deposited nanoparticle surface area. *Journal of Nanoparticle Research* 9 (1): 53–59.

23 Premnath, V., Oberreit, D., and Hogan, C.J. Jr., (2011). Collision-based ionization: bridging the gap between chemical ionization and aerosol particle diffusion charging. *Aerosol Science and Technology* 45 (6): 712–726.

24 Kolakowski, B.M., Grossert, J.S., and Ramaley, L. (2004). Studies on the positive-ion mass spectra from atmospheric pressure chemical ionization of gases and solvents used in liquid chromatography and direct liquid injection. *Journal of the American Society for Mass Spectrometry* 15 (3): 311–324.

25 Sabo, M. and Matejčík, S. (2012). Corona discharge ion mobility spectrometry with orthogonal acceleration time of flight mass spectrometry for monitoring of volatile organic compounds. *Analytical Chemistry* 84 (12): 5327–5334.

26 Sekimoto, K. and Takayama, M. (2009). Fundamental processes of corona discharge. *The Institute of Electrostatics Japan* 33 (1): 38–42.

27 Feng, S., Liu, H., Zhou, C. et al. (2025). Direct corona charging: a new strategy for enhancing sensitivity and stability in charged aerosol detection. *Analytical Chemistry* 97 (13): 7079–7088.

28 Jung, H. and Kittelson, D.B. (2005). Characterization of aerosol surface instruments in transition regime. *Aerosol Science and Technology* 39 (9): 902–911.

29 Kaufman, S., Medved, A., and Pöcher, A. (2002). An electrical aerosol detector based on the corona-jet charger. AAAR Conference (Poster).

30 Li, L., Chen, D.-R., and Tsai, P.-J. (2009). Evaluation of an electrical aerosol detector (EAD) for the aerosol integral parameter measurement. *Journal of Electrostatics* 67 (5): 765–773.

31 Woo, K.-S., Chen, D.-R., Pui, D.Y., and Wilson, W.E. (2001). Use of continuous measurements of integral aerosol parameters to estimate particle surface area. *Aerosol Science and Technology* 34 (1): 57–65.

32 Shin, W.G., Qi, C., Wang, J. et al. (2009). The effect of dielectric constant of materials on unipolar diffusion charging of nanoparticles. *Journal of Aerosol Science* 40 (5): 463–468.

33 Guiochon, G., Moysan, A., and Holley, C. (1988). Influence of various parameters on the response factors of the evaporative light scattering detector for a number of non-volatile compounds. *Journal of Liquid Chromatography* 11 (12): 2547–2570.

34 Mourey, T.H. and Oppenheimer, L.E. (1984). Principles of operation of an evaporative light-scattering detector for liquid chromatography. *Analytical Chemistry* 56 (13): 2427–2434.

35 Keller, B.O., Sui, J., Young, A.B., and Whittal, R.M. (2008). Interferences and contaminants encountered in modern mass spectrometry. *Analytica Chimica Acta* 627 (1): 71–81.

36 Zhang, T., Nguyen, D., and Franco, P. (2006). Use of evaporative light scattering detector in the detection and quantification of enantiomeric mixtures by HPLC. *Journal of Separation Science* 29 (10): 1517–1524.

37 Hazotte, A., Libong, D., Matoga, M., and Chaminade, P. (2007). Comparison of universal detectors for high-temperature micro liquid chromatography. *Journal of Chromatography. A* 1170 (1): 52–61.

38 Hutchinson, J.P., Remenyi, T., Nesterenko, P. et al. (2012). Investigation of polar organic solvents compatible with corona charged aerosol detection and their use for the determination of sugars by hydrophilic interaction liquid chromatography. *Analytica Chimica Acta* 750: 199–206.

39 Megoulas, N.C. and Koupparis, M.A. (2004). Enhancement of evaporative light scattering detection in high-performance liquid chromatographic determination of neomycin based on highly volatile mobile phase, high-molecular-mass ion-pairing reagents and controlled peak shape. *Journal of Chromatography. A* 1057 (1): 125–131.

40 Deschamps, F.S., Baillet, A., and Chaminade, P. (2002). Mechanism of response enhancement in evaporative light scattering detection with the addition of triethylamine and formic acid. *The Analyst* 127 (1): 35–41.

41 Lantz, M.D., Risley, D.S., and Peterson, J.A. (1997). Simultaneous resolution and detection of a drug substance, impurities, and counter ion using a mixed-mode HPLC column with evaporative light scattering detection. *Journal of Liquid Chromatography and Related Technologies* 20 (9): 1409–1422.

42 Kou, D., Manius, G., Zhan, S., and Chokshi, H.P. (2009). Size exclusion chromatography with corona charged aerosol detector for the analysis of polyethylene glycol polymer. *Journal of Chromatography. A* 1216 (28): 5424–5428.

43 Merle, C., Laugel, C., Chaminade, P., and Baillet-Guffroy, A. (2010). Quantitative study of the stratum corneum lipid classes by normal phase liquid chromatography: comparison between two universal detectors. *Journal of Liquid Chromatography and Related Technologies* 33 (5): 629–644.

44 Shaodong, J., Lee, W.J., Ee, J.W. et al. (2010). Comparison of ultraviolet detection, evaporative light scattering detection and charged aerosol detection methods for liquid-chromatographic determination of anti-diabetic drugs. *Journal of Pharmaceutical and Biomedical Analysis* 51 (4): 973–978.

45 Takahashi, K., Kinugasa, S., Senda, M. et al. (2008). Quantitative comparison of a corona-charged aerosol detector and an evaporative light-scattering detector for the analysis of a synthetic polymer by supercritical fluid chromatography. *Journal of Chromatography. A* 1193 (1): 151–155.

46 Vervoort, N., Daemen, D., and Török, G. (2008). Performance evaluation of evaporative light scattering detection and charged aerosol detection in reversed phase liquid chromatography. *Journal of Chromatography. A* 1189 (1): 92–100.

47 Wipf, P., Werner, S., Twining, L.A., and Kendall, C. (2007). HPLC determinations of enantiomeric ratios. *Chirality* 19 (1): 5–9.

48 Oppenheimer, L.E. and Mourey, T.H. (1985). Examination of the concentration response of evaporative light-scattering mass detectors. *Journal of Chromatography. A* 323 (2): 297–304.

49 Righezza, M. and Guiochon, G. (1988). Effects of the nature of the solvent and solutes on the response of a light-scattering detector. *Journal of Liquid Chromatography* 11 (9–10): 1967–2004.

50 Righezza, M. and Guiochon, G. (1988). Effect of the wavelength of the laser beam on the response of an evaporative light scattering detector. *Journal of Liquid Chromatography* 11 (13): 2709–2729.

51 Koropchak, J.A., Sadain, S., Yang, X. et al. (1999). Peer reviewed: nanoparticle detection technology for chemical analysis. *Analytical Chemistry* 71 (11): 386A–394A.

52 Magnusson, L.-E., Risley, D., and Koropchak, J. (2015). Aerosol-based detectors for liquid chromatography. *Journal of Chromatography. A* 1421: 68–81.

53 Koropchak, J., Heenan, C., and Allen, L. (1996). Direct comparison of evaporative light-scattering and condensation nucleation light-scattering detection for liquid chromatography. *Journal of Chromatography. A* 736 (1): 11–19.

54 Hering, S.V., Stolzenburg, M.R., Quant, F.R. et al. (2005). A laminar-flow, water-based condensation particle counter (WCPC). *Aerosol Science and Technology* 39 (7): 659–672.

55 Kupc, A., Bischof, O., Tritscher, T. et al. (2013). Laboratory characterization of a new nano-water-based CPC 3788 and performance comparison to an ultrafine butanol-based CPC 3776. *Aerosol Science and Technology* 47 (2): 183–191.

56 Liu, W., Kaufman, S.L., Osmondson, B.L. et al. (2006). Water-based condensation particle counters for environmental monitoring of ultrafine particles. *Journal of the Air & Waste Management Association* 56 (4): 444–455.

2

Charged Aerosol Detection—Application Overview

Ian N. Acworth[1] *and Paul H. Gamache*[2]

[1] *Consultant, Melrose, MA, USA*
[2] *Consultant, Nashua, NH, USA*

CHAPTER MENU

2.1 Summary

This chapter provides a selected overview of publications using charged aerosol detection (CAD) since mid-2016. Readers interested in reviewing articles from 2005 through mid-2016, are referred to the chapter by Acworth and Kopaciewicz [1]. Since mid-2016, more than 1300 articles have been published. For convenience, instead of reviewing articles by research areas [1] and to better harmonize with the different chapters in this book, publications are now grouped by analyte class and include sections on carbohydrates, drugs, excipients, lipids, and phytochemicals.

2.2 Introduction

The most common detection technique for HPLC is UV/vis absorbance, due to its sensitivity, wide linear dynamic range and gradient compatibility. However, this approach is somewhat selective and is only applicable if the analyte possesses a chromophore. For those that do not possess a chromophore, other "more universal" detectors are used including refractive index (RI), mass spectrometry (MS), evaporative light scattering, and charged aerosol. All these approaches have some limitations. For example, RI is easy to use but lacks sensitivity and is not gradient compatible, so it is limited to less complex samples and analytes with similar physicochemical properties. MS, depending on the type being used, can be difficult to operate, requires that the compound be capable of forming gas-phase ions and, without the availability of labeled standards, may not be quantitative. The evaporative light scattering detection (ELSD), is one of the most widely used universal detectors, but suffers from several limitations including its nonlinear behavior, relative insensitivity, narrow linear and dynamic ranges, and significant inter-analyte response variability [2, 3]. CAD is also nonlinear but overcomes many of the limitations of ELSD and related detection technologies (e.g., condensation nucleation light scattering detection (CNLSD) [4]). A comparison of performance characteristics for CAD, ELSD, CNLSD and RI is presented in Table 2.1. Table 2.2 reviews some publications directly comparing CAD, ELSD, and CNLSD performance. These and other studies comparing CAD and ELSD are further described throughout this book (e.g., Chapters 5 and 9).

2.3 CAD History and Background

CAD involves the nebulization of the liquid stream coming from the column to form droplets, which are then desolvated to produce dried particles. Particles are charged and the charge is measured using an electrometer. The amount of charge is proportional to the amount of analyte present (see Chapter 1 for greater detail of nebulization, particle formation, particle charging and charge measurement). CAD uses an electrical charging method based on TSI Inc.'s (St. Paul, MN) electrical aerosol detector (EAD) [5]. The charged aerosol detector was co-developed by ESA Inc. (Chelmsford, MA) and TSI Inc. and was introduced to the scientific community at the Pittsburgh Conference in 2005. It was awarded the Silver Pittcon Editor's award for best new product and, later in 2005, was selected for a prestigious R&D 100 award. Continued research and engineering improvements in product design led to the evolution of a range of charged aerosol detectors with ever increasing capabilities (Figure 2.1).

Figure 2.1 shows the evolution of commercial instrument models first introduced by ESA Inc. and, after corporate acquisitions, further developed as part of Dionex Corporation and now Thermo Fisher Scientific, Inc. The first four models use the same internal flow path design and can be described as a first generation (referred to collectively as UCAD in Chapter 1). Models 5 through 7 represent a second generation design, which is referred to collectively as VCAD in Chapters 1 and 3. The downstream electrical aerosol measurement technique is

Table 2.1 Comparison of CAD, ELSD, CNLSD, and RI performance characteristics.

	CAD	ELSD	CNLSD	RI
Analyte properties	Nonvolatile and many semi volatiles	Nonvolatile and many semi volatiles	Nonvolatile and many semi volatiles	No analyte restrictions
Response curve	Curvilinear response over a wide dynamic range	Sigmoidal	Curvilinear response over a wide dynamic range	Linear
Dynamic range (orders of magnitude)	Up to 4	2–3	2–3	2–5
Linear range (orders of magnitude)	1.5–2	<1	1.5–2	1–2
Response uniformity	Non-volatiles give similar response	Analyte dependent (e.g., particle refractive index, absorption can affect response)	Significant wide variation in response between analytes (depends upon differences in wettability and solubility of the analyte particle for a given condensing fluid (e.g., water, *n*-butanol))	Analyte dependent
LOD	High pg-low ng	Low to mid ng	High pg-low ng	High ng
Gradient compatibility	Yes[a]	Yes[a],[b]	Yes[a]	No
Ease of use	Simple	Can be complex	Can be complex	Simple

a) Gradient elution can affect analyte response. This can be minimized by using an inverse gradient (see Chapters 4 and 13 and Grosse et al. [6]).

b) An alternative approach for some ELSD models uses a nebulization gas flow gradient in an attempt to maintain a constant response. However, this approach appears to be somewhat limited and of questionable benefit.

Table 2.2 Comparison of detector performance reported in literature.

Application	Conclusions	References
Carbohydrates	CAD was more sensitive than ELSD, but ELSD was slightly better for precision, repeatability, and reproducibility	[7]
Column ranking	System for analysis of any combination of 26 sugar and sugar alcohols CAD showed superior performance to ELSD	[8]
Detector comparison	CAD showed superior performance to ELSD including higher sensitivity, wider linear dynamic range, better reproducibility, better response uniformity, and was reported to be easier to use	[3]
Detector comparison	Response uniformity for CAD was better than ELSD and CNLSD. Under aqueous conditions CAD and CNLSD sensitivity were comparable	[9]
Drug analysis	Using reversed-phase conditions CAD was much more sensitive than ELSD. Using HILIC conditions the ELSD was comparable to CAD in terms of linearity, sensitivity, precision, and accuracy	[10]
Drug analysis	CAD showed the most consistent response uniformity, best accuracy, and LOD compared to UV and ELSD	[11]
Drug analysis	CAD showed the lower LODs compared to ESLD and MS, and a comparable dynamic range to MS	[12]
Drug impurities	CAD was 3.6–42 times more sensitive than ELSD. The authors concluded that ELSD was not an alternative to the CAD for the intended purpose of the method	[13]
Drug impurities	The sensitivity of CAD was about 3–9 times better than ELSD	[14]
Drug impurities	CAD was up to six-fold more sensitive than ELSD	[15]
Drug library screening	ELSD could not be used for this study due to sigmoidal calibration curves. CAD showed potential for generalized application	[16]
Drug quality control	The CAD was ~10–37 times more sensitive than ELSD	[17]
Drug quality control	CAD showed better reproducibility, more uniform response, and was two to six-fold more sensitive than ELSD	[18]
Excipients	Analysis of cationic and anionic surfactants. CAD was approximately 1.4–2 times more sensitive than ELSD. CAD showed better linearity than ELSD	[19]

Table 2.2 (Continued)

Application	Conclusions	References
Excipients	CAD was found to provide more accurate impurity and polydispersity profiles of multiple lots of polyethylene glycol (PEG) to better determine their quality. RI lacked sensitivity and ELSD led to underestimating both impurity levels and polydispersity	[20]
Excipients	CAD was 10 times more sensitive than ELSD and showed a wider dynamic range	[21]
Excipients	Molar mass data by ELSD was 4.6% lower than the certified value of PEG 1000. Molar mass distribution by CAD was virtually the same as the certified value without any calibration	[22]
Excipients	CAD was superior to ELSD with respect to sensitivity, precision, recovery, and linearity	[23]
Excipients	CAD was more sensitive than ELSD and was able to quantify impurities below 0.05% (m/m)	[24]
Food	CAD LOQs were typically five times better than ELSD. CAD showed superior %RSD both at low and high levels, and better R2 values (both logarithmic and linearity fit)	[25]
Lipids	CAD offered better sensitivity, repeatability, precision, and accuracy than ELSD	[26]
Lipids	CAD was more sensitive and more precise than ELSD at the lower end of the calibration curve. CAD LODs and LOQs were typically three-fold lower than ELSD	[27]
Method development	CAD LOD was 2.5 lower than CNLSD	[28]
Method development	The nonlinear nature of ELSD caused issues when measuring enantiometric ratios. No issues were found when using CAD	[29]
Phyto-chemicals	Saikosaponins. CAD had a wider linear dynamic range, higher sensitivity, and better reproducibility than ELSD. The high temperature of the drift tube required by ELSD is an issue for thermally labile analytes	[30]
Phyto-chemicals	Escin. CAD and UV were complementary with the CAD measuring analytes with weak chromophores, and DAD providing peak purity data	[31]
Phyto-chemicals	CAD was found to be 30–55 times more sensitive than ELSD, enabling the measurement of low levels of reference compound impurities that could not be detected by ELSD	[32]
Phyto-chemicals	CAD was more sensitive than ELSD. The intra-day and inter-day precision by CAD was better than ELSD and closer to UV	[33]

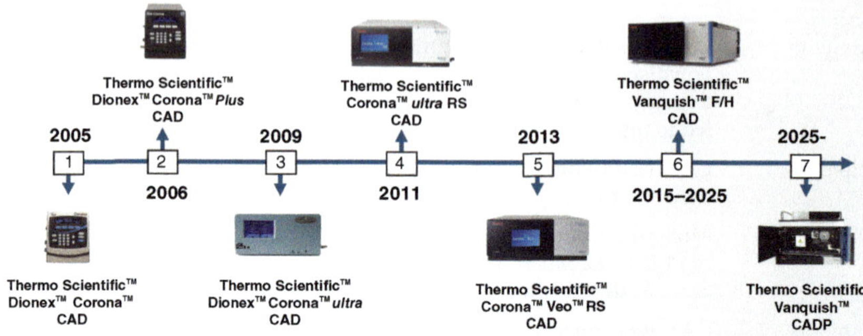

Figure 2.1 Evolution of CAD instruments.

effectively the same in all instruments while the upstream spray-drying design differs significantly between the two generations. The VCAD instruments all use the same concentric nebulizer/spray chamber design while the UCAD instruments all use the same cross-flow nebulizer/impactor design [34]. A recent paper describes a new, commercially available, instrument (SparkFlux2000-CAD, Lingfeng Instrument Co., Ltd., China). The most significant difference from the above designs is that the corona discharge is located directly within the analyte aerosol flow path, i.e., at the outlet of the evaporation tube [35]. This is described as a "direct corona charging" technique where the authors report high sensitivity and stability, although performance was mainly evaluated with relatively simple eluents and no mobile phase additives. Further evaluation (Chapter 1) suggests that, since the corona discharge needle maintains direct physical interface with aerosol residue particles, accumulation of contaminant species may result in nonuniform charging fields and compromised measurement reproducibility. It will be interesting to see additional, more thorough, evaluations of this device, which include its longer-term use in a wider range of applications.

Since all applications described within this book use either a UCAD or VCAD instrument, practical differences between the two main designs and the most relevant differences between specific models are summarized below.

2.3.1 Liquid Flow Range

The VCAD instrument design enables operation at lower liquid flow rates (0.01–2.0 mL/min) compared to the UCAD instruments (0.2–2.0 mL/min).

This facilitates its use for microbore LC separations and with post-column flow splitting configurations (Chapter 13). The concentric nebulizer also generates a more stable spray for eluents with low viscosity and surface tension [36], which provides a more stable, less noisy, chromatographic baseline for a variety of applications (e.g., normal phase, nonaqueous reversed phase, SFC).

2.3.2 Temperature Control

While some UCAD models included nebulizer temperature (T_n) control, this was mainly intended to prevent freezing of water due to rapid evaporative cooling that occurred with the cross-flow nebulizer for tetrahydrofuran (THF)–water solvent gradients. The temperature of the UCAD instrument evaporation tube is not controlled while that of the VCAD instrument design allows evaporation temperature (T_e) settings from ambient +5 to 100 °C. This provides a more robust control variable for method optimization than T_n, which facilitates use with a wider range of eluent compositions and flow rates and can be a useful variable to optimize "selectivity" between analytes and semi volatile eluent impurities. The newest Thermo Scientific™ Vanquish CAD™ P model allows T_e and the downstream detection process temperature to be controlled separately, which is particularly beneficial when using low-temperature evaporation for analysis of semi volatile analytes. Also, when using a high T_e, it is recommended to use the option to couple the two temperatures in order to prevent the creation of a decreasing-temperature gradient. This can result in unwanted downstream recondensation of aerosol vapor.

2.3.3 Power Function Value

The spray-drying technique of the VCAD design produces a lower relative proportion of small (<10 nm diameter) residue particles than first generation CADs. This has the effect of simplifying response curves, extending the quasi-linear and overall dynamic response range, and increasing sensitivity to low concentrations of both nonvolatile and semi volatile analytes (Chapter 1). As detailed in Chapter 3, the intrinsic response of VCAD for analytes that behave as nonvolatiles remains sublinear throughout its dynamic range. To provide approximately linear response for typical methods and quantitation ranges, the instrument's digital signal processing algorithm applies a fixed exponent or internal *Power Function* (PF) to the instantaneous signal. For many methods, this internal PF provides sufficiently linear response to allow accurate quantitation with linear regression using default conditions; the signal output can then be further optimized with a user-defined *Power Function Value* (PFV). Chapter 3 describes several options to determine an optimal PFV that requires only experimentation at the default

PFV setting of 1 and can be used with any chromatographic system. For any application, it is highly recommended to use an optimal PFV, as it can extend the linear range and enable more straightforward quantitation while often improving accuracy, signal to noise ratios, precision, and peak shapes. The newest VCAD P model uses an updated internal PV (a modified PFV) that was chosen, based on user input, to provide an optimal signal output "out of the box" for a greater majority of methods than previous models. It is hoped that this and enhanced tools to more efficiently determine a method's optimal PFV will lead to its more widespread use as a data acquisition parameter.

2.3.4 The Use of CAD with Different Chromatographic Techniques

CAD can be used with a variety of different HPLC/UHPLC approaches (including reversed-phase, normal-phase, HILIC, mixed-mode, chiral, size-exclusion, and 2D-LC) as well as supercritical-fluid chromatography. The newer VCAD devices further facilitate use with microbore chromatography and with flow-splitting configurations even for high organic-containing eluents. CAD is very often coupled with UV-diode array detection and with other complementary detectors, especially MS and is used, with fraction collection, in semipreparative or analytical-scale preparative approaches. Practical considerations for use of CAD in multi-detector configurations are detailed in Chapter 13, and many examples of its use with various separation techniques are described throughout this book.

A prerequisite for CAD, ELSD, and CNLSD, like MS, is that the mobile phase must be volatile. A wide variety of solvents can be used including but not limited to water, acetonitrile, methanol, isopropanol, acetone, tetrahydrofuran, heptane, and chloroform. With CAD, CNLSD, and ELSD, the best performance is usually obtained when the concentration of less volatile impurities in the eluent is minimized by using ultra-high purity water, solvents with low evaporation residue, columns and conditions that produce minimal bleed, LC systems and operational practices that minimize contamination, and low concentrations of low molar mass additives. (See Chapter 3 and Muellner et al. [37] for details.)

The response of nebulizer-based detectors can be affected by the organic content of the mobile phase entering the detector during gradient operation, a result of changes in nebulization efficiency due to changes in surface tension and viscosity of the eluent. While CAD response depends on particle size, independent of particle composition, solvent gradients affect the particle size distribution. This results in nonuniform response between analytes that elute in different parts of the gradient. For applications requiring analyte quantitation in the absence of individual standards, inverse gradient compensation can be used to normalize response in gradient methods. This topic is detailed in Chapters 4 and 13.

2.4 CAD Publication Overview

The annual number of CAD publications is presented in Figure 2.2. The marked increase from 2021 onward may be partly due to the impact of COVID and the use of CAD in COVID-19 vaccine development and production. Additional factors contributing to the observed growth of CAD publications may include: an increased understanding of its capabilities, especially compared to ELSD, and best practices for its use; adoption of standard methods (e.g., in pharmacopeias, Chapters 8 and 10); advances in column technologies and separation techniques that use volatile eluents compatible with both CAD and MS; hot topic areas such as polysorbate analysis in therapeutic protein formulations (Chapter 12); characterization and quality control of traditional medicines (Chapter 9); and development of non-COVID-related lipid-based drug delivery systems (Chapter 11).

When examining CAD publications by research area, the Pharmaceutical/Biopharmaceutical industries are the most prolific producing ~30% of the total number of publications (Figure 2.3) covering all aspects of drug discovery, development, and production (Swartz et al. [38]), and involving measurement of a broad range of analyte classes. Many papers have also been published for food and beverages (13%) and natural products and traditional medicine (16%) applications. Most food and beverage applications involve the analysis of carbohydrates (Table 2.3, Chapter 7) or lipids (Table 2.6, Chapter 5) while those

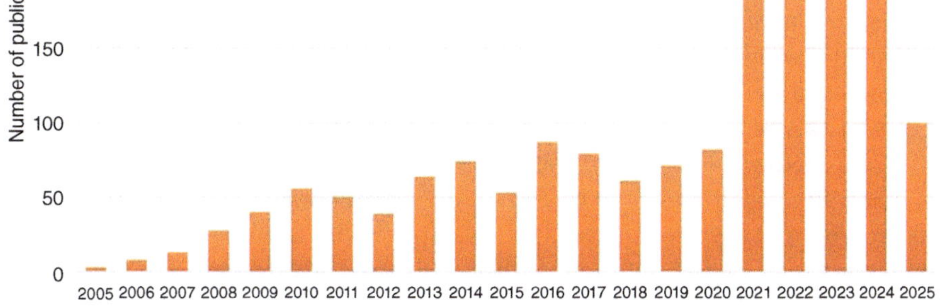

Figure 2.2 Publications by year (data obtained from Google Scholar and PubMed). Results for 2025 are though July.

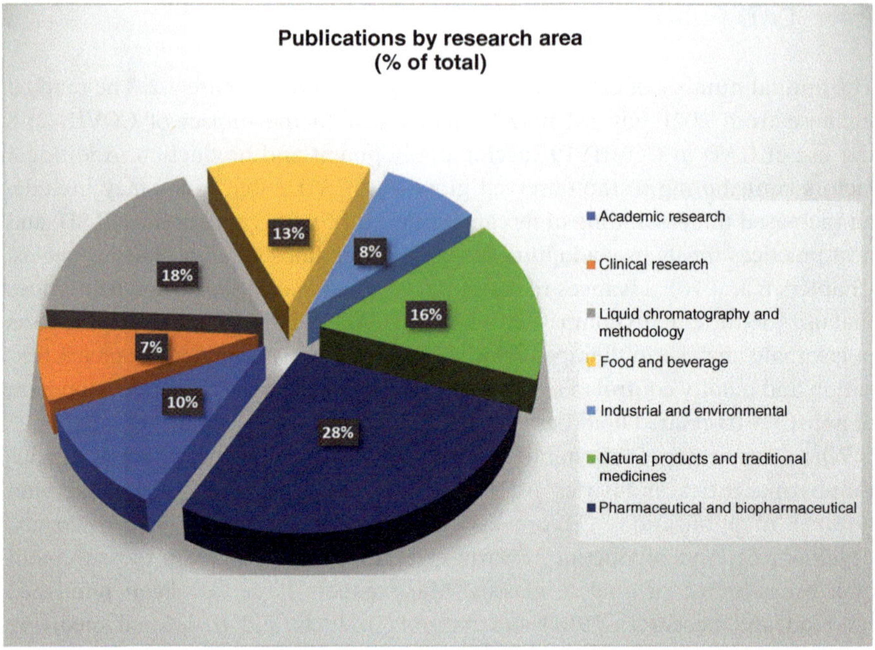

Figure 2.3 Publications by research area.

of natural products and traditional medicines encompass a very broad range of analyte classes (Table 2.7, Chapter 9).

While the CAD instrument is often described as the detector of choice for analysis of weak chromophores, its use in drug discovery and in natural products and herbal medicines demonstrates its advantages for quantitation of strong chromophores alike. Gradient compatibility, dynamic range, response uniformity, precision, and sensitivity enable quantitation of chemically diverse and unknown species that are often encountered in drug libraries and natural products. Chapter 14 highlights its particular usefulness in high-throughput microgram-scale library compound purification and describes best practices for optimization, calibration, routine maintenance and performance tracking. Likewise, several papers describe the use of the CAD instrument in generic or platform methods. These approaches are applicable to a wide range of analytes and samples, significantly increasing efficiency for various processes, especially in the pharmaceutical industry. One example is the analysis of 25 commonly used pharmaceutical counterions with a single validated method (Chapter 6) that requires minimal sample preparation for a diverse range of drug substances and sample matrices. The method is described as having the required accuracy to control stoichiometry and the sensitivity for purity control and to establish mass balance for use in pharmaceutical development and manufacturing. Further examples of comprehensive methods include analysis of extractables and leachables from packaging materials and safety assessments of consumer products (Chapter 13); profiling of complex excipients including surfactants and polymers (Chapters 8 and 11); and several methods for carbohydrate and glycoconjugate based therapeutics (Section 2.5.1 and Chapter 7). Many publications describe CAD as providing a more accurate and complete picture of what is in a sample than other techniques. One example is its use as part of a comprehensive characterization scheme for new chemical entities (NCEs), where it is necessary to find and quantify every component and arrive at mass balance and true purity of a drug substance (Chapter 6). Here, an NCE was analyzed using DAD, RI, MS, and CAD using various orthogonal separation methods. Out of all these analyses, two new unknown peaks were found only by a mixed-mode LC–CAD method. Chapters 8 and 10 stress the importance of using a comprehensive approach for impurity analysis as part of quality control (QC) to ensure drug product safety. Examples are described where CAD is shown to uncover and enable quantitation of unexpected impurities that go unnoticed with other methods. This includes the detection of fatty acids in pharmaceutical excipients, which were not previously found by existing compendial methods. For several other QC methods, the combination of UV and CAD allows for more robust approach to routine drug product testing.

The universal nature of CAD can also be a limitation since its selectivity is only based on volatility. In most cases, analytes must be separated chromatographically, which often makes CAD less suitable than other techniques (e.g., LC–MS/MS) for trace quantitation in complex matrices. This may explain its relatively low use in clinical research where biological matrices (e.g., plasma, tissue) are very common. It should be noted, however, that CAD is used quite extensively for lipid analysis in these matrices. The high pg low ng sensitivity of CAD is also a factor in this regard where more sensitive and selective techniques like MS or fluorescence detection (FLD), after derivatization, may be more suitable. CAD is often described as having: a relatively narrow (~2 orders of magnitude) quasi-linear response range, a response that varies with eluent composition; and a lack of analyte specific information such as with diode array or MS. CAD also requires the use of volatile mobile phases and is limited to detection of analytes that behave as nonvolatiles or semi volatiles with the latter having poorer sensitivity. With the newer CAD technology, use of an optimal PFV and optimal T_e along with IGC for gradient methods can address some of these limitations. Also, the combination of CAD with UV-DAD (a very common configuration) and with MS can be used for the detection of more volatile species. For most analyses with CAD, it is recommended to use the lowest T_e setting that consistently produces the required sensitivity limits and where the ideal goal is for all analytes to behave as nonvolatiles. This generally provides more universal detection, more uniform response, and enables more accurate quantitation. The ability to use a low T_e in order to realize these benefits mainly depends on the cleanliness of the eluent (Chapter 3). The development of new column technologies for use with volatile mobile phases along with more advanced separation techniques such as 2D-LC, appear to be primary reasons for the greatly expanded use of CAD in recent years. Stationary phases for HILIC, mixed mode columns and porous graphitic columns along with a greater understanding of the variables that affect their performance are among the key areas of development. This has expanded the applications of both CAD and MS. Significant opportunities remain, however, to further develop columns with low bleed and to understand the variables that affect not only chromatographic selectivity but MS and CAD detector performance (Chapter 3).

2.5 CAD Publications by Analyte Classes

It is outside the purview of this chapter to provide a comprehensive review of every CAD publication. Rather, this section will concentrate on the main classes of analytes based on the number of publications (Figure 2.4). These classes include carbohydrates, drugs, excipients, lipids, and phytochemicals.

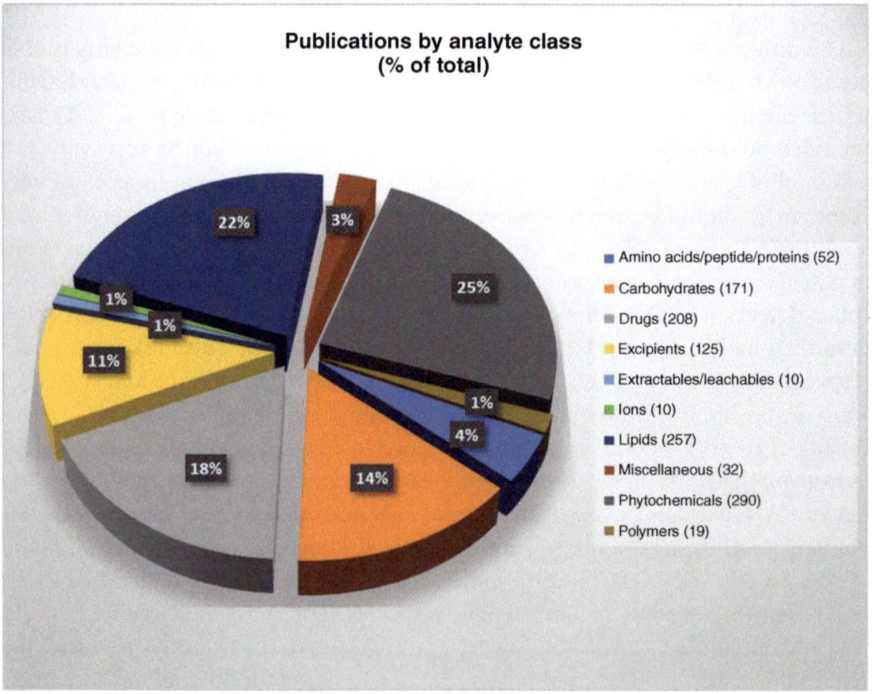

Figure 2.4 Publications by analyte class. The number of publications for each is given in parenthesis.

2.5.1 Carbohydrates

Carbohydrate analysis is now one of the main application areas of CAD (Figure 2.4) and its use includes many fields of study (Table 2.3). Representative methods are described in Chapter 7 where Table 7.1 summarizes the separation conditions, key detector settings and reported limits of quantitation (LOQs). A majority of early applications involve the adoption of CAD in methods that traditionally employed refractive index detection (RID) and ELSD. This includes its use to monitor enzyme reactions, for measurement of carbohydrate excipients and impurities in pharmaceuticals and for quantitation in beverages, herbal medicines, and plants. More recently, there has been an increase in the number of papers that describe the use of CAD for analyses that have historically employed pre-column derivatization for HPLC with fluorescence detection (HPLC-FLD) and/or direct analysis by high performance anion exchange chromatography with pulsed amperometric detection (HPAEC-PAD). While HPLC-FLD provides very high sensitivity, the derivatization process is often tedious, time-consuming and variable. HPAEC-PAD allows direct, label free, analysis of carbohydrates, and provides high sensitivity but requires a dedicated system with limited flexibility in the choice of eluents and detectors compared to HPLC. Examples of the use of HPLC-CAD for direct label free analyses include quantification of: N-linked

Table 2.3 Carbohydrates.

Area	Examples
Biological/clinical research	• Digestion [39] • Disease/biomarkers [40] • Enzymology [41, 42] • Human milk oligosaccharides [43–46]
Food	• Adulteration/authentication [47] • Contamination [48] • Functional foods [49] • Maillard [50] • Safety [51–53] • Sweeteners/artificial sweeteners [54–56]
Industrial (including agriculture; environmental; oils/fuels)	• Biofuels [57, 58] • Biomass [59] • Bioreactors [60] • Polymers [61, 62] • Plant Growth [63, 64]
Methodology	• Analysis [65–70] • Column evaluation [71] • Multidetector platforms/2D systems [72]
Pharma/biopharma	• Glycans [73–75] • Biomimetics [76]
Phytochemicals	• Herbal medicines/Traditional medicines [77–80]

glycan moieties in therapeutic monoclonal antibodies (mAbs) [73]; total sialic acids in mAbs [75]; polysaccharide glycoconjugate antigens for quality evaluation of vaccines used against bacterial infections [66]; precursors in the biosynthesis of therapeutic heparin [81]; compositional analysis of diverse monosaccharides derived from carbohydrate-based therapeutics [62]. These and other applications have been consistently described as accurate, faster, simpler, more comprehensive, and more versatile. A key advantage of these and other CAD methods over HPAEC-PAD is its greater compatibility with MS, which is often required at some point for component identification or peak confirmation. Since HPLC-CAD is not as sensitive as HPAEC-PAD or HPLC-FLD, some of the methods were adapted to analyze higher sample amounts. While this was not found to be an issue in these studies, it may be a limitation for other applications. Key considerations with HPLC-CAD or HPLC–MS for direct analysis of carbohydrates is column selection and choice of mobile phase additives. In general, higher LOQs and poorer precision are observed for methods that utilize columns or conditions that produce column bleed. This includes many silica-based columns used for HILIC or normal phase separations and the use of higher pH eluents where dissolution of the silica support is a well-described cause of stationary phase instability. Higher LOQs are also commonly observed when higher molar mass additives are used, especially when an oppositely charged additive is also present within the column eluent. The topics of column bleed, choice of additive are discussed in general within Chapter 3 and in relation to specific analyte classes in Chapter 5 (Lipids), Chapter 6 (Ions), and Chapter 7 (Carbohydrates).

2.5.2 Drugs

HPLC combined with CAD and UV, and sometimes MS, has now become a standard configuration within the pharmaceutical industry [1], but especially during the earlier stages of drug discovery and development (Chapter 14). Here CAD plays a primary role in the quantitation of drug substances and impurities regardless of whether the drug, library compound, or newly synthesized material possesses a strong chromophore. This is mainly due to the lack of individual standards and the requirement for comprehensive characterization. Since many drugs have a sufficiently strong chromophore, UV typically becomes the primary detector for upstream applications, where suitable calibration standards are usually available for the active pharmaceutical ingredient and most impurities. In these cases, CAD is often used as a complementary detector to find and quantify impurities and contaminants, which may not have a chromophore or well-characterized standards. For drug substances without a strong chromophore, CAD is a primary tool in drug development and in drug product quality control. Examples of this are detailed in Chapter 10 and Table 2.4, which

Table 2.4 Drugs.

Area	Examples
Biological/clinical research	Anticonvulsants [82] Disease [83] Drug absorption [84] Enzymology [85] Toxicology [86]
Food	Antibiotic contamination [87]
Industrial	Soil contamination [88]
Pharmaceuticals/ biopharmaceuticals	Drug discovery and development [38, 89–91] Drug measurement [92–94] Impurity testing [95–97]

also includes drug analysis in biological matrices, food, and the environment. CAD is also a primary tool for the analysis of drug substances that are obtained from natural sources (e.g., paclitaxel) or biosynthetic processes (amino acid derivatives, antibiotics, and heparin) since their impurity profiles are usually more complex and diverse.

2.5.3 Excipients

CAD is very widely used for analysis of pharmaceutical excipients, many of which do not have strong chromophores and are complex mixtures often derived from natural materials. This includes polymers and surfactants, carbohydrates (Chapter 7), lipids (Chapter 5), and various other components. These materials can differ in composition and impurity content from different manufacturers, between lots from the same manufacturer, and their stability may depend on the specific drug formulation. It is therefore important to characterize and monitor their composition from raw material to final product. Polymer analysis typically includes quantitation, determination of average molar mass, measurement of impurities including aggregation products and polydispersity profiling, and utilizing SEC, RPLC, and SFC. Earlier methods using CAD for analysis of polyethylene glycols (PEGs), PEG reagents, PEGylated molecules, and surfactants were described by Kou et al. [98] in the first edition of this book. More recent methods are described in Chapter 8. The analysis and characterization of polysorbate, which is present as a surfactant in most therapeutic protein formulations and the analysis of lipid-based pharmaceutical formulations (e.g., emulsions, liposomes, lipid nanoparticles) are two of the most significant application areas for CAD. The latter

Table 2.5 Excipients.

Area	Examples
Biological/clinical research	• Biosurfactants [99] • Metabolic engineering [100]
Food	• Emulsifiers [101]
Industrial (*including agriculture; environmental; oils/fuels*)	• Detergents/soaps [98, 102] • Environmental contamination [103, 104]
Methodology	• Analysis [105–113]
Pharma/biopharma	• Drug delivery/formulations [114–125]

is discussed in detail in a dedicated chapter (Chapter 11), while the analysis of polysorbate is detailed in Chapter 12. Polymers and surfactants are also important components of food, soaps, and cosmetics and are commonly present in the environment and many industrial processes. Selected examples are listed in Table 2.5. Chapter 15 discusses the analysis of complex industrial polymers. This includes the use of solvents and separation techniques that are less frequently used in other fields of study. This includes gradient polymer elution chromatography and liquid chromatography at critical conditions along with the use of solvents such as hexafluoroisopropanol and trichlorobenzene. For these applications, the combined use of CAD and MS is a common configuration, which also may include NMR. It should be noted that the examples described in Chapter 15 only use first-generation UCAD instruments where nonlinear response may lead to inaccurate measurement of polymer distributions. With the VCAD instruments, use of an optimal PFV may provide significant benefits in this regard.

2.5.4 Lipids

The use of CAD for quantitative analysis of lipids is now quite common across many fields of study and includes its coupling with both LC and SFC (See Table 2.6). Chapter 5 provides an overview of methods for the study of lipids primarily in biological (animal, plant, microbial) sources. This includes the quantification of individual lipid molecules, within-class molecular species, and diverse lipid classes. Representative methods and their reported limits of quantitation (LOQs) and precision are summarized in Chapter 5, Table 5.1. This shows the use of a variety of separation modes including HILIC, normal phase, and reversed phase and a broad range of mainly organic eluents. Most separations involve gradient elution with complex multi-step ternary and quaternary gradients often used for lipid-class separations. Reversed-phase separations

Table 2.6 Lipids.

Area	Examples
Biological/clinical research	• Health/Disease [126, 127] • Intestinal absorption [128] • Lipotoxicity [129] • Metabolic changes [130] • Mitochondrial activity [131–133]
Food	• Authentication/adulteration [134–140] • Contamination [141] • Lipidomics [142] • Storage/stability [143, 144] • Supplements [145, 146]
Industrial	• Biofuels [147, 148] • Pesticides [149]
Methodology	• Analysis [150–158] • Detector comparison [159] • 2D/Multidetector platforms [160]
Pharma/biopharma	• Drug delivery (e.g., adjuvants; cationic lipids; lipid nanoparticles; liposomes) [161–173]
Phytochemicals	• Herbal medicines/Traditional medicines [174–179]

using low molar mass additives and simple (e.g., binary) gradient elution techniques tend to provide the best precision and sensitivity. This includes the analysis of fatty acids (FAs) and triacylglycerols (TAGs) with typically reported LOQs of <5 ng, the main exceptions being the shorter chain (<C16) FAs. These analytes exhibit significant volatility resulting from the detector's spray-drying process, where the effects are most prominent for low analyte levels. Interestingly, C18 FAs and, in some cases, monoacylglycerols are also shown to exhibit some level of volatility. This highlights the importance of optimizing the evaporation temperature (T_e), while making sure to evaluate this parameter for analyte levels near the targeted LOQ. Many of the described studies involve quantitation over a wide mass-concentration range. Especially in these cases, it is important to optimize the PFV-instrument setting, which directly optimizes the signal output with respect to linearity. Many lipid analyses also involve quantitation of species, either chromatographically separated or eluting as a group, for which specific calibration standards are unavailable. Quantitative accuracy may therefore depend on factors that affect response uniformity (see Chapter 4). This includes consideration of T_e, choice of mobile phase additives, solvent gradient effects, and use of an optimal PFV. Chapter 11 discusses the use of CAD for analyses related to lipid-based pharmaceutical formulations (e.g., emulsions, liposomes, lipid nanoparticles).

2.5.5 Phytochemicals

Phytochemicals are commonly defined as non-nutrient substances naturally occurring in plants which, when consumed (e.g., as a supplement or traditional medicine), may have potential health benefits but may also have adverse effects. Phytochemicals represent an extremely broad and chemically diverse group of compounds (e.g., alkaloids, carbohydrates, flavonoid glycosides, ginsenosides, lipids, polyphenols, saponins, and many other compounds) which are often present in very complex matrices with many unknown components and at widely varying abundance. Many phytochemicals lack a chromophore making them ideal candidates for measurement by CAD—see Table 2.7 (and Acworth and Kopaciewicz [1]). Chapter 9 provides a comprehensive review of CAD applications for traditional herbal medicines. This includes various chromatographic approaches, multidetector configurations, analytical-scale preparative techniques with fraction collection, and the use of IGC to normalize CAD response.

Table 2.7 Phytochemicals.

Area	Examples
Biological/clinical research	• Bioactivity [180–182] • Disease [183] • Enzymology [184] • Physiology [185]
Food	• Dietary supplements [186] • Functional Foods [187] • Nutraceuticals [188] • Processing [189] • Quality control [190] • Safety [191] • Sweeteners [192]
Industrial (including agriculture; environmental)	• Metal recovery [193] • Methanogenesis [194, 195] • Pest control [196, 197] • Salinity impact on fruit quality [198]
Methodology	• Analysis [175, 199–202] • Column evaluation [203] • Metabolomics [204] • Multidetector platforms [205, 206]
Pharma/biopharma	• Drug activity [207] • Impurity testing [208]
Natural products/traditional medicines	• Adulteration/authentication/fingerprinting [209–211] • Composition [212, 213] • Disease treatment [214] • Pharmacopoeia [215] • Quality [216, 217]

References

1 Acworth, I.N. and Kopaciewicz, W. (2017, Chapter 2). Charged aerosol detection: a literature review. In: *Charged Aerosol Detection for Liquid Chromatography and Related Techniques* (ed. P. Gamache), 67–162. Wiley and Sons, Inc.

2 Cohen, R.D. and Liu, Y. (2014). Advances in aerosol-based detectors. *Advances in Chromatography* 52: 1–53.

3 Thomas, D., Bailey, B., Plante, M., and Acworth, I.N. (2014). Charged aerosol detection and evaporative light scattering detection – fundamental differences affecting analytical performance. Thermo Scientific PN70990.

4 Allen, L.B. and Koropchak, J.A. (1993). Condensation nucleation light scattering: a new approach to development of high-sensitivity, universal detectors for separations. *Analytical Chemistry* 65: 841–844.

5 Kaufman, S.L. (2003). Evaporative Electrical Detector. US Patent # US 6,568,245 B2.

6 Grosse, S., Muellner, T., Lovejoy, K. et al. (2024). Why use charged aerosol detection with inverse gradient? Thermo Scientific Technical Note 73449.

7 Márquez-Sillero, I., Cárdenas, S., and Valcárcel, M. (2013). Comparison of two evaporative universal detectors for the determination of sugars in food samples by liquid chromatography. *Microchememical Journal* 110: 629–635.

8 Hetrick, E.M., Kramer, T.K., and Risley, D.S. (2017). Evaluation of a hydrophilic interaction liquid chromatography design space for sugars and sugar alcohols. *Journal of Chromatography A* 1489: 65–74.

9 Hutchinson, J.P., Li, J., Farrell, W. et al. (2011). Comparison of the response of four aerosol detectors used with ultra-high pressure liquid chromatography. *Journal Chromatography A* 1218: 1646–1655.

10 Jia, S., Park, J.H., Lee, J., and Kwon, S.W. (2011). Comparison of two aerosol-based detectors for the analysis of gabapentin in pharmaceutical formulations by hydrophilic interaction chromatography. *Talanta* 85: 2301–2306.

11 Shaodong, J., Lee, W.J., Ee, J.W. et al. (2010). Comparison of ultraviolet detection, evaporative light scattering detection and charged aerosol detection methods for liquid-chromatographic determination of anti-diabetic drugs. *Journal Pharmaceutical Biomedical Analysis* 51: 973–978.

12 Pistorino, M. and Pfeifer, B.A. (2008). Polyketide analysis using mass spectrometry, evaporative light scattering, and charged aerosol detector systems. *Analytical Bioanalytical Chemistry* 390: 1189–1193.

13 Holzgrabe, U., Nap, C.J., and Almeling, S. (2010). Control of impurities in L-aspartic acid and L-alanine by high-performance liquid chromatography coupled with a corona charged aerosol detector. *Journal Chromatography A* 1217: 294–301.

14 Ilko, D., Neugebauer, R.C., Brossard, S. et al. (2017, Chapter 11). Impurity control in topiramate with high performance liquid chromatography. Validation and comparison of the performance of evaporative light scattering detection and charged aerosol detection. In: *Charged Aerosol Detection for Liquid Chromatography and Related Techniques* (ed. P. Gamache), 379–392. Wiley and Sons, Inc.

15 Vervoort, N., Daemen, D., and Török, G. (2008). Performance evaluation of evaporative light scattering detection and charged aerosol detection in reversed phase liquid chromatography. *Journal Chromatography A* 1189: 92–100.

16 Loughlin, J., Phan, H., Wan, M. et al. (2007). Evaluation of charged aerosol detection (CAD) as a complementary technique for high-throughput LC-MS-UV-ELSD analysis of drug discovery screening libraries. *American Laboratory* 39: 24–27.

17 Holzgrabe, U., Nap, C.-J., Beyer, T., and Almeling, S. (2020). Alternatives to amino acid analysis for the purity control of pharmaceutical grade L-alanine. *Journal of Separation Sciences* 33: 2402–2410.

18 Schilling, K. and Holzgrabe, U. (2020). Recent applications of the charged aerosol detector for liquid chromatography in drug quality control. *Journal of Chromatography A* 1619: 460911.

19 Kim, B.H., Jang, J.B., and Moon, D.C. (2013). Analysis of ionic surfactants by HPLC with evaporative light scattering detection and charged aerosol detection. *Journal Liquid Chromatography and Related Technologies* 36: 1000–1012.

20 Kou, D., Manius, G., Zhan, S., and Chokshi, H.P. (2009). Size exclusion chromatography with corona charged aerosol detector for the analysis of polyethylene glycol polymer. *Journal of Chromatography A* 1216: 5424–5428.

21 Lobback, C., Backensfeld, T., Funke, A., and Weitschies, W. (2007). Quantitative determination of nonionic surfactants with CAD. *Chromatography Techniques* 11: 18–20.

22 Takahashi, K., Kinugasa, S., Senda, M. et al. (2008). Quantitative comparison of a corona-charged aerosol detector and an evaporative light-scattering detector for the analysis of a synthetic polymer by supercritical fluid chromatography. *Journal Chromatography A* 1193: 151–155.

23 Nair, L.M. and Werling, J.O. (2009). Aerosol based detectors for the investigation of phospholipid hydrolysis in a pharmaceutical suspension formulation. *Journal Pharmaceutical Biomedical Analysis* 49: 95–99.

24 Sperber, B., Gutmann, M., Kenrein, J. et al. (2025). Impurity profiling of PEGylated myristoyl diglyceride, DMG-PEG 2000, a functional excipient used in mRNA lipid nanoparticle formulations. *European Journal of Pharmaceutics and Biopharmaceutics* 214: 114762.

25 Wang, H., Song, S., Shao, M. et al. (2019). Determination of bisphenol analogues in food-contact plastics using diode array detector, charged aerosol detector and evaporative light-scattering detector. *Ecotoxicology and Environmental Safety* 186: 109778.

26 Merle, C., Laugel, C., Chaminade, P., and Baillet-Guffroy, A. (2010). Quantitative study of the stratum corneum lipid classes by normal phase liquid chromatography: comparison between two universal detectors. *Journal Liquid Chromatography* 33: 629–644.

27 Ramos, R.G., Libong, D., Rakotomanga, M. et al. (2008). Comparison between charged aerosol detection and light scattering detection for the analysis of *Leishmania* membrane phospholipids. *Journal Chromatography A* 1209: 88–94.

28 Cohen, R.D., Liu, Y., and Gong, X. (2012). Analysis of volatile bases by high performance liquid chromatography with aerosol-based detection. *Journal Chromatography A* 1229: 172–179.

29 Wipf, P., Werner, S., Twining, L.A., and Kendall, C. (2007). HPLC determinations of enantiomeric ratios. *Chirality* 19: 5–9.

30 Eom, H.Y., Park, S.Y., Kim, M.K. et al. (2010). Comparison between evaporative light scattering detection and charged aerosol detection for the analysis of saikosaponins. *Journal Chromatography A* 1217: 4347–4354.

31 Filip, K., Grynkiewicz, G., Gruza, M. et al. (2014). Comparison of ultraviolet detection and charged aerosol detection methods for liquid-chromatographic determination of protoescigenin. *Acta Poloniae Pharmaceutica* 71: 933–938.

32 Long, Z., Guo, Z., Acworth, I.N. et al. (2016). A non-derivative method for the quantitative analysis of isosteroidal alkaloids from Fritillaria by high performance liquid chromatography combined with charged aerosol detection. *Talanta* 151: 239–244.

33 Wang, L., He, W.S., Yan, H.X. et al. (2009). Performance evaluation of charged aerosol and evaporative light scattering detection for the determination of ginsenosides by LC. *Chromatographia* 70: 603–608.

34 Gamache, P.H. and Kaufman, S.L. (2017, Chapter 1). Principles of charged aerosol detection. In: *Charged Aerosol Detection for Liquid Chromatography and Related Techniques* (ed. P. Gamache), 3–65. Wiley and Sons, Inc.

35 Feng, S., Liu, H., Zhou, C. et al. (2025). Direct corona charging: a new strategy for enhancing sensitivity and stability in charged aerosol detection. *Analytical Chemistry* 97: 7079–7088.

36 Bailey, B., Plante, M., Thomas, D. et al. (2017, Chapter 3). Practical use of CAD: achieving optimal performance. In: *Charged Aerosol Detection for Liquid Chromatography and Related Techniques* (ed. P. Gamache), 163–189. Wiley and Sons, Inc.

37 Muellner, T., Acworth, I., and Gamache, P. (2021). Getting the most out of your charged aerosol detector: factors influencing charged aerosol detector performance Thermo Scientific Technical Note 73914.

38 Swartz, M., Emanuele, M., and Awad, A. (2017, Chapter 10). Charged aerosol detection in pharmaceutical analysis. In: *Charged Aerosol Detection for Liquid Chromatography and Related Techniques* (ed. P. Gamache), 357–377. Wiley and Sons, Inc.

39 Song, Q., Zou, J., Cheng, S. et al. (2025). Insights into metabolic signatures and regulatory effect of Dendrobium officinale polysaccharides in gut microbiota: a comparative study of healthy and diabetic status. *Food Science and Nutrition* 13: e4651.

40 Makino, E., Klodnitsky, H., Leonard, J. et al. (2018). Fast, sensitive method for trisaccharide biomarker detection in mucopolysaccharidosis type 1. *Scientific Reports* 8: 3681.

41 Cerutti, G., Gugole, E., Montemiglio, L.C. et al. (2021). Crystal structure and functional characterization of an oligosaccharide dehydrogenase from *Pycnoporus cinnabarinus* provides insights into fungal breakdown of lignocellulose. *Biotechnology and Biofuels* 14: 161.

42 Westereng, B., Arntzen, M.O., Ostby, H. et al. Analyzing activities of lytic polysaccharide monooxygenases by liquid chromatography and mass spectrometry. In: *Carbohydrate-Protein Interactions. Methods in Molecular Biology*, vol. 2657 (ed. D.W. Abbott and W.F. Zandberg), 27–51. New York, NY: Humana.

43 Hansen, M.E., Sakanaka, M., Jensen, M. et al. (2025). Uptake of fucosylated type I human milk oligosaccharide blocks by *Bifidobacterium longum subsp. Infantis. mBio* 16: e00368–e00325.

44 Kosmerl, E., Gonzalez-Orozco, B.D., Garcia-Cano, I. et al. (2023). Milk phospholipids protect *Bifidobacterium longum subsp. infantis* during in vitro digestion and enhance polysaccharide production. *Frontiers in Nutrition* 10: 1194945.

45 Lloyd, J.E. (2023). Human milk oligosaccharides: bioactive sugars to exploit the bacterial sweet tooth. PhD Thesis. The University of Manchester (United Kingdom). 30730210.

46 Machida, S., Saito, K., Nishimoto, M., and Kitaoka, M. (2022). Production of lacto-N-biose I using crude extracts of bifidobacterial cells. *Journal of Applied Glycoscience* 69: 15–21.

47 Perini, M., Nardin, T., Camin, F. et al. (2018). Combination of sugar and stable isotopes analyses to detect the use of nongrape sugars in balsamic vinegar must. *Journal of Mass Spectrometry* 53: 772–780.

48 Skinner, M.M., Fong, M.A., Rimkus, T.P. et al. (2025). Pulsed electric field treatment of sweet potatoes to reduce oil and acrylamide in kettle chips. *Foods* 14: 577.

49 Sato, K., Nagai, N., Yamamoto, T. et al. (2019). Identification of a functional oligosaccharide in maple syrup as a potential alternative saccharide for diabetes mellitus patients. *International Journal of Molecular Science* 20: 5041.

50 Lee, J., Roux, S., Le Roux, E. et al. (2022). Unravelling caramelization and Maillard reactions in glucose and glucose + leucine model cakes: formation and degradation kinetics of precursors, α-dicarbonyl intermediates and furanic compounds during baking. *Food Chemistry* 376: 131917.

51 EFSA Panel on Nutrition, Novel Foods and Food Allergens (2023). Safety of 3-fucosyllactose (3-FL) produced by a derivative strain of Escherichia coli K-12 DH1 as a novel food pursuant to Regulation (EU) 2015/2283. *EFSA Journal* 21: 8026.

52 EFSA Panel on Nutrition, Novel Foods and Food Allergens (2023). Safety of lacto-N-fucopentaose I/2′-fucosyllactose (LNFP-I/2′-FL) mixture as a novel food pursuant to Regulation (EU) 2015/2283. *EFSA Journal* 21: 8026.

53 Pitsch, J., Hoglinger, O., and Weghuber, J. (2020). Roasted rye as a coffee substitute: methods for reducing acrylamide. *Foods* 9: 925.

54 Cheng, S., Wang, S., Zheng, M. et al. (2024). Simultaneous analysis of natural and artificial sweeteners in sugar-free drinks and urine samples by column-switching UHPLC-charged aerosol detection method. *Journal of Chromatography A* 1713: 464533.

55 Leng, Z.-S., Guo, H.-M., Cai, H.-Q., and Zhang, J. (2023). Research on simultaneous determination of eight artificial sweeteners in food by high performance liquid chromatography with a charged aerosol detector. *Chinese Journal of Applied Chemistry* 40: 436–440.

56 Ma, K., Li, X., Zhang, Y., and Liu, F. (2020). Determining high-intensity sweeteners in white spirits using an ultrahigh performance liquid chromatograph with a photo-diode array detector and charged aerosol detector. *Molecules* 25: 40.

57 Ortiz-Tena, J.G., Rühmann, B., Schieder, D., and Sieber, V. (2016). Determination of carbohydrates using liquid chromatography with charged aerosol detection. *Algal Research* 17: 227–235.

58 Tahar, I.B., Fickers, P., Gorka, A. et al. (2021). Higher thermal and ethanol tolerance of a yeast strain isolated from oral cavity. *Indian Journal of Experimental Biology* 59: 798–803.

59 Sakuragi, K. and Otaka, M. (2021). Milling characteristics of coal and torrefied biomass blends in a roller mill. *ACS Omega* 6: 29814–29819.

60 Karunasinghe, G., Qian, D., Meier, M.S. et al. (2024). Development of an effective Au : Pd bimetallic heterogeneous catalyst for oxidative lignin depolymerization to low molecular weight aromatics. *Chemistry Select* 9: e202401532.

61 Egan, J., Wang, S., Shen, J. et al. (2023). Enzymatic textile fiber separation for sustainable waste processing. *Resources, Environment and Sustainability* 13: 100118.

62 Ghosh, R. and Kline, P. (2019). HPLC with charged aerosol detector (CAD) as a quality control platform for analysis of carbohydrate polymers. *BMC Research Notes* 12: 268.

63 Malovrh, K., Ravnjak, B., Bavcon, J., and Krizman, M. (2024). Nectar production and three main sugars in nectar of Salvia pratensis and *Salvia glutinosa* in correlation with abiotic factors. *Agriculture* 14: 668.

64 Mouliom-Ntapnze, A.M., Fangue-Yapseu, G.Y., and Missihoun, T.D. (2025). Developmental stages of bell pepper influence the response to far-red light supplements in a controlled environment. *Agronomy* 15: 732.

65 Asthana, C., Peterson, G.M., Shastri, M.D., and Patel, R.P. (2020). A novel and sensitive HILIC-CAD method for glucosamine quantification in plasma and its application to a human pharmacokinetic study. *Journal of Pharmaceutical and Biomedical Analysis* 178: 112954.

66 Corrado, A., De Martino, M., Bordoni, V. et al. (2023). A universal UHPLC-CAD platform for the quantification of polysaccharide antigens. *Scientific Reports* 13: 10646.

67 Molnar-Gabor, D., Lengyel, M., and Krongaard, T. (2024). Rapid method for quantitation of seven human milk oligosaccharides in infant formula and premix. *Carbohydrate Research* 541: 109149.

68 Grembecka, M., Lebiedzinska, A., and Szefer, P. (2024). Application of HPLC coupled with a charged aerosol detector to the evaluation of fructose, glucose, sucrose, and inositol levels in fruit juices, energy drinks, sports drinks, and soft drinks. *Beverages* 10: 94.

69 Miyoshi, M., Kimura, I., Inazu, T., and Izumori, K. (2019). High sensitivity analysis and food processing stability of rare sugars. *Food Science and Technology Research* 25: 891–901.

70 Pitsch, J. and Weghuber, J. (2019). Hydrophilic interaction chromatography coupled with charged aerosol detection for simultaneous quantitation of carbohydrates, polyols and ions in food and beverages. *Molecules* 24: 4333.

71 Fu, X., Cebo, M., Ikegami, T., and Lämmerhofer, M. (2020). Separation of carbohydrate isomers and anomers on poly-N-(1H-tetrazole-5-yl)-methacrylamide-bonded stationary phase by hydrophilic interaction chromatography as well as determination of anomer interconversion energy barriers. *Journal of Chromatography A* 1620: 460981.

72 Wang, H., Jin, H., Chai, R. et al. (2024). An analysis of polysaccharides from eight plants by a novel heart-cutting two-dimensional liquid chromatography method. *Foods* 13: 1173.

73 Knihtila, R., Chemmalil, L., Sawant, P. et al. (2022). Development of a novel, label-free N-glycan method using charged aerosol detection. *Journal of Chromatography B* 1212: 123502.

74 Boßmann, D., Bartling, B., de Vries, I. et al. (2019). Charged aerosol detector HPLC as a characterization and quantification application of biopharmaceutically relevant polysialicacid from *E. coli* K1. *Journal of Chromatography A* 1599: 85–94.

75 Wang, Y., Xu, Y., Peters, B.-H. et al. (2025). Development and validation of a label-free HPLC-CAD method to determine total sialic acid in therapeutic proteins. *Journal of Pharmaceutical and Biomedical Analysis* 263: 116937.

76 Lenhart, D., Tischhofer, M.-T., Gruber, R. et al. (2023). Chemical and biological differences between original and mimetic pentosan polysulfates. *Carbohydrate Polymers* 319: 121201.

77 Chen, Q.-L., Chen, Y.-J., Zhoua, S.-S. et al. (2018). Laser microdissection hyphenated with high performance gel permeation chromatography-charged aerosol detector and ultra performance liquid chromatography-triple quadrupole mass spectrometry for histochemical analysis of polysaccharides in herbal medicine: ginseng, a case study. *International Journal of Biological Macromolecules* 107: 332–342.

78 Hu, D., Han, B., Chen, C. et al. (2020). Determination of seven oligosaccharides and sucrose in *Pseudostellaria heterophylla* by pressurized liquid extraction and ultra-high performance liquid chromatography with charged aerosol detector and tandem mass spectrometry. *Journal of Chromatography A* 1609: 460441.

79 Lan, J., Wu, G., Wu, L. et al. (2023). Development of a quantitative chromatographic fingerprint analysis method for sugar components of xiaochaihu capsules based on quality by design concept. *Separations* 10: 13.

80 Li, L.-F., Zhang, Q.-W., and Han, Q.-B. (2022). Recent advances in qualitative and quantitative analysis of polysaccharides in natural medicines: a critical review. *Journal of Pharmaceutical and Biomedical Analysis* 220: 115016.

81 Ji, X., Hu, G., Zhang, Q. et al. (2016). Simultaneous analysis of heparosan oligosaccharides by isocratic liquid chromatography with charged aerosol detection/mass spectrometry. *Carbohydrate Polymers* 152: 337–342.

82 Pańczyk, K., Żelaszczyk, D., Koczurkiewicz, P. et al. (2018). Synthesis and anticonvulsant activity of phenoxyacetyl derivatives of amines, including aminoalkanols and amino acids. *Medicinal Chemistry Communications* 9: 1933–1948.

83 Kavouris, J.A., McCall, L.-I., Giardini, M.A. et al. (2023). Discovery of pyrazolopyrrolidinones as potent, broad-spectrum inhibitors of *Leishmania* infection. *Frontiers in Tropical Diseases* 3: 1011124.

84 Torres-Teran, I., Venczel, M., Stieler, T. et al. (2023). Prediction of subcutaneous drug absorption - characterization of subcutaneous interstitial fluids as a basis for developing biorelevant in vitro models. *International Journal of Pharmaceutics* 638: 122906.

85 Robert-Hazotte, A., Faure, P., Neiers, F. et al. (2019). Nasal mucus glutathione transferase activity and impact on olfactory perception and neonatal behavior. *Scientific Reports* 9: 3104.

86 Kriikku, P., Pelkonen, S., Kaukonen, M., and Ojanpera, I. (2023). Propranolol and metoprolol: two comparable drugs with very different post-mortem toxicological profiles. *Journal of Analytical Toxicology* 47: 615–622.

87 Kiszkiel-Taudul, I., Starczewska, B., and Wierzbowska, M. (2023). Development of chromatographic techniques connected with corona and tandem mass spectrometry detection systems for determination of amoxicillin in bovine milk. *Food Control* 144: 109342.

88 Jayashree, R., Prasher, S.O., and Dinesh, G.K. (2024). Column study to assess the mobility of salinomycin in soil environment. *International Journal of Environment and Climate Change* 14: 773–783.

89 Barnhart, W.W., Handlovic, T.T., Lanman, B.A., and Haidar Ahmad, I.A. (2025). Use of charged aerosol detection in drug discovery chemistry. Chapter 14. This Book. In:

90 Goyon, A., Wang, S., Hofmann, K. et al. (2024). Unified and versatile multiplex platform for expedited method development and comprehensive characterization of therapeutic peptides. *Analytical Chemistry* 96: 14011–14019.

91 Hrovat, F., Zimmer, M., Fu, M. et al. (2025). Multi-attribute characterization of pharmaceutical materials by HPLC coupled with charged aerosol detection. *Journal of Chromatography A* 1763: 466424.

92 Hemida, M., Barrientos, R.C., Singh, A.N. et al. (2024). Automated multicolumn screening workflow in ultra-high pressure hydrophilic interaction chromatography for streamlined method development of polar analytes. *Journal of Chromatography A* 1733: 465266.

93 Holzgrabe, U. (2025). Analysis of small molecule pharmaceuticals with charged aerosol detection. Chapter 10. This Book. In:

94 Soliven, A., Haidar Ahmad, I.A., Tam, J. et al. (2017). A simplified guide for charged aerosol detection of non-chromophoric compounds-analytical method development and validation for the HPLC assay of aerosol particle size distribution for amikacin. *Journal of Pharmaceutical and Biomedical Analysis* 143: 68–76.

95 Chin, S., Lin, X., Santarra, B. et al. (2021). Multiplexed small molecule impurity monitoring in antibody-based therapeutics by mixed-mode chromatography paired with charged aerosol detection. *Journal of Pharmaceutical and Biomedical Analysis* 197: 113952.

96 Long, Z., Guo, Z.M., Liu, X.D. et al. (2016). A sensitive non-derivatization method for apramycin and impurities analysis using hydrophilic interaction liquid chromatography and charged aerosol detection. *Talanta* 146: 423–429.

97 Thakkar, H., Jain, S., Kumar, S. et al. (2023). Hyphenated liquid chromatography - diode array detection - charged aerosol detection - high resolution - multistage mass spectrometry with online hydrogen/deuterium exchange: one stop solution for pharmaceutical impurity profiling. *Journal of Chromatography A* 1689: 463725.

98 Kou, D., Manius, G., Zhan, S., and Chokshi, H.P. (2009). Size exclusion chromatography with corona charged aerosol detector for the analysis of polyethylene glycol polymer. *Journal of Chromatography A* 1216: 5424–5428.

99 Sivapuratharasan, V., Lenzen, C., Michel, C. et al. (2022). Metabolic engineering of *Pseudomonas taiwanensis* VLB120 for rhamnolipid biosynthesis from biomass-derived aromatics. *Metabolic Engineering Communications* 15: e00202.

100 Tiso, T., Zauter, T., Tulke, H. et al. (2017). Designer rhamnolipids by reduction of congener diversity: production and characterization. *Microbial Cell Factories* 16: 225.

101 Plante, M., Bailey, B., and Acworth, I.N. (2016). Analysis of emulsifiers in foods by high pressure liquid chromatography and corona charged aerosol detection. Thermo Scientific. PN-70995.

102 Lazana, P., García-Mayoral, M.-F., Lamothe, B., and Pena-Abaurrea, M. (2021). Comprehensive ethoxymer characterization of complex alcohol ethoxy sulphate products by mixed-mode high-performance liquid chromatography coupled to charged aerosol detection. *Journal of Chromatography A* 1639: 461927.

103 Choyke, S. and Ferguson, P.L. (2019). Molecular characterization of non-ionic surfactant components of the Corexit® 9500 oil spill dispersant by high-resolution mass spectrometry. *Rapid Communications in Mass Spectrometry* 33: 1683–1694.

104 Paun, I., Iancu, V.I., Cruceru, L. et al. (2018). Simultaneous determination of anionic, amphoteric and cationic surfactants mixtures in surface water. *Extraction* 7: 9.

105 Carnes, K.A., Oliver, L.D. Jr., Brown, T.A. et al. (2022). A platform analytical method for intact polysorbates in protein-containing biopharmaceutical products via HPLC-CAD. *Journal of Liquid Chromatography and Related Technologies* 45: 259–270.

106 Cloutier, M., Prévost, M.-J., Lavoie, S. et al. (2021). Total synthesis, isolation, surfactant properties, and biological evaluation of ananatosides and related macrodilactone-containing rhamnolipids. *Chemical Science* 12: 7533–7546.

107 Holgrabe, U. (n.d.). Analysis of pharmaceutical excipients with charged aerosol detection. In: Chapter 8. This book.

108 Schilling, K., Pawellek, R., Lovejoy, K. et al. (2018). Influence of charged aerosol detector instrument settings on the ultra-high-performance liquid chromatography analysis of fatty acids in polysorbate 80. *Journal of Chromatography A* 1576: 58–66.

109 Theiss, C. and Holzgrabe, U. (2018). Characterization of polydisperse macrogols and macrogol-based excipients via HPLC and charged aerosol detection. *Journal of Pharmaceutical and Biomedical Analysis* 160: 212–221.

110 Wang, J., Tang, S., and Zhang, K. (2024). Quantitation of polyethylene glycol by size exclusion chromatography with charged aerosol, differential refractive index, and multi-angle light scattering detectors. *Journal of Pharmaceutical and Biomedical Analysis* 238: 115854.

111 Wang, S.-Q., Zhao, X., Zhang, L.-J. et al. (2024). Discrimination of polysorbate 20 by high-performance liquid chromatography-charged aerosol detection and characterization for components by expanding compound database and library. *Journal of Pharmaceutical Analysis* 14: 100929.

112 Zegota, M.M., Achenbach, J., Schuster, G. et al. (2025). Direct comparison of single peak and gradient chromatographic methods for routine analysis of surfactants in biopharmaceuticals. *European Journal of Pharmaceutical Sciences* 209: 107065.

113 Zhang, S., Xiao, H., and Li, N. (2022). The characterization of polysorbate. *Current Protocols* 2: e489.

114 Hackbush, S., White, S., and Du, M. (2024). Characterization of DSPE-mPEG raw materials from different vendors reveals differences in impurity profiles and polymer chain length polydispersity. *bioRxiv*.

115 He, Y., Cook, K.S., Littlepage, E. et al. (2015). Ion-pair reversed phase liquid chromatography with ultraviolet detection for analysis of ultraviolet transparent cations. *Journal Chromatography A* 1408: 261–266.

116 Huang, J.C., Zongyun, S.L., Huang, Z., and Bolgar, M.S. (2017). Enablement of the direct analysis of excipients in monoclonal antibody formulations through the incorporation of a wide pore C18 protein trap with hydrophilic interaction liquid chromatography. *Journal of Chromatography B* 1068–1069: 131–135.

117 Jahn, M. (2023). Chapter 6 - Analytical methods for surfactant characterization in biologic drug products. In: *Surfactants in Biopharmaceutical Development* (ed. A.V. Koulov and S.K. Singh), 119–136. Academic Press.

118 Kurmi, M., Suryavanshi, V., Panduranga, N.S. et al. (2021). Development of HPLC-CAD stability indicating assay method for polyethylene glycol-conjugated phospholipid (DMPE-PEG 2000) and identification of its degradation products. *Journal of Pharmaceutical and Biomedical Analysis* 198: 113967.

119 Lebar, B., Lekic, T., Kosir, P. et al. (2024). Polysorbate stability: effects of packaging materials, buffers, counterions, and pH. *International Journal of Pharmaceutics* 665: 124598.

120 Martos, A., Koch, W., Jiskoot, W. et al. (2017). Trends on analytical characterization of polysorbates and their degradation products in biopharmaceutical formulations. *Journal of Pharmaceutical Sciences* 106: 1722–1735.

121 Nguyen, T., Sosic, Z., and Xu, C.-F. (2024). Direct quantitation of poloxamer 188 in adeno-associated virus products using reversed-phase chromatography with charged aerosol detection. *Separation Science Plus* 7: e202400045.

122 Shi, L., Shen, G., Chai, R. et al. (2025). A review of polysorbate quantification and its degradation analysis by liquid chromatography coupled with charged aerosol detection. *Journal of Chromatography A* 1742: 465651.

123 Tomlinson, A., Zarraga, I.E., and Demeule, B. (2020). Characterization of polysorbate ester fractions and implications in protein drug product stability. *Molecular Pharmaceutics* 17: 2345–2353.

124 Webster, G.K., Chang, J.C., and Heflin, J.L. (2021). Stability indicating method for polysorbate 80 in protein formulations. *Journal of Chromatographic Science* 59: 706–713.

125 Zhang, H., Hong, S., Tan, S.S.K. et al. (2022). Polysorbates versus hydroxypropyl beta-cyclodextrin (HPβCD): comparative study on excipient stability and stabilization benefits on monoclonal antibodies. *Molecules* 27: 6497.

126 Brun, A., Denis, P., Rambeau, M. et al. (2024). Polyunsaturated fatty acids prevent myosteatosis and lipotoxicity. *The Journal of Nutritional Biochemistry* 134: 109722.

127 de Waal, T., Brouers, J., Berbern, P. et al. (2023). Characterization of aspirated duodenal fluids from Parkinson's disease patients. *Pharmaceutics* 15: 1243.

128 Effinger, A., McAllister, M., Tomaszewska, I. et al. (2021). Investigating the impact of Crohn's disease on the bioaccessibility of a lipid-based formulation with an in vitro dynamic gastrointestinal model. *Molecular Pharmaceutics* 18: 1530–1543.

129 King, H.M., Cosslett, A.G., Thomas, C.P., and Price-Davies, R. (2018). A HPLC method to monitor the occurrence of lipid peroxidation in intravenous lipid emulsions used in parenteral nutrition using in-line UV and charged aerosol detection. *Clinical Nutrition ESPEN* 28: 96–102.

130 Swanepoel, A., Truter, C., Viljoen, F. et al. (2024). Temporal dynamics of plasma catecholamines, metabolic and immune markers, and corticosterone: DHEA ratio in farmed crocodiles before and after an acute stressor. *Animals* 14: 2236.

131 Khadhraoui, N., Prola, A., Vandestienne, A. et al. (2023). Hacd2 deficiency in mice leads to an early and lethal mitochondrial disease. *Molecular Metabolism* 69: 101677.

132 Lahtvee, P.-J., Kumar, R., Hallström, B.M., and Nielsen, J. (2016). Adaptation to different types of stress converge on mitochondrial metabolism. *Molecular Biology of the Cell* 27: 2505–2514.

133 Prola, A., Blondelle, J., Vandestienne, A. et al. (2021). Cardiolipin content controls mitochondrial coupling and energetic efficiency in muscle. *Science Advances* 7: eabd6322.

134 Green, H.S., Li, X., De Pra, M. et al. (2020). A rapid method for the detection of extra virgin olive oil adulteration using UHPLC-CAD profiling of triacylglycerols and PCA. *Food Control* 107: 106773.

135 Ismaiel, L., Fanesi, B., Kuhalskaya, A. et al. (2023). The determination of triacylglycerols and tocopherols using UHPLC–CAD/FLD methods for assessing the authenticity of coffee beans. *Foods* 12: 4197.

136 Jiménez-Carvelo, A.M., Pérez-Castaño, E., González-Casado, A., and Cuadros-Rodríguez, L. (2024). Synergistic use of three complementary detectors coupled to reverse phase liquid chromatography for olive oil authentication based on the phospholipid fraction. SSRN.

137 Lopez-Ruiz, R., Carvelo, A.M.J., and Rodriguez, L.C. (2024). The phospholipid chromatographic fingerprint: an analytical cutting-edge strategy in the distinguished characterization of edible vegetable oils. SSRN.

138 Martín-Torres, S., Jiménez-Carvelo, A.M., González-Casado, A., and Cuadros-Rodríguez, L. (2019). Differentiation of avocados according to their botanical variety using liquid chromatographic fingerprinting and multivariate classification tree. *Journal of the Science of Food and Agriculture* 99: 4932–4941.

139 Obisesan, K.A., Jiménez-Carvelo, A.M., Cuadros-Rodriguez, L. et al. (2017). HPLC-UV and HPLC-CAD chromatographic data fusion for the authentication of the geographical origin of palm oil. *Talanta* 170: 413–418.

140 Vera, D.N., Jiménez-Carvelo, A.M., Cuadros-Rodríguez, L. et al. (2019). Authentication of the geographical origin of extra-virgin olive oil of the Arbequina cultivar by chromatographic fingerprinting and chemometrics. *Talanta* 203: 194–202.

141 Skinner, M.M., Fong, M.A., Rimkus, T.P. et al. (2025). Pulsed electric field treatment of sweet potatoes to reduce oil and acrylamide in kettle chips. *Foods* 14: 577.

142 Tietel, Z., Hammann, S., Meckelmann, S.W. et al. (2023). An overview of food lipids toward food lipidomics. *Comprehensive Reviews in Food Science and Food Safety* 22: 4302–4354.

143 Pattaraudomchok, P., Winotapun, C., Thanomjitr, D. et al. (2024). Glycolipid-coated mangoes in microperforated film packaging for shelf-life-extended application. *ACS Food Science & Technology* 4: 2857–2866.

144 Zeng, X.-B., Yin, F.-W., Zhao, G.-H. et al. (2024). Mechanism of color change in Antarctic krill oil during storage. *Food Chemistry* 444: 138583.

145 Al Hamimi, S., Heyman-Lindén, L., Plaza, M. et al. (2017). Alterations in the plasma metabolite profile associated with improved hepatic function and glycemia in mice fed lingonberry supplemented high-fat diets. *Molecular Nutrition and Food Research* 61: 1600442.

146 Szterk, A., Zmysłowski, A., and Bus, K. (2018). Identification of cis/trans isomers of menaquinone-7 in food as exemplified by dietary supplements. *Food Chemistry* 243: 403–409.

147 Bogevik, A.S., Seppanen-Laakso, T., Samuelsen, T.A., and Thoresen, L. (2022). Fractionation of oil from black soldier fly larvae (*Hermetia illucens*). *European Journal Lipid Science and Technology* 124: 2100252.

148 Nordblad, M., Pedersen, A.K., Rancke-Madsen, A., and Woodley, J.M. (2016). Enzymatic pretreatment of low-grade oils for biodiesel production. *Biotechnology and Bioengineering* 113: 754–760.

149 Bredenbruch, S., Muller, C., Atemnkeng, H. et al. (2024). The biological activity of bacterial rhamnolipids on Arabidopsis thaliana and the cyst nematode *Heterodera schachtii* is linked to their molecular structure. *Pesticide Biochemistry and Physiology* 204: 106103.

150 Abreu, S., Solgadi, A., and Chaminade, P. (2017). Optimization of normal phase chromatographic conditions for lipid analysis and comparison of associated detection techniques. *Journal of Chromatography A* 1514: 54–71.

151 Barry, K.M., Dinan, T.G., Murray, B., and Kelly, P.M. (2016). Comparison of dairy phospholipid preparative extraction protocols in combination with analysis by high performance liquid chromatography coupled to a charged aerosol detector. *International Dairy Journal* 56: 179–185.

152 Causevic, A., Olofsson, K., Adlercreutz, P., and Grey, C. (2021). Non-aqueous reversed phase liquid chromatography with charged aerosol detection for quantitative lipid analysis with improved accuracy. *Journal of Chromatography A* 1652: 462374.

153 Libong, D., Heron, S., Tchapla, A., and Chaminade, P. (2017, Chapter 5). Lipid analysis with the corona CAD. In: *Charged Aerosol Detection for Liquid Chromatography and Related Separation Techniques* (ed. P.H. Gamache), 223–287. Wiley.

154 Lucci, P., Moret, S., Buchini, F. et al. (2018). Improved analysis of olive oils triacylglycerols by UHPLC-charged aerosol detection. *Journal of Food Composition and Analysis* 66: 230–236.

155 Pawellek, R., Krmar, J., Leistner, A. et al. (2021). Charged aerosol detector response modeling for fatty acids based on experimental settings and molecular features: a machine learning approach. *Journal of Cheminformatics* 13: 53.

156 Takeda, H., Takahashi, M., Hara, T. et al. (2019). Improved quantitation of lipid classes using supercritical fluid chromatography with a charged aerosol detector. *Journal of Lipid Research* 60: 1465–1474.

157 Walther, R., Krmar, J., Leistner, A. et al. (2023). Analytical quality by design: achieving robustness of an LC-CAD method for the analysis of non-volatile fatty acids. *Pharmaceuticals* 16: 478.

158 Zhao, X., He, Y., Chen, J. et al. (2021). Identification and direct determination of fatty acids profile in oleic acid by HPLC-CAD and MS-IT-TOF. *Journal of Pharmaceutical and Biomedical Analysis* 204: 114238.

159 Fibigr, J., Šatínský, D., and Solich, P. (2017). A UHPLC method for the rapid separation and quantification of phytosterols using tandem UV/Charged aerosol detection - a comparison of both detection techniques. *Journal of Pharmaceutical and Biomedical Analysis* 140: 274–280.

160 Takahashi, R., Nakaya, M., Kotaniguchi, M. et al. (2018). Analysis of phosphatidylethanolamine, phosphatidylcholine, and plasmalogen molecular species in food lipids using an improved 2D high-performance liquid chromatography system. *Journal of Chromatography B* 1077–1078: 35–43.

161 Ai, L., Li, Y., Zhou, L. et al. (2023). Lyophilized mRNA-lipid nanoparticle vaccines with long-term stability and high antigenicity against SARS-CoV-2. *Cell Discovery* 9: 9.

162 Beckert, N., Dietrich, A., and Hubbuch, J. (2024). RP-CAD for lipid quantification: systematic method development and intensified LNP process characterization. *Pharmaceuticals* 17: 1217.

163 Broset, E., Lampaya, V., Larraga, A. et al. (2025). A complete approach for circRNA therapeutics from purification to lyophilized delivery using novel ionizable lipids. *International Journal of Molecular Sciences* 26: 5138.

164 Clogston, J.D., Crist, R.M., Dobrovoslaia, M.A., and Stern, S.T. (2024). *Characterization of Nanoparticles Intended for Drug Delivery*. Humana Press.

165 Fan, Y., Marioli, M., and Zhang, K. (2021). Analytical characterization of liposomes and other lipid nanoparticles for drug delivery. *Journal of Pharmaceutical and Biomedical Analysis* 192: 113642.

166 Ferey, L., Slabi, S.A., Roy, C.-E. et al. (2018). Chromatographic study of nucleoside-lipids by RP-UHPLC-DAD/CAD. *Analytical and Bioanalytical Chemistry* 410: 7711–7721.

167 Gerhardt, A., Voigt, E., Archer, M. et al. (2022). A flexible, thermostable nanostructured lipid carrier platform for RNA vaccine delivery. *Molecular Therapy Methods & Clinical Development* 25: 205–214.

168 Kim, K.H., Lee, J.E., Lee, J.C. et al. (2023). Optimization of HPLC single CAD method for simultaneous analysis of different lipids in lipid nanoparticles with analytical QbD. *Journal of Chromatography A* 1709: 464375.

169 Matsumoto, T., Komori, T., Yoshino, Y. et al. (2021). A Liposomal gemcitabine, ff-10832, improves plasma stability, tumor targeting, and antitumor efficacy of gemcitabine in pancreatic cancer xenograft models. *Pharmaceutical Research* 38: 1093–1106.

170 Rysin, A., Paal, M., Lokerse, W. et al. (2021). Evaluation of release and pharmacokinetics of hexadecylphosphocholine (miltefosine) in phosphatidyldiglycerol-based thermosensitive liposomes. *Biochimica et Biophysica Acta – Biomembranes* 1863: 183698.

171 Su, W.-C., Lieu, R., Fu, Y. et al. (2025). A platform method for simultaneous quantification of lipid and nucleic acid components in lipid nanoparticles. *Journal of Chromatography A* 1746: 465788.

172 Valachovič, M. and Hapala, I. (2017). Biosynthetic approaches to squalene production: the case of yeast. In: *Vaccine Adjuvants. Methods in Molecular Biology*, vol. 1494 (ed. C.B. Fox), 95–106. Springer Science.

173 Phan, T., Devine, C., Laursen, E.D. et al. (2020). Squalene emulsion manufacturing process scale-up for enhanced global pandemic response. *Pharmaceuticals* 13: 168.

174 Abuzaytoun, R., Budge, S.M., Xia, W., and MacKinnin, S. (2022). Unusual ether lipids and branched chain fatty acids in sea cucumber (*Cucumaria frondosa*) viscera and their seasonal variation. *Marine Drugs* 20: 435.

175 Bony, N.F., Libong, D., Champy, P. et al. (2016). Establishing the non-aqueous liquid chromatographic profile of non-polar metabolites of herbal medicinal products. *Chemistry Reports* 19: 863–875.

176 Dianat, M., Straaten, S., Maritato, A. et al. (2024). Exploration of in situ extraction for enhanced triterpenoid production by *Saccharomyces cerevisiae*. *Microbial Biotechnology* 17: e70061.

177 Kim, H.S., Moon, B.C., Sungyu, Y. et al. (2019). Determination of fatty acids in the seeds of *Lepidium apetalum Willdenow, Descurainia sophia*

(L.) Webb ex Prantl, and Draba nemorosa L. by ultra-high-performance liquid chromatography equipped with a charged aerosol detector. *Journal of Liquid Chromatography & Related Technologies* 42: 128–136.

178 Kim, M.-O., Kang, M.-J., Lee, S.-U. et al. (2021). Polyacetylene (9Z,16S)-16-hydroxy-9,17-octadecadiene-12,14-diynoic acid in *Dendropanax morbifera* leaves. Food. *Bioscience* 40: 100878.

179 Yang, R., Wu, Z., Pu, Y. et al. (2019). Fast and non–derivative method based on high–performance liquid chromatography–charged aerosol detection for the determination of fatty acids from *Agastache rugosa (Fisch. et Mey.) O. Ktze*. Seeds. *Natural Product Research* 33: 1969–1974.

180 Kowalczyk, M., Rolnik, A., Adach, W. et al. (2021). Multifunctional compounds in the extract from mature seeds of *Vicia faba* var. *minor*: phytochemical profiling, antioxidant activity and cellular safety in human selected blood cells in in vitro trials. *Biomedicine and Pharmacotherapy* 139: 111718.

181 Krzyżanowska-Kowalczyk, J., Kowalczyk, M., Ponczek, M.B. et al. (2021). *Pulmonaria obscura* and *Pulmonaria officinalis* extracts as mitigators of peroxynitrite-induced oxidative stress and cyclooxygenase-2 inhibitors–in vitro and in silico studies. *Molecules* 26: 631.

182 Plaza, M., Batista, A.G., Cazarin, C.B.B. et al. (2016). Characterization of antioxidant polyphenols from *Myrciaria jaboticaba* peel and their effects on glucose metabolism and antioxidant status: a pilot clinical study. *Food Chemistry* 211: 185–197.

183 Sowa, I., Moldoch, J., Paduch, R. et al. (2023). Polyphenolic composition of *Carlina acaulis L.* extract and cytotoxic potential against colorectal adenocarcinoma and cervical cancer cells. *Molecules* 28: 6148.

184 Cruz de Carvalho, T., de Oliveira Silva, E., Soares, G.A. et al. (2020). Fungal biocatalysts for labdane diterpene hydroxylation. *Bioprocess and Biosystems Engineering* 43: 1051–1059.

185 Puchalska, J., Szumacher-Strabel, M., Patra, A.K. et al. (2021). The effect of different concentrations of total polyphenols from *Paulownia* hybrid leaves on ruminal fermentation, methane production and microorganisms. *Animals* 11: 2843.

186 Singh, J., Tareq, F.S., and Luthria, D.L. (2025). Comparative investigation of untargeted and targeted metabolomics in turmeric dietary supplements and rhizomes. *Foods* 14: 7.

187 Saber, F.R., Elosaily, A.H., Mahrous, E.A. et al. (2023). Detailed metabolite profiling and in vitro studies of *Urospermum picroides* as a potential functional food. *Food Chemistry* 427: 136677.

188 Park, M.H., Kim, D.-Y., Arbianto, A.D. et al. (2021). Characterization of compounds and quantitative analysis of oleuropein in commercial olive leaf extracts. *Journal of Applied Biological Chemistry* 64: 113–119.

189 Plaza, M., Oliveira, D., Nilsson, A., and Turner, C. (2017). Green and efficient extraction method to determine polyphenols in cocoa and cocoa products. *Food Analytical Methods* 10: 2677–2691.

190 Shi, W., Wu, Z., Wu, J. et al. (2023). A comprehensive quality control evaluation for standard decoction of *Smilax glabra Roxb* based on HPLC-MS-UV/CAD methods combined with chemometrics analysis and network pharmacology. *Food Chemistry* 410: 135371.

191 Waidyanatha, S., Collins, B.J., Cristy, T. et al. (2024). Advancing botanical safety: a strategy for selecting, sourcing, and characterizing botanicals for developing toxicological tools. *Food and Chemical Toxicology* 186: 114537.

192 Hollá, M., Šatínský, D., Švec, F., and Sklenarova, H. (2022). UHPLC coupled with charged aerosol detector for rapid separation of steviol glycosides in commercial sweeteners and extract of *Stevia rebaudiana*. *Journal of Pharmaceutical and Biomedical Analysis* 207: 114398.

193 Doydora, S.A., Baars, O., Cubeta, M.A., and Duckworth, O.W. (2024). Using manganese oxidizing fungi to recover metals from electronic waste. *Minerals* 14: 111.

194 Huang, H., Szumacher-Strabel, M., Patra, A.K. et al. (2021). Chemical and phytochemical composition, in vitro ruminal fermentation, methane production, and nutrient degradability of fresh and ensiled *Paulownia* hybrid leaves. *Animal Feed Science and Technology* 279: 115038.

195 Yanza, Y.R., Szumacher-Strabel, M., Lechniak, D. et al. (2022). Dietary *Coleus amboinicus Lour.* decreases ruminal methanogenesis and biohydrogenation and improves meat quality and fatty acid composition in longissimus thoracis muscle of lambs. *Journal of Animal Science and Biotechnology* 13: 5.

196 Musa, T., Kagi, A., Gutlin, H. et al. (2025). Potato late blight control with a botanical product and reduced copper applications. *Plant Disease* 109 (11): 2309–2320. e-ISSN: 1943-7692.

197 Olishevska, Z., Nickzad, A., Restieri, C. et al. (2023). *Bacillus velezensis* and *Paenibacillus peoriae* strains effective as biocontrol agents against Xanthomonas bacterial spot. *Applied Microbiology* 3: 1101–1119.

198 Tareq, F.S., Singh, J., Ferreira, J. et al. (2024). A targeted and an untargeted metabolomics approach to study the phytochemicals of tomato cultivars grown under different salinity conditions. *Journal of Agricultural and Food Chemistry* 72: 7694–7706.

199 Kanbayashi, Y., Tsutsumi, Y., Mizukoshi, T., and Yamaguchi, H. (2020). Development of an LC-NMR system using a diode array detector and a charged aerosol detector. *Bunseki Kagaku* 69: 151–155.

200 Liu, G., Zhu, B., Ren, X., and Wanga, J. (2021). Universal response method for the quantitative analysis of multi-components in josamycin and midecamycin using liquid chromatography coupled with a charged aerosol detector. *Journal of Pharmaceutical and Biomedical Analysis* 192: 113679.

201 Plante, M., Acworth, I.N., Bailey, B.A. et al. (2016). Improved universal approach to measure natural products in a variety of botanical and supplements. Thermo Scientific PN-70543-EN.

202 Wu, L., Gong, X., and Qu, H. (2024). An optimization strategy for charged aerosol detection to linearize the detector response in the multicomponent quantitative analysis of Qishen Yiqi dripping pills. *Journal of Separation Science* 47: e2300784.

203 Jin, G., Ding, J., Zhou, Y. et al. (2021). Synthesis and chromatographic evaluation of pyrazinedicarboxylic anhydride bonded stationary phase. *Journal of Chromatography A* 1638: 461825.

204 Kirchhoffer, O.A., Quiros-Guerrero, L., Nitschke, J. et al. (2025). Prioritization of novel anti-infective stilbene derivatives by combining metabolomic data organization and a stringent 3R-infection model in a knowledge graph J-L. RSC. *Advances* 15: 13010–13030.

205 Rutz, A. and Wolfender, J.-L. (2023). Automated composition assessment of natural extracts: untargeted mass spectrometry-based metabolite profiling integrating semiquantitative detection. *Journal of Agricultural and Food Chemistry* 71: 18010–18023.

206 Wang, R., Liu, Y., Zhou, H. et al. (2020). Integration of micro-fractionation, high-performance liquid chromatography-ultraviolet detector-charged aerosol detector-mass spectrometry analysis and cellular dynamic mass redistribution assay to accelerate alkaloid drug discovery. *Journal of Chromatography A* 1616: 460779.

207 Gunia-Krzyżak, A., Bareyre, F.M., Marona, H., and Waszkielewicz, A.M. (2019). Four N-(E)-cinnamoyl (cinnamamide) derivatives of aminoalkanols with promising anticonvulsant and analgesic activity. *Bioorganic & Medicinal Chemistry Letters* 29: 1298–1303.

208 Zhu, F., Zhang, X., Du, B.-Y. et al. (2023). Using UPLC-LTQ-Orbitrap-MS and HPLC-CAD to identify impurities in cycloastragenol, which is a pre-clinical candidate for COPD. *Molecules* 28: 6382.

209 Wu, L., Gong, X., Pan, J., and Qu, H. (2022). Establishing a chromatographic fingerprint using tandem UV/charged aerosol detection and similarity analysis for Shengmai capsule: a novel method for natural product quality control. *Phytochemical Analysis* 33: 460–472.

210 Yang, G., Zhao, X., Wen, J. et al. (2017). Simultaneous fingerprint, quantitative analysis and anti-oxidative based screening of components in *Rhizoma Smilacis Glabrae* using liquid chromatography coupled with charged aerosol and coulometric array detection. *Journal of Chromatography B* 1049-1050: 41–50.

211 Zhu, B., Hu, D., Zhao, J., and Li, S. (2024). Rapid identification and quantification of *Pseudostellaria heterophylla* with its adulterants by HPLC-CAD fingerprint combined with improved quantitative analysis of multi-components by single marker (QAMS). *Journal of Pharmaceutical and Biomedical Analysis* 247: 116205.

212 Liu, L., Fan, W., Zhang, H. et al. (2024). A versatile economic strategy by HPLC–CAD for quantification of structurally diverse markers in quality control of Shengmai Formula from raw materials to preparations. *Phytomedicine* 129: 155625.

213 Ye, X.-Y., Guo, Q., Guo, B. et al. (2019). Qualitative and quantitative analysis of cyclovirobuxine D and related substances by HPLC-CAD in the active pharmaceutical ingredient of huangyangning tablets. *Journal of Pharmaceutical Sciences* 54: 2303–2307.

214 Pak, M.E., Yang, H.J., Li, W. et al. (2022). Yuk-Gunja-Tang attenuates neuronal death and memory impairment via ERK/CREB/BDNF signaling in the hippocampi of experimental Alzheimer's disease model. *Frontiers in Pharmacology* 13.

215 Xu, X., Xu, H., Shang, Y. et al. (2021). Development of the general chapters of the Chinese *Pharmacopoeia* 2020 edition: a review. *Journal of Pharmaceutical Analysis* 11: 398–404.

216 Liu, R., Shi, S., Xiong, S. et al. (2022). Quality markers of *Dendrobium officinale* by "oligosaccharide-spectrum-effect" relationships. *Frontiers in Nutrition* 9: 914380.

217 Zhang, C.-E., Liang, L.-J., Yu, X.-H. et al. (2019). Quality assessment of *Astragali Radix* from different production areas by simultaneous determination of thirteen major compounds using tandem UV/charged aerosol detector. *Journal of Pharmaceutical and Biomedical Analysis* 165: 233–241.

3

Practical Use of CAD—Achieving Optimal Performance

Paul H. Gamache[1], Ian N. Acworth[2], Troy T. Handlovic[3], and Imad A. Haidar Ahmad[3]

[1] *Consultant, Nashua, NH, USA*
[2] *Consultant, Melrose, MA, USA*
[3] *One Amgen Center Drive, AMGEN Inc., Thousand Oaks, CA, USA*

CHAPTER MENU

3.1 Summary

A primary factor affecting the performance of charged aerosol detection is the concentration of impurities in the column eluent. This can be minimized through various means including use of solvents specified as having low evaporation residue, columns and conditions that produce minimal bleed, LC systems and operational practices that minimize contamination, and low concentrations of low-molar-mass additives. While the default settings of newer instruments are suitable for many applications, evaporation temperature (T_e) and power function value (PFV) are key parameters for optimization. Use of a low T_e generally results in more universal detection and more uniform response. This may, however, lead to increased baseline noise, which is mainly a function of eluent impurity concentration. When optimizing T_e, it is important to examine its effects on analyte levels near the targeted limit of quantitation. PFV can be used to optimize the signal output for a given method and range of quantitation. An optimal PFV extends the linear range and enables straightforward quantitation while often improving accuracy, signal-to-noise ratios, precision, and peak shapes. Three different techniques are described that facilitate the determination of an optimal PFV.

3.2 Introduction

CAD can be used with a range of chromatographic techniques where several variables can influence performance. This chapter provides guidance to help analysts achieve optimal results and includes consideration of analyte properties, chromatographic variables, detector settings and laboratory practices. The following concepts from Chapter 1 are highly relevant to this discussion:

- CAD and other evaporative aerosol detectors (EADs) distinguish analytes from eluent components based on differences in their extent of vaporization that results from the spray-drying process
- High concentrations of nonvolatile and semi volatile eluent impurities produce high baseline signal, drift and noise, thus limiting performance
- Analytes that behave as nonvolatiles should produce similarly shaped response curves and with a constant solvent composition, should produce nearly uniform response
- Especially at low mass levels, analytes that behave as semi volatiles exhibit a more supralinear curve shape with lower overall sensitivity than those that behave as non-volatiles
- In the spray-drying process, volatile ionic mobile phase additives can combine with other ionic solutes (e.g., analytes, impurities, other additives) to form less-volatile salts. The choice of additive type and concentration is therefore an important consideration
- The PFV is a key instrumental setting that can be used to optimize and extend the linear response range of CAD. An optimal setting can simplify calibration, improve peak shapes and the accuracy of both quantitation and determination of detection limits
- T_e is a useful parameter to optimize signal-to-noise ratio

The first part of this chapter describes the main factors that affect results and best practices in developing methods and in maintaining and monitoring performance during routine use. Detailed recommendations for use of the PFV and for calibration will also be provided. Technical considerations for inverse gradient compensation and for combining other detectors with CAD in multi-detector configurations will be covered in Chapters 4 and 13, respectively. The discussions in this chapter, while focused on VCAD instrument design as described in Chapter 1, also generally apply to other EADs.

3.3 Factors Affecting Performance

With EADs, it is often stated that eluents must be volatile and that all nonvolatile and some semi volatile analytes are detected. This is certainly an oversimplification when considering the specific process of spray-drying, the results of which depend not only on solute properties (e.g., vapor pressure) but also experimental conditions (e.g., T_e, influence of mobile phase additives) and, to some extent, instrument (e.g., spray chamber) design. For these reasons, it is appropriate to think in terms of either a solute's expected behavior for a range of conditions or its observed behavior within a given analytical method. The following section will discuss eluent impurities, which are often unknown species, but their behavior as semi volatile and nonvolatile solutes results in an unwanted contribution to EAD signal.

3.3.1 Eluent Impurities

After droplet evaporation, eluent impurities, together with any non- and semi volatile analyte present, contribute to the size of each dried aerosol particle in proportion to their combined mass concentration. Higher impurity concentrations lead to elevated baseline levels, baseline noise, and, especially with solvent gradients, baseline drift [1–4]. Prolonged exposure to impure eluents may also result in accumulation of contaminants within the instrument, which can degrade longer term performance and require more frequent maintenance. Figure 3.1 compares chromatograms obtained with purposefully contaminated and "clean" eluents under gradient conditions. The clean eluent shows a background current (i.e., baseline level of CAD without autozero) of <1 pA, which is generally indicative of a relatively clean eluent. The contaminated eluent represents a rather extreme example with background current >20 pA. With such a high level of impurities, it is not surprising to see higher baseline noise, more gradient drift and late-eluting "ghost" peaks. These ghost peaks are attributed to impurities from the weaker eluent that are retained and concentrated on-column before elution with a stronger eluent. The relatively poor analyte response seen here is another consequence of this rather prohibitive level of background impurities.

While eluent impurities cannot be completely eliminated, their levels should be kept to a minimum. Doing so can ensure good quantitative performance especially in terms of detection limits, precision, reproducibility and accuracy. There can be multiple sources of impurities including the chromatographic flow path, column bleed, general lab equipment, and mobile phase components. Common sources and recommended practices to minimize impurity concentration are further discussed below.

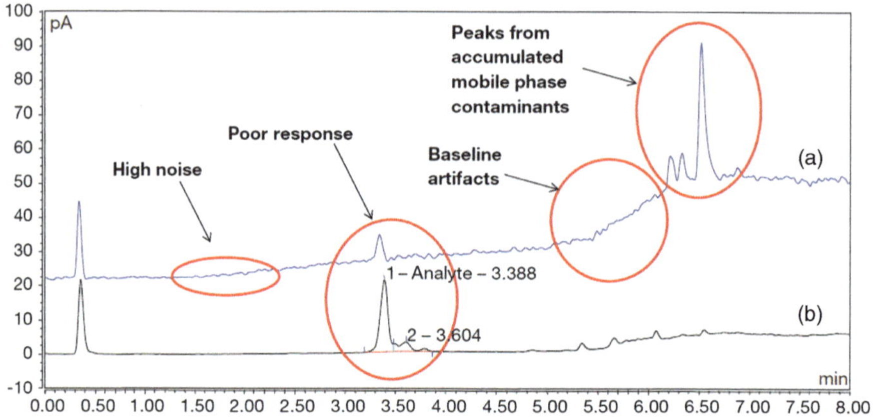

Figure 3.1 HPLC-CAD gradient analysis of a nonvolatile analyte (2840 ng injected): (a) Mobile phase with purposefully added nonvolatile impurity, (b) Freshly prepared clean mobile phase (no impurity added). *Source:* Reproduced with permission of Thermo Fisher Scientific, Inc.

Water, a very common solvent, can be a main source of impurities. A best practice with CAD is to use ultrapure water from a well-maintained point-of-use purification system, ideally where the feed water is purified by reverse osmosis. Typical ultrapure water specifications are 18.2 MΩ·cm resistivity at 25 °C and <5 parts per billion total organic carbon. Since impurities can accumulate upon storage and in a stagnant system, the purification system should be thoroughly flushed before use. Commercially available HPLC-grade bottled water is not recommended as it is often found to produce a significantly higher background level than that obtained from point-of-use systems.

The impurity concentration of **organic solvents** is especially important to consider since these solvents typically have lower viscosity and surface tension than water, which leads to much higher aerosol mass transport. An organic solvent with the same impurity concentration as water should therefore produce much higher background current and noise. The most relevant specification for an organic solvent is its residue after evaporation. As expected, solvents with low residue after evaporation, as commonly specified with LC–MS grades, typically produce lower background and noise. Several reports have shown that better limits of detection can be obtained with methods that use eluents with higher organic content [e.g., hydrophilic interaction liquid chromatography (HILIC)] than those that use mainly aqueous eluents [e.g., reversed phase liquid chromatography (RPLC)]. This may be attributed to higher mass-transport efficiency but since this efficiency applies equally to both impurities and analytes, any gains in signal-to-noise ratio should significantly depend on ensuring that the organic solvents are pure [5–7].

3.3.2 Mobile Phase Additives

While the choice of mobile phase additives (i.e., pH modifiers, buffers and ion-pairing [IP] reagents) is often primarily based on chromatographic objectives, it is also very important to consider their influence on detection. A basic requirement with EADs is that additives must, to the extent possible, behave as volatile solutes. In practice, however, an increase in background current and noise is usually observed whenever an additive is utilized. Table 3.1 lists common additives and typical concentrations used with CAD and LC–MS.

Formic acid (FA), as the sole additive, or ammonium formate (AmF) buffers are often the first choices when using CAD and LC–MS due to their relative volatility and availability as pure reagents. As mentioned, volatile ionic mobile phase additives can, during spray-drying, combine with other ionic solutes (e.g., analytes, impurities, sample matrix components, and other oppositely charged additives) to form less-volatile salts. These effects can be significant especially considering that the additive concentrations listed in Table 3.1 should, in most cases, be several orders of magnitude higher than typical analyte concentrations within the eluent [8]. This can have beneficial and/or adverse effects depending on the nature of the interacting species. For example, chloride normally behaves as a volatile solute in the absence of a basic mobile phase additive. However, if

Table 3.1 Volatile additives and typical concentrations.

Acidic additive	Typical conc.	mM	mg/mL	pK_a	pH	
Acetic acid	**0.1% v/v**	17.4	1.04	4.8	3.27	
Formic acid	**0.1% v/v**	23.6	1.09	3.8	2.7	
Trifluoroacetic acid	**0.1% v/v**	13	1.48	0.0	1.9	
Buffer					**~Buffer range**	**Buffer species**
Ammonium acetate	**10 mM**	10	0.77	4.8	3.8–5.8	Acetic acid–acetate
Ammonium formate	**10 mM**	10	0.63	3.8	2.8–4.8	Formic acid–formate
Ammonium carbonate	**10 mM**	10	0.96	10.3	7–11	Ammonia–ammonium
				9.3		Hydrogen carbonate–carbonate
				7.8		Carbonic acid–hydrogen carbonate

ammonium is used as an additive, then a detectable nonvolatile chloride salt can be formed. This can be beneficial if chloride is an analyte of interest but detrimental if it is an eluent impurity or a high-level sample matrix component.

The higher background currents observed when any additive is used may be partly attributed to salt formation with common ionic eluent impurities. As discussed in Chapter 1, the extent to which additives contribute to CAD signal significantly depends on their molar mass where larger additive molecules, through salt formation, contribute more mass to the aerosol particles. For mobile phases with only an acidic modifier, the trend in background current and baseline noise is expected to be FA < Acetic Acid (HAc) ≪ Trifluoroacetic Acid (TFA) and may be attributed to their salt formation with common cationic impurities (e.g., Na^+, K^+). The expected trend when including buffers is FA < HAc < AmF < AmAc, where salt formation may occur with both cationic and anionic (e.g., Cl^-) impurities. The lowest molar mass additives are therefore preferred since any salt formed with impurities should be more volatile and of lower mass, therefore contributing less to the background. It is also recommended to restrict the concentration of additives to the minimum required to achieve the desired separation since higher additive concentrations typically produce higher background signal.

The influence of additive volatility is most significant when using IP reagents such as TFA or triethylamine since, by their nature, they tend to form nonvolatile salts with impurities. It is recommended to avoid the use of TFA or longer chain acids (e.g., heptafluorobutyric acid [HFBA]) together with ammonium or other basic additives. Likewise avoid using higher molar-mass basic additives (e.g., butylamine) together with formic or other acids. These combinations may, by themselves, form salts often leading to prohibitively high background currents and baseline noise. Additives may also form salts with ionic sample matrix components which, if present at sufficiently high levels, can lead to a large solvent front, ghost-peaks or other baseline artifacts. As with background currents and baseline noise, these issues can be quite significant when using IP reagents and are minimized by using low molar-mass additives at low concentrations. For these and other reasons it is sometimes preferable to use a mixed-mode column instead of IP reagents. Several mixed-mode stationary phases, for example, with RP + ion-exchange retention characteristics or ion-exchange + normal phase (via hydrophilic interaction) characteristics, have become available to address the many challenging separations faced by chromatographers. These columns may avoid several problems associated with the use of IP reagents. In particular, mixed-mode columns that exhibit low-column bleed (see below) and provide good separation with minimal additive concentration can provide an alternative or complement to IP-RPLC for use with CAD and LC–MS. Examples of this are described in Chapters 8 and 12.

Elevated background current and noise may also be observed when using higher pH (> ~7) eluents. In some cases, higher background current and noise can be attributed to column bleed (see below). Another potential cause is the formation, through spray-drying, of non-volatile carbonate salts. While several labs have observed prohibitively high background currents when using ammonium carbonate buffers, this issue can also occur with AmAc buffers. Freshly prepared AmAc buffers typically produce low background currents and noise. However, this is often found to significantly increase over a relatively short period of time (e.g., within a day or so) and is attributed to ammonium carbonate salt formation that results from absorption of atmospheric CO_2 by the eluent. In either case, the use of an optimally higher T_e (e.g., 50 °C) can be very effective at reducing background signal and noise [8].

3.3.3 Column Bleed

Column bleed, in the context of EADs, refers to the addition of nonvolatile impurities into the eluent by the column itself. This is commonly attributed to degradation (e.g., cleavage of bonded phase, dissolution of support material) or dissolution of impurities (e.g., metals in lower purity silica, prior exposure to non-volatiles) from the stationary phase. Some columns (e.g., silica-based amino, cyano, zwitterionic) exhibit significant bleed throughout their operating range [9]. However, for many columns, bleed is observed mainly when operated near the extremes of their compatibility range. Most reports of column bleed with silica-based columns appear to be when operated toward the lower or higher ends of their pH compatibility range (typically 2–8) and/or when operated at higher T. In these cases, alternative columns developed for wider pH compatibility can be used including those with polymeric, protected silica, or other more pH-stable supports. As observed by Teutenberg et al. [10], columns designed for wide pH compatibility also seem to exhibit low bleed when operated at high T. In any case, it is recommended to use the lowest column T that has a positive impact on the chromatographic separation while minimizing column bleed. Figure 3.2 compares the baseline levels (background currents), of four HILIC columns over a range of pH 7 aqueous buffer/acetonitrile fractions [11]. For columns with the same bonded phase, much higher currents and noise, attributed to column bleed, were observed with the silica (ZIC-HILIC) compared to the polymeric (ZIC-pHILIC) based column. Also, while two of the silica-based columns exhibited high currents, low currents were observed with the silica-based Luna HILIC column and is attributed to protection of the silica by a unique cross-linked diol bonded phase. Since some columns may exhibit significant bleed while still maintaining acceptably stable retention times and peak shapes, it is recommended to screen columns specifically for use with CAD. A simple approach is to

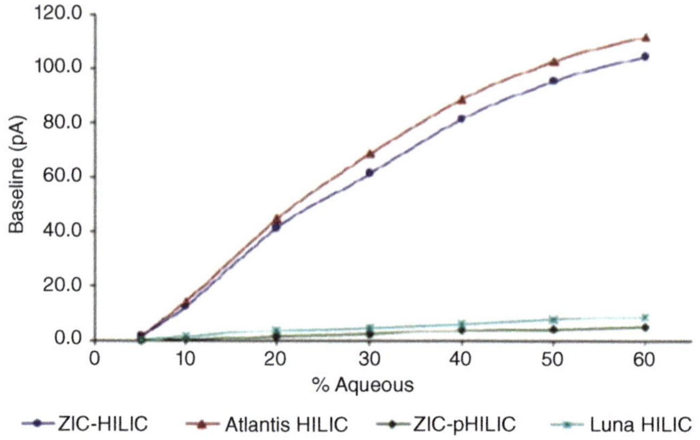

Figure 3.2 Plots of CAD baseline level as a function of aqueous fraction of mobile phase with different HILIC columns: ZIC-HILIC = zwitterionic. *Source:* Huang et al. [11] / with permission of Elsevier.

compare background currents with, and without, a column. It is recommended to avoid using columns that have been previously exposed to nonvolatile buffers or IP reagents. Also, when using a column for the first time, it is recommended that it be thoroughly washed before connecting to CAD, ideally with a stronger solvent at slightly higher operating temperature. If using a gradient elution method, allow the column to experience several gradient cycles prior to connecting to CAD.

3.3.4 Mobile Phase Preparation

Recommended practices for mobile phase preparation to minimize impurities with EADs are similar to those for LC–MS. This includes the use of glassware that is dedicated for use with volatile mobile phases and is thoroughly rinsed with suitable solvents like ultrapure water just prior to use. Use of detergents to clean glassware should, in most cases, be avoided since residual surfactants will likely be detected. It is also highly recommended to prepare any buffers gravimetrically rather than to titrate to the desired pH with a pH electrode. Gravimetric preparation of buffers where the amount of reagent needed is calculated and measured usually results in improved batch-to-batch repeatability of retention times [12]. pH electrodes can also be a significant source of contamination as they are usually stored and calibrated in highly concentrated nonvolatile buffers. The practice of filtering the mobile phase can also add impurities. Ultrapure water and HPLC-grade or better solvents are "pre-filtered," and so it is better to avoid additional refiltering. Buffer salts are often best prepared as concentrated solutions in water, which are then filtered prior to dilution to the desired concentration. TFA and other modifiers may degrade over time, and therefore single-use vials are preferred for mobile phase preparation [13].

3.3.5 Chromatographic System

Before connecting CAD to any chromatographic system, it is important to thoroughly flush with suitable solvents. If the system has been previously used with nonvolatile mobile phases, then more extensive preparation is required. As with columns, many long-term issues with system components can be attributed to "memory effects" after extended exposure to nonvolatile solutes (e.g., phosphate buffers). Commonly reported sources include in-line filters (e.g., solvent reservoir filters), online degassers (membrane with vacuum), and pulse dampers (membrane with solvent reservoir). In some cases, the component can be flushed with a suitable solvent, but in others, it should be replaced with a part that is then dedicated for use with CAD and LC–MS.

3.4 Evaporation Temperature

Evaporation temperature (T_e) is a useful instrument parameter to optimize SNR by altering the extent of vaporization of solutes. This is often a compromise between maximizing the vaporization of eluent impurities to reduce baseline noise and minimizing vaporization of target analytes. An increase in T_e may reduce noise but possibly at the expense of response for semi volatile analytes. Conversely, a decrease in T_e should increase signal for semi volatile analytes often at the expense of higher baseline noise. When optimizing T_e, it is important to assess sensitivity by actual analysis of low-level standards [e.g., near the targeted limit of quantitation (LOQ)]. This is exemplified in Figure 3.3 where the effect of increasing T_e is much more pronounced for low analyte levels. A good practice is to optimize T_e by testing small incremental changes in T_e (e.g., $\leq 5\,°C$). This type of testing is also useful to investigate method robustness and relative temperature stability of each analyte focusing on levels near the limit of detection (LOD). It is recommended to use the lowest T_e setting that consistently produces the required sensitivity limits and where the ideal goal is for all analytes to behave as nonvolatiles. This generally provides more universal detection, more uniform response, and enables more accurate quantitation. The ability to use a low T_e in order to realize these performance benefits mainly depends on the cleanliness of the eluent as discussed above.

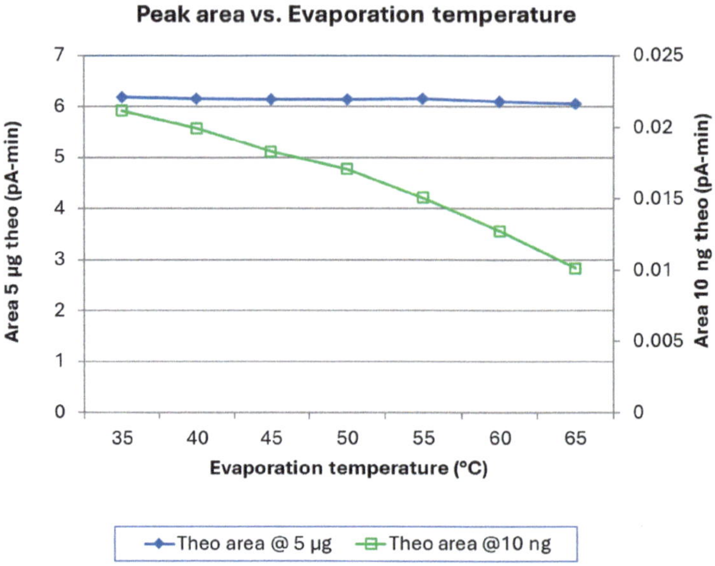

Figure 3.3 **Peak** area response for low and high amounts of theophylline (Theo) as a function of evaporation temperature.

3.5 Calibration and Sensitivity Limits

This section provides recommended approaches for calibrating and determining sensitivity limits with CAD and will begin with an overview of the relationship between CAD's response and analyte mass, which is detailed in Chapter 1. A primary focus will be on the PFV, which is a key instrumental setting that is used to optimize and extend the linear response range. Guidance for PFV optimization and how to evaluate calibration curve-fitting models to achieve the most accurate quantitative results will be discussed.

3.5.1 Overview of CAD Response Curves

CAD response can be described by a "power law" equation, which relates instrument response (represented here by peak height, h in pA) to the analyte's injected mass (m_{inj}),

$$h = a(m_{\mathrm{inj}})^b \tag{3.1}$$

where the coefficient a indicates the instrument's sensitivity for an analyte, and the variable exponent b describes the curve shape. When a concentration versus response curve is fit with Equation (3.1) and b is 1 the instrument response for that mass range can be said to be "linear," when less than 1 the response is

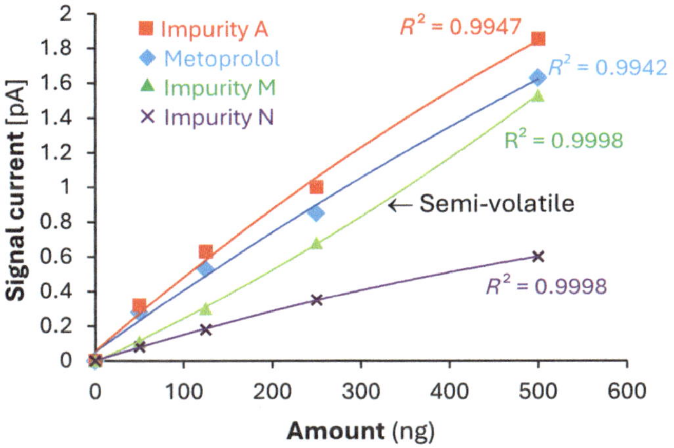

Figure 3.4 Response curves for non-volatile and semi volatile analytes. Note: Differences in response magnitude for these analytes are attributed to use of a solvent gradient method without inverse compensation and, for impurity M, volatility. *Source:* Reproduced with permission of Thermo Fisher Scientific, Inc.

"sublinear," and when greater than 1 "supralinear." For analytes that behave as non-volatiles, b tends to be less than one across CAD's entire dynamic range. Importantly, non-volatiles generally exhibit similarly shaped response curves where the changes in b that occur across the full dynamic range are similar. However, if the analyte shows some volatility, b can vary from greater than 1 toward the LOQ to less than 1 for higher levels. Response curves in Figure 3.4 show three analytes with a slight downward curving or sublinear shape, and one analyte (impurity M) with an upwards curving or supralinear shape. The similarity in curve shape for three of these analytes is highly characteristic of their nonvolatile behavior while the different, supralinear shape of impurity M indicates semi volatile behavior.

The variable exponent b for any nonvolatile or semi volatile analyte is always maximal at the lowest m_{inj} values and gradually decreases with increasing m_{inj} values. This trend is attributed to changes in the fraction of smaller aerosol residue particles that have a higher power law exponent of response and are more prevalent with low m_{inj} (e.g., toward the LOQ) and the low concentration that occurs near the edges of any peak. The contribution of these smaller particles to the shape of CAD's response curve was first detailed in 2017 [14] and is reiterated in Chapter 1. It should be noted that descriptions of CAD response as simply being proportional to particle surface area (d^2) are not fully consistent with this updated theory.

The spray-drying technique used in the newer VCAD design, first introduced in 2013, (e.g., Corona™ Veo™ and Vanquish™ series CAD, Thermo Fisher Scientific Inc.) produces a lower relative proportion of smaller residue particles than first-generation CADs. As a result, VCAD's intrinsic response for analytes that behave as nonvolatiles remains sublinear ($b < 1$) throughout its dynamic range and varies somewhat more closely in proportion to the expected cube root of m_{inj}. To address this intrinsic nonlinearity, the CAD has a user defined PFV that can be used to extend the linear range of the instrument. To provide approximately linear response for typical methods and quantitation ranges, the instrument's digital signal processing (DSP) algorithm applies a fixed exponent or internal *Power Function* (PF) to the instantaneous signal. For many methods, this internal PF provides sufficiently linear response to allow accurate quantitation with linear regression using the default PFV = 1. In other instances, adjustment of the PFV enables one to further optimize the range over which response is sufficiently linear for a given method and its range of quantitation.

3.5.2 Approaches for Linearization of the CAD Signal in the Literature

Prior to discussing the mathematical implications of the power function, a variety of methods used in the literature (Figure 3.5) to handle the instrument's nonlinearity are first discussed. Linearization of the CAD signal enables straightforward quantitation, improves high-throughput analysis, and often increases the precision and accuracy of the results. Since the CAD signal follows a power law (Equation 3.1) trend, a log–log plot is one of the easiest methods to improve the dynamic range of the CAD detector without optimizing PFV. The use of log–log plots (Figure 3.5a) is accompanied with some drawbacks including the loss of sensitivity at low concentrations, more complicated data analysis, and added complexity in validation/implementation for quality compliance labs [15]. Log–log plots can also be visually deceptive, as small deviations on the plot can represent large absolute differences. Also, linear regression calculations, such as correlation coefficients, are valid only when the experimental uncertainty of replicate y values is not related to the values of x or y. This assumption may not be fully valid after data have been transformed [16].

Several approaches have been proposed in the literature to optimize the PFV both experimentally and computationally to result in a linear calibration curve. One approach includes adjusting PFV while monitoring the variability in the response factor as a function of concentration (or mass injected), as shown in Figure 3.5b. The PFV corresponding to the lowest RSD of response factors across a given mass range provides the best linearity [17]. Another approach to visually determine the optimum PFV is to plot the response factor as a function of concentration range, at various PFVs (Figure 3.2c) [17]. The approaches in Figure 3.5b/c were originally proposed with experimental data collection at each evaluated PFV to determine the optimum PFV. Therefore, they were time-consuming

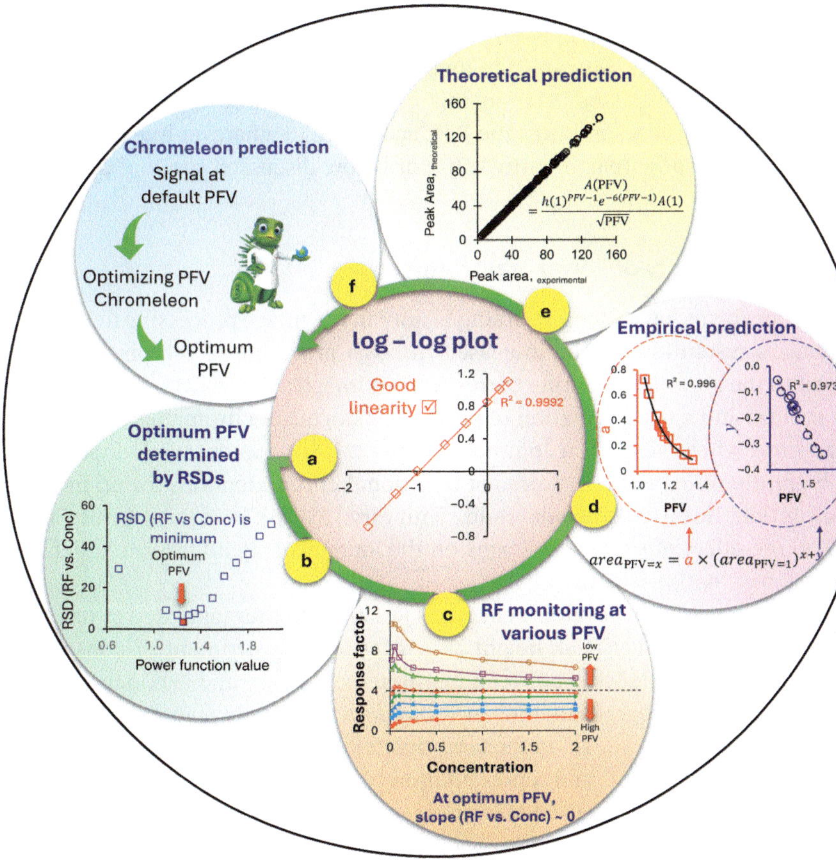

Figure 3.5 Various methods to handle nonlinearity, determine linearity, and predict the CAD response based on limited experimentation at different PFVs.

and required significant resources. Moreover, the optimum PFV may no longer provide sufficiently linear response if there is a change in the required mass range of quantitation or a significant change in the chromatographic conditions, which means that the process may need to be repeated. This slow process lowered analyst's motivation to optimize their CAD's performance.

Later, approaches were developed to determine the optimum PFV based on predictions from limited experimentation. The first proposed empirical model (Figure 3.5d) fit the CAD's response at multiple PFVs and used these relationships to predict the CAD's response at different PFVs [18]. Thereafter, a theoretical model which takes into consideration the peak height, and area was proposed (Figure 3.5e) [19]. This theoretical model will be explained explicitly in the next section. A final method of prediction is the "power law" feature in Chromeleon™ (Thermo Fisher Scientific) software (Figure 3.5f) [20]. Through automatic data transformation in the software, the analyst can easily find the PFV leading to best linearity, as discussed later in this chapter. All three of these prediction methods can be used to optimize the PFV according to whatever linearity criteria fits the given purpose (e.g., Figure 3.5b/c or below discussed residual plots/back calculations).

3.5.3 Brief Theory of Power Transforms

Power transforms have a long-standing history in the image processing field where an image's resolution can be increased through powering of the pixels. This is related to the so-called "gamma correction" in image science [21]. This work led to a natural and early adaptation of power transforms in chromatography for 2D separations whose signal is a common image [22]. Subsequent work appeared in the literature on power transforms of traditional chromatograms for an increase in resolution, linearity, efficiency, and symmetry [23–26]. Below is a brief discussion on the effect of these transforms on the analytical signal prior to a detailed discussion on optimizing the PFV.

Analysts in research, development, and QC labs, whether regulated or not, tend to prefer peak area over peak height as a measure of instrument response. This complicates the relationship between instrument response and exponentiation as powering a chromatogram to a given value *does not simply power the peak area* to the same value. To better understand this relationship, the Gaussian function is used as the ideal functional representation of a chromatographic peak. When a Gaussian function $G(t)$ is raised to a power n (belonging to real positive numbers) the peak's area $A(n)$ scales as,

$$A(n) = \int_{-\infty}^{\infty} G(t)^n \mathrm{d}t = h_0^n \frac{\sigma_0}{\sqrt{n}} \sqrt{2\pi} \tag{3.2}$$

where t is time, h_0 is the peak's original unprocessed height, and σ_0 is the peak's original unprocessed standard deviation as elucidated by Dasgupta and later Wahab [23, 26]. Due to properties of exponential functions, a powered Gaussian peak is still a Gaussian peak in form, but a powered exponentially modified Gaussian peak (EMG, often used to model asymmetric chromatographic peaks) will approach a Gaussian peak in form [23]. After powering to n, the height of the peak is equal to h_0^n, the width is equal to $\sigma_0/\sqrt{n}$, and the area is equal to $A_0 h_0^{n-1}/\sqrt{n}$ all compared to the original peak [26]. Applying a power to a chromatogram does differentially impact peaks of variable heights, shapes, and widths so the use of power transforms should be done with care [24]. This is especially important if quantitation is based on a non-authentic standard in CAD which may have different peak/retention characteristics compared to the analyte.

3.5.4 Internal Power Function and User-Defined PFV

The default signal for the VCAD instruments is reported at a power of 1.5, or more recently 1.8, as the older model's default signal was reported at a power of 1. The user defined PFV can then be used to adjust the default signal by applying an additional scaled power transform. Optimizing the PFV is recommended to improve the linearity of the instrument, make quantitation more accurate, and simplify calibration. The general equation for the power function is,

$$h(\mathrm{PFV}) = h(1)^{\mathrm{PFV}} \times z(\mathrm{PFV}) \tag{3.3}$$

where $h(\mathrm{PFV})$ is the signal output after the power function, PFV is the power function value, $h(1)$ is the default signal at PFV = 1, and $z(\mathrm{PFV})$ is a scaling function of PFV. Two scaling functions have been used to model the PFV being,

$$z(\mathrm{PFV}) = 1/500^{\mathrm{PFV}-1} \tag{3.4A}$$

or,

$$z(\mathrm{PFV}) = e^{-6(\mathrm{PFV}-1)} \tag{3.4B}$$

whose outputs match closely at all PFV values. Due to the better agreement of Equation (3.4B) with recently collected experimental data, the second scaling factor is used for the following discussion [19]. The purpose of the scaling factor in the power function is not to linearize the data, but rather to ensure that the analog output of the detector does not exceed the instrument's full-scale range of 500 pA regardless of the PFV. If peak height is used to represent instrument response, the preceding equations can be applied to compute response at any given PFV from the data collected at PFV = 1. However, if peak area is desired additional equations are required to calculate response at a given PFV from PFV = 1 data.

Referring to the CAD response model summarized above and discussed in detail in Chapter 1, the exponent of response (b in Equation (3.1)) is changing across the peak's profile. The edges of the peak have the lowest analyte concentration, the largest proportion of small particles, and the highest exponent of response. The apex of the peak has the highest analyte concentration, the smallest proportion of small particles, and the lowest exponent of response. Therefore, there is not, by definition, a PFV that always produces a perfectly linear response. Rather, the PFV in which the most linear response is produced is a "best fit" value for your specific method. To compute the peak area ($A(\mathrm{PFV})$) at any PFV from the data at PFV = 1 the following equation can be used,

$$A(\mathrm{PFV}) = \frac{h(1)^{\mathrm{PFV}-1} e^{-6(\mathrm{PFV}-1)} A(1)}{\sqrt{\mathrm{PFV}}} \tag{3.5}$$

where $A(1)$ is the peak's area at PFV = 1, and the scaling factor provided in Equation (3.4B) is used (see [19] for derivation). Using this equation, peak areas

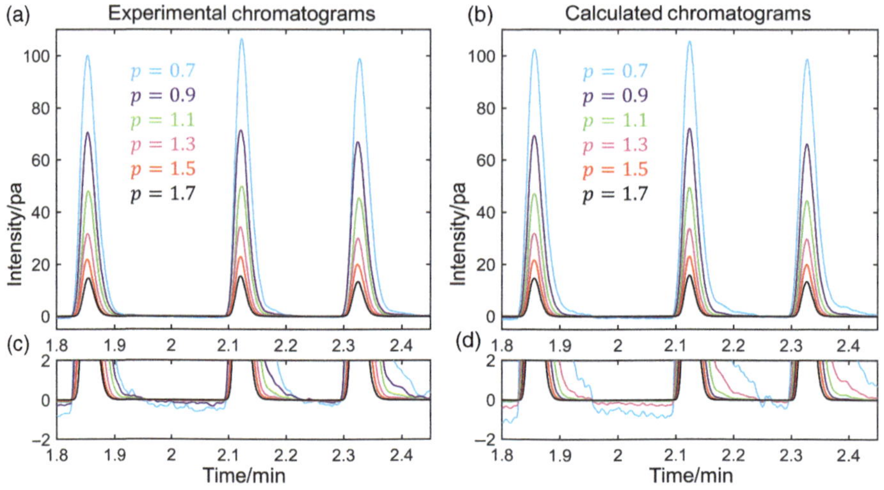

Figure 3.6 Experimentally collected chromatograms (a and c) at six different PFVs compared to those calculated using Equations (3.3), (3.4A) and (3.4B) based on the data collected at PFV = 1 (b and d). c and d are shown at higher amplification to illustrate the effect of PFV setting on baseline noise. *Source:* Handlovic et al. [19] / with permission of the American Chemical Society.

at any PFV can be calculated if the data was collected at PFV = 1. Alternatively, the entire chromatogram can be computed at each PFV using Equations (3.3), (3.4A) and the data at PFV = 1 if each data point in the chromatogram is treated as a discrete "height" signal. An example of calculating entire chromatograms at any PFV from the data collected at PFV = 1 is provided in Figure 3.6.

The peaks in the calculated chromatograms can then be integrated to obtain PFV-specific peak areas. A similar approach can be completed for those who operate their CAD using the Chromeleon software using the "power law" feature [27]. Calculating entire chromatograms is advantageous compared to Equation (3.5) if the peaks of interest are noticeably asymmetric as Equation (3.5) is designed for Gaussian-shaped peaks [19].

3.5.5 Optimizing the PFV

An example calibration used to establish the CAD's response for microgram purification quantitation in Amgen's Drug Discovery program (see Chapter 14) is shown in Figure 3.7. This calibration includes eight mass levels (12.5–2000 ng) for three pharmaceutically relevant analytes (amcinonide, bendroflumethiazide,

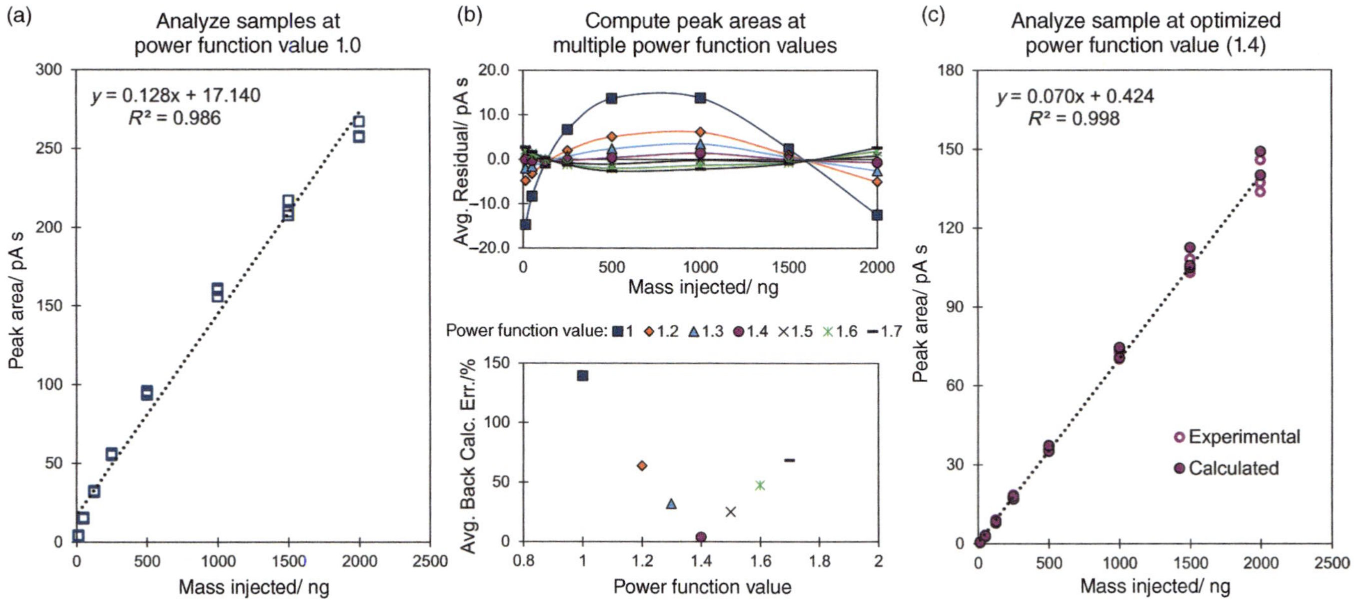

Figure 3.7 Global optimization of the PFV to linearize the response for three non-volatile analytes (amcinonide, bendroflumethiazide, and tolnaftate) across greater than two orders of magnitude of mass injected (12.5–2000 ng) using a three-step process. (a) Eight calibration standards analyzed at PFV one for all three compounds. (b) Calculated peak areas at a variety of PFVs using the data at PFV of one and Equation (3.5). Peak areas for each PFV were individually regressed with a linear function to calculate the average residuals and back calculated error rate (see below). Using both the residual plots and back calculated error rate the PFV of 1.4 was chosen. (c) Comparison of experimentally collected peak areas to calculated peak areas at the optimized PFV (1.4). For experimental details see: *Anal. Chem.* 2024, 96, 40, 16045–16052.

and tolnaftate). A three-step optimization procedure is proposed to quickly find the PFV that produces the most linear response:

1) Collection of the data at PFV = 1. This establishes the default instrument response for the desired method and mass range.
2) Computation of the peak areas at a variety of PFVs. Each dataset corresponding to a PFV can then be regressed and inspected without the need for timely physical experimentation and data analysis. The PFV that produces the desired results, using criterion such as the ones below or in Figure 3.5b/c, can readily be identified.
3) Collection of the data at the desired PFV to be used for the final calibration and for future data collection.

This entire approach is pragmatic and mathematically simple allowing for the analysis process to be done using Chromeleon, MATLAB® or even Excel® in a matter of minutes. An automated version of this optimization process in the MATLAB environment is provided free of charge in the supporting information of [19].

Figure 3.7a shows the first step of the optimization process where the data is collected at PFV = 1. Notice that the peak area versus concentration curve is not linear indicated by the poor fit of a linear trendline. This is not inherently bad, as nonlinear functions can be used to model calibration data; however, we have found that a power law equation (Equation (3.1) or nearly analogous Log–Log plot) tends to better fit the bottom portion of this mass range, but not the top and a second order polynomial fits the top range of the mass range but not the bottom. Since neither of these nonlinear functions completely describes the area-based response, the data is linearized by optimizing the PFV providing a more accurate calibration. Figure 3.7b shows the second step of the optimization procedure where the peak areas at PFV 1.0–1.7 are computed using Equation (3.5) and the data in Figure 3.7a. The range and interval of PFVs accessed can be adjusted across the entire PFV range of 0.67–2.0 depending on the scenario. This data could also be obtained by using Equations (3.3), (3.4A) and (3.4B) and then integrating the peaks after the power function is applied to the entire chromatogram or by using the power law function in Chromeleon.

Analytical chemists have long argued on the "proper" definition of linearity, and there is no universally agreed on metric [28, 29]. In this chapter, four methods are shown: (i) minimization of relative standard deviation for the response functions (Figure 3.5b), (ii) visual monitoring of the slope of response function versus concentration plots (Figure 3.5c), (iii) visual monitoring of residual plots (Figure 3.7b), and (iv) average back calculation error rate (Figure 3.7b). In general, the use of aggregate measures such as the total sum of square residuals (TSS), the correlation coefficient (R^2), the F-statistic, RSD of response factors, or average back calculation error should not be used as stand-alone criterion for goodness-of-fit. Instead, a holistic approach should be applied combining visual inspection of the residual plots with analysis of aggregate measures which fit the desired purpose.

In Figure 3.7b two methods are shown being residual plots and back calculation. Each set of peak areas representing a PFV was regressed versus mass injected using a linear equation. The average residuals between that linear function and the computed area are shown on the residual plot (top). In this plot, a convex pattern represents a sublinear response, as a concave pattern represents a supralinear response. The PFV which produces residuals near the zero value is the ideal linear response, here represented best by the PFV = 1.4 curve. The back-calculation approach provided at the bottom of Figure 3.7b takes the same computed data and fits linear equations that was discussed above. Here, the linear equation parameters are used to back calculate the mass values at each mass level [30]. The average error between the true mass level and the back-calculated mass level is assessed. This method also agrees that 1.4 produces the least amount of error when a linear function is used for this method and mass range. Finally, Figure 3.7c shows the experimentally collected data at the optimal PFV of 1.4 versus the computed values. There is excellent agreement between the two datasets, and the linear function shows a strong fit across the entire mass range indicating that 1.4 is indeed the ideal PFV here for all three analytes.

3.5.6 Sensitivity Limits

From the plots in Figure 3.3c, it is clear that the greater deviations from linearity are toward the low end of the response range. The deviations are similar in nature to those observed with ELSD but much less pronounced (Chapter 1). It is common to calculate the LOD and LOQ using a linear function such as signal-to-noise ratio (SNR) of 3 and 10, respectively. However, since the response may be nonlinear especially near these limits, an accurate estimation of LOD and LOQ usually requires actual analysis of appropriately low analyte levels rather than estimation from higher analyte levels. The latter can lead to very misleading results especially when assuming linear response. In general, it is best to optimize the PFV to obtain a sufficiently linear response before calculating detection limits. As mentioned, for analytes that behave as non-volatiles, CAD (PFV = 1.0) is sub-linear through most of its dynamic range, where the power law exponent b (Equation 3.1) is <1 but is maximal and closest to 1 for the lowest m_{inj}. As described in Chapter 1, sublinear response causes artificial peak broadening and for supralinear response causes peak shaving. Use of an optimal PFV setting, as described above, in addition to its benefits for calibration can also help to avoid these common pitfalls.

3.5.6.1 Recommended Practices for Calibration

The choice of PFV should be made using a robust evaluation of calibration models for "quality of fit." The following are recommended practices for any quantitative method [31–34] and are especially important to use with CAD and other EADs.

1) Limit calibration to as small a range as possible above and below the expected sample concentration.
2) Use the simplest curve-fitting model that adequately describes the response-amount relationship over the required range of interest. A linear curve fit can be used for initial evaluation and, especially when using an optimal PFV, is often adequate for many CAD methods that require quantitation over mass ranges of 10^2 to ~10^3. For quantitation over wider ranges, quadratic (i.e., second-order polynomial), log–log, and point-to-point are often used.
3) Evaluate quality of fit by using a sufficient number of levels and replicates over the entire range with special consideration to the upper and lower limits. The curve should be fit to the individual data points as opposed to first averaging the response between replicates at each level.
4) Do not use aggregate measures [e.g., coefficient of determination (r^2)] as the only metrics for assessing quality of fit. As widely described [29, 34], a curve fit with an r^2 of 0.9990 may still be poor, with the largest error typically near the low end. Reliance only on r^2 can be especially problematic when using linear regression on log-transformed data.
5) Consider using weighted regression (e.g., 1/amount, 1/amount^2) as it often provides a better fit to the lower amounts [35].

3.5.7 Summary of Calibration and Sensitivity Limits

The default signal output of newer CAD instruments provides approximately linear response over an analyte mass range that is suitable for many applications. The signal output can then be further optimized for a given method and range of quantitation with a user-defined PFV. An optimal PFV extends the linear range and enables straightforward quantitation while often improving accuracy, detection limits, precision, and peak shapes. The process of experimentally determining an optimal PFV, however, requires significant time and resources. In this section, three techniques are described that greatly facilitate this process and require only data collection at a PFV of 1.0.

References

1 Aruda WO, Walfish S, Krull IS. Review and optimization of linearity and precision in quantitative HPLC-ELSD with chemometrics. *LC GC North America*. 2008;26(10) 1032-1041.

2 Magnusson, L.-E., Risley, D., and Koropchak, J. (2015). Aerosol-based detectors for liquid chromatography. *Journal of Chromatography. A* 1421: 68–81.

3 Reilly, J., Everatt, B., and Aldcroft, C. (2008). Implementation of charged aerosol detection in routine reversed phase liquid chromatography methods. *Journal of Liquid Chromatography and Related Technologies* 31 (20): 3132–3142.

4 Young, C.S. and Dolan, J.W. (2003). Success with evaporative light-scattering detection. *LC-GC The Magazine of Separation Science* 21 (2): 120–128.

5 Hutchinson, J.P., Li, J., Farrell, W. et al. (2011). Comparison of the response of four aerosol detectors used with ultra high pressure liquid chromatography. *Journal of Chromatography. A* 1218 (12): 1646–1655.

6 Hutchinson, J.P., Remenyi, T., Nesterenko, P. et al. (2012). Investigation of polar organic solvents compatible with Corona Charged Aerosol Detection and their use for the determination of sugars by hydrophilic interaction liquid chromatography. *Analytica Chimica Acta* 750: 199–206.

7 Russell, J.J., Heaton, J.C., Underwood, T. et al. (2015). Performance of charged aerosol detection with hydrophilic interaction chromatography. *Journal of Chromatography. A* 1405: 72–84.

8 Bailey, B., Plante, M., Thomas, D. et al. (2017). Practical use of CAD: achieving optimal performance. In: *Charged Aerosol Detection for Liquid Chromatography and Related Separation Techniques* (ed. P.H. Gamache), 163–189. John Wiley and Sons.

9 Cohen, R.D. and Liu, Y. (2014). 1 Advances in aerosol-based detectors. *Advances in Chromatography* 52: 1.

10 Teutenberg, T., Tuerk, J., Holzhauser, M., and Kiffmeyer, T. (2006). Evaluation of column bleed by using an ultraviolet and a charged aerosol detector coupled to a high-temperature liquid chromatographic system. *Journal of Chromatography. A* 1119 (1): 197–201.

11 Huang, Z., Richards, M., Zha, Y. et al. (2009). Determination of inorganic pharmaceutical counterions using hydrophilic interaction chromatography coupled with a Corona® CAD detector. *Journal of Pharmaceutical and Biomedical Analysis* 50 (5): 809–814.

12 Stoll DR, Makey DM. Mobile phase buffers in LC: effect of buffer preparation method on retention repeatability LC-GC North America 2019 37(7), 444-449.

13 Lantz, M.D., Risley, D.S., and Peterson, J.A. (1997). Simultaneous resolution and detection of a drug substance, impurities, and counter ion using a mixed-mode HPLC column with evaporative light scattering detection. *Journal of Liquid Chromatography and Related Technologies* 20 (9): 1409–1422.

14 Gamache, P.H. and Kaufman, S.L. (2017). Principles of charged aerosol detection. In: *Charged Aerosol Detection for Liquid Chromatography and Related Separation Techniques* (ed. P.H. Gamache), 1–65. John Wiley and Sons.

15 Haidar Ahmad, I.A., Blasko, A., Wang, H. et al. (2021). Charged aerosol detection in early and late-stage pharmaceutical development: selection of regression models at optimum power function value. *Journal of Chromatography. A* 1641: 461997.

16 Motulsky, H.J. and Ransnas, L.A. (1987). Fitting curves to data using nonlinear regression: a practical and nonmathematical review. *The FASEB Journal* 1 (5): 365–374.

17 Soliven, A., Haidar Ahmad, I.A., Tam, J. et al. (2017). A simplified guide for charged aerosol detection of non-chromophoric compounds—analytical method development and validation for the HPLC assay of aerosol particle size distribution for amikacin. *Journal of Pharmaceutical and Biomedical Analysis* 143: 68–76.

18 Haidar Ahmad, I.A., Blasko, A., Tam, J. et al. (2019). Revealing the inner workings of the power function algorithm in charged aerosol detection: a simple and effective approach to optimizing power function value for quantitative analysis. *Journal of Chromatography. A* 1603: 1–7.

19 Handlovic, T.T., Roy, D., Barnhart, W.W., and Haidar Ahmad, I.A. (2024). Role of the power function value in linearity and universality for charged aerosol detectors: theoretical elucidations from a validated model. *Analytical Chemistry* 96 (40): 16045–16052.

20 Gamache, P., Muellner, T., Eggart, B. et al. (n.d.). Charged aerosol detection – use of the power function and robust calibration practices to achieve the best quantitative results, Thermo Fisher Scientific Technical Note. 73299.

21 Farid, H. (2001). Blind inverse gamma correction. *IEEE Transactions on Image Processing* 10 (10): 1428–1433.

22 Shalliker, R.A., Stevenson, P.G., Shock, D. et al. (2010). Application of power functions to chromatographic data for the enhancement of signal to noise ratios and separation resolution. *Journal of Chromatography. A* 1217 (36): 5693–5699.

23 Dasgupta, P.K., Chen, Y., Serrano, C.A. et al. (2010). Black box linearization for greater linear dynamic range: the effect of power transforms on the representation of data. *Analytical Chemistry* 82 (24): 10143–10150.

24 Hellinghausen, G., Wahab, M.F., and Armstrong, D.W. (2018). Improving visualization of trace components for quantification using a power law based integration approach. *Journal of Chromatography. A* 1574: 1–8.

25 Wahab, M.F., Berthod, A., and Armstrong, D.W. (2019). Extending the power transform approach for recovering areas of overlapping peaks. *Journal of Separation Science* 42 (24): 3604–3610.

26 Wahab, M.F., Gritti, F., O'Haver, T.C. et al. (2019). Power law approach as a convenient protocol for improving peak shapes and recovering areas from partially resolved peaks. *Chromatographia* 82 (1): 211–220.

27 Pawellek, R., Muellner, T., Gamache, P., and Holzgrabe, U. (2021). Power function setting in charged aerosol detection for the linearization of detector response – optimization strategies and their application. *Journal of Chromatography. A* 1637: 461844.

28 Raposo, F. (2016). Evaluation of analytical calibration based on least-squares linear regression for instrumental techniques: a tutorial review. *TrAC, Trends in Analytical Chemistry* 77: 167–185.

29 Raposo, F. and Barceló, D. (2021). Assessment of goodness-of-fit for the main analytical calibration models: guidelines and case studies. *TrAC, Trends in Analytical Chemistry* 143: 116373.

30 Jurado, J.M., Alcázar, A., Muñiz-Valencia, R. et al. (2017). Some practical considerations for linearity assessment of calibration curves as function of concentration levels according to the fitness-for-purpose approach. *Talanta* 172: 221–229.

31 Dolan, J.W. (2009). Calibration curves, part I: to b or not to b? *LC GC North America* 27 (3): 224–230.

32 Dolan JW. Calibration curves, part II: what are the limits? *LC GC North America*. 2009; 27(4) 306-12.

33 Dolan, J.W. (2009). Calibration curves, part 3: a different view. *LC GC Europe* 22: 304.

34 Kiser, M.M. and Dolan, J.W. (2004). Selecting the best curve fit. *LCGC North America* 22 (2): 112–117.

35 Almeida, A.M., Castel-Branco, M.M., and Falcao, A. (2002). Linear regression for calibration lines revisited: weighting schemes for bioanalytical methods. *Journal of Chromatography B* 774 (2): 215–222.

4

Practical Realization of Response Uniformity and its Benefits for Global Quantification

Frank Steiner[1], Katherine S. Lovejoy[1], and Paul H. Gamache[2]

[1] *Thermo Fisher Scientific, Product Applications, Germering, Germany*
[2] *Consultant, Nashua, NH, USA*

CHAPTER MENU

4.1 Summary

Uniform detector response can be defined as equivalent response curves for a range of analytes independent of their physicochemical properties. This provides the potential to quantify multiple analytes with one calibration curve. This is very desirable for many applications where it is impractical to use individual standards, such as drug libraries, drug impurities, and natural products. Charged aerosol detection (CAD) is often observed, or assumed, to provide uniform response. However, significant variability in response factors (RFs) (i.e., signal per unit mass) has been observed with some analytes and methods. CAD is based on spray-drying a column effluent to form an aerosol residue where each particle size is proportional to its initial droplet size and its mass concentration of less volatile solutes. Signal from the downstream electrical aerosol detector is primarily attributed to residue particle sizes $\gg$10 nm, below which the detector response exponentially drops. It then follows that, except for very large molecules at mass levels near the detection limit, most of the signal from CAD arises from measurement of particles each consisting of many (e.g., 10 to $>10^6$) analyte molecules. The electrical aerosol measurement technique of CAD is based on a mechanism called unipolar diffusion charging of aerosols, which is well known to impart a size-dependent charge that is effectively independent of particle composition. This strongly suggests that deviations in CAD response uniformity are then mainly related to the upstream spray-drying process where analyte volatility, solvent composition, salt formation between ionizable analytes and eluent additives and bulk particle density are known to be the primary factors. This is supported by several studies using flow injection analysis under isocratic conditions where very similar response has been demonstrated for chemically diverse analytes that behave as nonvolatiles. This includes inorganic, organic, neutral, polar, non-polar, hydrophobic, hydrophilic, small and large molecules and, after correction for salt formation, ionic species. Many studies have demonstrated the ability of CAD to provide accurate single calibrant or surrogate standard quantitation when combined with various liquid and supercritical fluid separations methods. However, significant limitations have also been observed. This chapter will examine variables that may influence the observed response uniformity of these methods including those related to the chromatographic method, detector settings, and pre-analytical considerations. The main focus will be on practical strategies toward achieving accurate quantification in the absence of individual standards.

4.2 Introduction

Analytical separation science is generally concerned with separating, identifying, quantifying, and providing chemical structural information on components in complex mixtures. These analyses can be time-consuming and resource-intensive and often require technologies that are able to detect all separated components. The detector should also produce a linear, or easily calibrated, response for sample components present over a wide concentration range with sufficient sensitivity. Moreover, the ideal characteristics of a detector are that it shows universal detection and uniform response and can be interfaced to a separation technique that employs a wide variety of mobile phases/separation media and conditions. A universal detector is a device that detects every component within the eluent, except the mobile phase. Uniform response can be defined as equivalent response curves for a range of analytes independent of their physicochemical properties. Zhang et al., in a 2019 review article entitled "Seeking universal detectors for analytical characterizations," described the characteristics of commonly used detectors for gas and liquid chromatography focusing on universal detection and uniform response [1]. Table 4.1 summarizes some of the most relevant characteristics of liquid chromatography detectors derived from Zhang's review.

Table 4.1 Universal response characteristics of commonly used HPLC detector.

Detector	Detection requirement		Response curves	Comments
	Analyte	*Mobile phase*		
UV	Sufficient UV chromophore	UV transparent	Highly dependent on analyte structure	Sensitive, compatible with LC, wide linear range
RI	RI different than mobile phase		Analyte and mobile phase dependent	Limited sensitivity and stability, incompatible with gradient elution
CAD	Low volatility in spray-drying process	Volatile in spray-drying process	Similar among nonvolatiles but depends on solvent composition, salt formation for ionic solutes and dried particle bulk density	Better accuracy, sensitivity, reproducibility, and wider linear range than ELSD
ELSD	Low volatility in spray-drying process	Volatile in spray-drying process	Same as CAD with additional dependencies on particle optical properties	Complex nonlinear response curves with limited sensitivity
CNLSD	Low volatility in spray-drying process	Volatile in spray-drying process	Same as CAD with additional dependencies on particle wettability and solubility	Similar sensitivity, linear range and reproducibility to CAD
CLND	Nitrogen containing	Nitrogen-free	Equimolar response for nitrogen-containing compounds except those with adjacent nitrogen atoms	Incompatible with nitrogen-containing eluents e.g., acetonitrile (ACN), ammonium, high maintenance
MS	Ability to form gas phase ion, dependent on analyte, ion source type, and conditions	Volatile, facilitates formation of gas phase ion	Highly dependent on analyte properties, matrix effects can lead to ion suppression or enhancement	Highly sensitive and specific, invaluable qualitative information, many instrumental options provide wide applicability
qNMR	Universal for chosen nuclide (e.g., ^1H)	Free of chosen nuclide	Same for all molecules	Poor sensitivity and very limited compatibility with LC

The "Detection Requirements" column in Table 4.1 clearly indicates that there must be a way to differentiate between analyte and mobile phase components, either through some form of selectivity (e.g., relative volatility) or specificity (e.g., NMR). This underlies the limitations of any of these devices to provide a truly universal response (see also Chapter 1). The "Response Curves" column represents response uniformity. Here the analyte and condition dependencies of ultraviolet (UV), mass spectrometric (MS), and refractive index (RI) detectors are known to severely limit their use for quantitation in the absence of individual standards. RI is additionally limited by its incompatibility with gradient elution and relatively poor sensitivity. While UV is the most commonly used LC detector, having many ideal characteristics, it is unable to directly provide a measurable response for poor chromophores and the response for chromophores varies widely with analyte molar absorptivity. MS is a very powerful technique for LC analysis with high sensitivity and specificity along with the ability to provide, often critical, qualitative information. MS is sometimes described as universal, since the technology in all of its implementations can be applied to a very broad range of analytes and matrices. However, universal detection and uniform response is typically not achieved with a given MS ion source, instrument and method. Quantitative nuclear magnetic resonance (qNMR) has been described as a gold-standard device for single-calibrant quantitation [2], but it has poor sensitivity and very limited compatibility with LC, requires significant expertise, and is often cost prohibitive. Chemiluminescent nitrogen detection (CLND) can also be very effective for quantitation in the absence of standards, but this requires knowledge of the analytes' molecular structure. This requirement and its applicability to only nitrogen-containing analytes, its incompatibility with commonly used mobile phases and its reported lack of ruggedness are significant limitations. Another device, based on an adaptation of flame ionization detection (FID) for use with LC eluents, was introduced in 2021 as a universally quantitative detector by Activated Research Company [3]. This was based on the use of a heated rotating disc to evaporate the mobile phase, prior to reduction to methane and detection by FID. There is currently very little information about this device in literature.

Amongst all LC detectors, evaporative aerosol detectors (EADs) are the ones that are most commonly referred to as universal. This includes CAD, evaporative light scattering detection (ELSD) and, to a lesser extent, condensation nucleation light scattering detection (CNLSD). As described in Chapter 1, all EADs involve the same successive steps of primary spray droplet formation from an eluent stream, conditioning by inertial impaction to remove droplets too large to evaporate during passage through the instrument, and evaporation of remaining droplets to form residue particles each comprised of species with sufficiently low volatility (i.e., background impurities and analyte). Detection of the residue particles produces the detector signal. Detection selectivity for

EADs is therefore based on differences in vaporization, or relative volatility, of components within an eluent. Accordingly, these techniques are expected to have very similar detection scope, eluent requirements, and solvent dependency of response. Their detection scope includes any analyte with sufficiently low volatility to substantially contribute to residue particles produced in the spray-drying process and is therefore less selective or more universal than UV and MS. For these reasons, CAD or ELSD are very often combined with UV for more comprehensive analysis. The use of EADs with UV and MS is also quite common and is facilitated by the shared requirement of EADs and MS to use volatile eluents. Practical considerations for multidetector combinations with CAD are detailed in Chapter 13 and key examples are described throughout this book, especially in Chapters 12 through 15.

4.3 Response Uniformity of EADs

The influence of solvent gradients, analyte volatility, salt formation, and bulk particle density (related to analyte density) on overall detector performance is expected to be similar among all EADs (Chapter 1). When comparing the response uniformity of these devices, a main differentiating factor may therefore be the relative dependency of the downstream aerosol detector's response on residue particle material (i.e., primarily determined by analyte properties). Aerosol charging by diffusion mechanisms is well known to have only a minor dependence on the dielectric constant of the particle material (Chapter 1). Accordingly, several studies have shown that, after accounting for other variables (e.g., solvent dependency, volatility and salt formation), CAD response exhibits a minor dependence on analyte properties [4–8] and consistently provides more uniform response than ELSD [9–16]. ELSD makes use of photometers to measure the combined angular light scattering of all aerosol particles within the detection flow path. Angular light scattering intensity is a very complex function of the light source intensity, scattering angle of measurement, aerosol particle concentration, RI of the particle material, particle shape, and ratio of particle diameter to the wavelength of incident light (d_p/λ) [17, 18]. The dependence of ELSD response on analyte RI has been previously described [17, 19]. Oppenheimer et al. [18] estimated that ELSD response may vary by a factor of 2 or more for analytes encompassing an RI range of 1.4–1.6. The RI of light-absorbing analytes also depends on their absorption coefficient [4]. Analyte-specific absorption and fluorescence have also been reported to influence ELSD response where variations are more pronounced when using monochromatic (e.g., laser diode) light sources [19, 20].

The technique of CNLSD is similar to ELSD but uses the condensation of liquid vapor onto the dried aerosol particles to amplify particle size into a region of higher light scattering efficiency [21]. CNLSD commonly provides better sensitivity than ELSD, but studies have shown that the signal significantly depends on analyte nature [9, 21]. Figure 1.17 in Chapter 1 compares CAD and CNLSD response for 24 chemically diverse nonvolatile species analyzed by flow injection analysis (FIA) with a constant solvent. After accounting for the counterion content of a few analytes that were obtained as salts, the variation in response (%RSD) among analytes was found to be ~8.0% for CAD and ~39.0% for CNLSD. The much larger variation observed with CNLSD is attributed to the downstream aerosol measurement technique where wettability and solubility of the particle material for a given condensing fluid (e.g., water, n-butanol) result in variable, analyte-dependent, response. This is supported by several studies of atmospheric aerosols [22–24].

Most detector comparison studies (e.g., Chapters 2 and 5) have shown significant advantages of CAD over ELSD. While both techniques allow universal detection of less volatile analytes, the response of CAD for nonvolatiles, as discussed above, is fundamentally less dependent on analyte properties. In addition, CAD is consistently observed to provide a wider linear and dynamic range, better sensitivity, better precision and better quantitative accuracy. These advantages are consistent with theory as discussed in Chapter 1. Briefly, a fundamental limitation of the downstream detector for both techniques is related to the exponential signal drop that occurs for smaller aerosol particles. These particles are more prevalent with low analyte levels and the low concentration that occurs near the edges of any peak. For ELSD this corresponds to Rayleigh light scattering for particle diameters typically <50 nm where a 2-fold change in diameter results in a ~ 64-fold drop in light scattering intensity. With CAD, a signal drop occurs for particle diameters < ~9 nm where a 2-fold change in diameter results in a ~5-fold drop in signal current. For CAD, the lower d transition and more gradual signal change underlies the widely observed lower detection limits, wider linear and dynamic range, and less complex response curves than ELSD. All of these properties are highly relevant toward achieving accurate quantitation, especially in the absence of individual standards. The following text will more closely examine variables that may influence the observed response uniformity of CAD-based methods and practical strategies toward achieving accurate quantification in the absence of individual standards.

4.4 Factors Affecting the Response Uniformity of CAD

The main factors affecting the response uniformity of CAD are related to the upstream spray-drying process. This includes analyte volatility, solvent composition, salt formation between ionizable analytes and eluent additives, and bulk particle density, each discussed in more detail in this section.

Beyond these detector-specific factors, some variables are essential to any analysis where uniform response is expected. These factors include known purity of calibration standards (certified reference standards should ideally be used and each standard should also be injected into the CAD to ensure no unexpected peaks appear), solubility of the standards in the mobile phase of choice, stability of the standards in the autosampler in solution, recovery from the column (if a column is being used), chromatographic resolution (complete separation of all peaks if the sample contains multiple substances), and peak shape (broad peaks, split peaks or peaks with riders interfere with quantitation). The linearity and precision for the application over the mass range of interest should be established using a system suitability test. Analyte interactions with wetted surfaces of the HPLC system should also be ruled out by comparing the likely chemical properties of the substances in the application and the wetted surfaces list from the HPLC manufacturer. Examples of such interactions affecting CAD measurements are detailed in Chapter 6.

4.4.1 Analyte Volatility

Analyte volatility has the greatest impact on detector response, because the effect of analyte evaporation during the spray-drying process can significantly decrease response and is the major root cause for non-uniformity of CAD. CAD is often described as capable of detecting all nonvolatile and some semivolatile analytes. This is certainly an oversimplification when considering the specific process of spray-drying, the results of which depend not only on solute properties but also experimental conditions. For these reasons, it is appropriate to think in terms of either a solute's expected behavior for a range of conditions or its observed behavior within a given analytical method. Uniform response with CAD is only obtained for analytes that behave as nonvolatiles. One way to determine if an unknown or impurity is above or below the volatility limit is by testing detector response at different evaporation temperatures, as described in Chapter 13 and in Eckhardt et al. [25]. An analyte that behaves as a semivolatile produces both an overall lower response and a characteristic exponential signal loss at lower mass levels (e.g., Figure 3.4, Chapter 3). This is attributed to evaporation rate being proportional to surface area (d^2), where, for example, a twofold smaller droplet or particle would evaporate four times faster. The signal drop is therefore more noticeable at low analyte levels, which compromises detection limits, precision and quantitative accuracy. A given analyte may, however, be observed as a nonvolatile if only measured at high levels. As noted above, an exponential signal drop is observed with ELSD for lower mass levels even for nonvolatiles and this drop is further exacerbated for semivolatiles.

Several studies have examined response uniformity for a diverse range of chemicals in relation to common indicators of volatility. This includes molar mass (M), vapor pressure (P_s), boiling point (B_p), and enthalpy of vaporization (ΔH_v). Most studies have shown an approximate value above or below which solutes are reliably detected, and these are typically described as nonvolatile (e.g., $B_p > {\sim}400\,°C$ [6, 10], or $M > 350\,g/mol$ and $\Delta H_v > 65\,kJ/mol$ [26]). Near these limits (e.g., B_p from 300 to 400 °C, [7]), weaker correlation (e.g., lower sensitivity or no response) has been observed. Solutes within this range are typically considered semivolatile and those beyond considered volatile. The lines between these categories, however, are certainly blurred, and there are unexpected outliers. Additional insight is provided from studies [26–29] that have examined the influence of variables related to solvents, additives (e.g., pH modifiers, ion-pairing agents), conditions (e.g., liquid flow rate, CAD evaporation temperature T_e), and solute (e.g., pK_a, charge state). The effect of mobile phase additives on the response of ionizable species has been shown to be highly relevant to response uniformity (see Section 4.4.2).

T_e is a key instrumental setting with regard to response uniformity. Here, it is recommended to use the lowest T_e setting that consistently produces the required sensitivity limits and where the ideal goal is for all analytes to behave as nonvolatiles. This generally provides more universal detection, more uniform response and enables more accurate quantitation.

The ability to use a lower T_e is, however, limited as it can lead to higher baseline noise due to more sensitive detection of less volatile impurities that are likely to be present in most experiments. The ability to use a low T_e in order to realize these performance benefits mainly depends on the cleanliness of the eluent as detailed in Chapter 3. A further limitation is the requirement to ensure that all droplets are fully evaporated prior to detection. This is especially challenging with aqueous eluents where the aerosol flow path is, in many conditions, saturated with water vapor [i.e., 100% relative humidity (RH)]. Some ELSD instruments allow the use of sub-ambient T_e, where a post-nebulizer gas flow is used to reduce %RH. However, this may be counteracted by the resulting higher gas flow velocity, which decreases the time for evaporation. Most literature examples of improved response for semivolatiles at sub-ambient T_e are observed for eluents with relatively low aqueous content [30].

4.4.2 Salt Formation with Ionizable Analytes

Evaporation of aerosol droplets produced by nebulizing an LC eluent leads to a rapid increase in the concentration of less volatile components thereby favoring solute-solute interactions. Several studies have shown that volatile acidic and/or basic mobile phase additives can be used to increase detection scope and influence sensitivity for ionizable analytes. This can be attributed to salt formation and therefore greater "combined" mass in the residue particles [27–29]. This can be advantageous by enabling better detection of volatile analytes (e.g., volatile basic analyte + volatile acidic modifier) or disadvantageous leading to higher background signal and noise (e.g., volatile basic impurity + volatile acidic modifier) or decreased response uniformity among analytes (e.g., differential interaction between analytes and a given volatile modifier). While the choice of mobile phase additive is usually driven by requirements of the chromatographic separation technique (i.e., pH modification, pH buffering, ion pairing), possible influences on aerosol detection should also be considered. The above references provide useful insight to the main factors influencing response for ionizable solutes. For example, Cohen et al. [27] described the main physicochemical properties influencing detectability and sensitivity of volatile basic analytes as: (i) pK_a's of both the acidic modifier and conjugate acid of basic analyte; (ii) number of ionizable functional groups on a given molecule; and (iii) steric effects (i.e., accessibility of ionizable site). While this study addressed basic analytes the same principles should apply to acidic analytes. Russell et al. [28] studied a diverse set of acidic, basic and neutral analytes spanning a range of properties related to volatility (e.g., M, B_P, melting point). For nonvolatile analytes, somewhat higher response was obtained for basic compared to neutral analytes when using volatile buffers such as ammonium formate and ammonium acetate or acidic modifiers formic acid (FA), trifluoroacetic acid (TFA), and heptafluorobutyric acid (HFBA). The response for basic analytes increased monotonically with molar mass of the acidic modifier (i.e., FA < TFA < HFBA). Evidence suggests that the likelihood that ionizable additives will contribute to the mass of a stable aerosol particle increases with both the additive concentration and extent of ionization (i.e., related to solution pH and strength of acidic and/or basic additives) and decreases with the volatility of the resulting salt [27, 31–33].

Figure 4.1 shows an example from FIA where uniform response was obtained for most analytes, but lower response was observed for oxalic acid, which is attributed to its semivolatile behavior. Higher response was observed for 2 of the 3 compounds whose starting materials were either Na or Cl salts. Since there was no chromatographic separation, diclofenac was detected as its Na salt while dopamine and guanidine were detected as their Cl salts. The observed increase in response for all three injected salts directionally corresponds to the molar mass (M) ratios of the respective salt and free analytes forms, M_{salt}/M_{free} (i.e., 1.60 for guanidine, 1.23 for dopamine and 1.08 for diclofenac). The lowest increase in response was therefore observed when the counterion Na ($M = 22.9$ g/mol) had the lowest mass contribution to the diclofenac-Na salt ($M = 296.15 + 22.9$ g/mol). Conversely, the highest increase was observed for guanidine ($M = 59.07$ g/mol) with its Cl ion ($M = 35.45$ g/mol). When such analytes are separated chromatographically

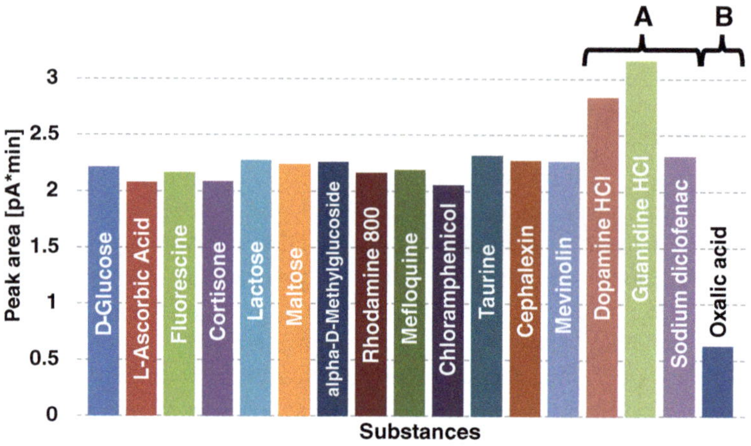

Figure 4.1 Factors that can affect the response uniformity of CAD (A: salt formation, B: volatilit). Flow injection analysis, 0.5 µg injection (diclofenac as free acid, dopamine and guanidine as free base), isocratic 80% aqueous ACN.

as ionized solutes, they can form salts with mobile phase additives. As illustrated with diclofenac, the use of low *M* additives (i.e., FA, ammonium formate) at low concentrations should minimize the influence of salt formation and generally produce the most uniform response for diverse analytes. Stronger and/or higher *M* additives may also be chosen to enable better detection of volatile ionizable analytes. An example is shown in Table 4.2, where the response of oxalic acid, which behaved as a semivolatile with no additive, was increased by the addition of 0.01% triethylamine (TEA). Further examples of the use of higher *M* additives (e.g., triethylamine TEA, TFA, HFBA) to enable or improve detection of otherwise volatile ionizable analytes include the use of TFA additive to detect ammonium, a very common mobile phase additive [27, 35]. While these approaches are quite useful for specific, targeted analyses, use of higher *M* additives can, more generally, have adverse effects on response uniformity and other method performance characteristics as detailed in Chapter 3.

The effect of salt formation on response uniformity can be corrected mathematically if one knows the extent of ionization and number of ionization sites of the analytes and additives. This can be inferred from the eluent pH and the pK_a's of acidic species or conjugate acid of basic species. Considering that additive concentrations, in most cases, should be several orders of magnitude higher than typical analyte concentrations within the column eluent [36], it is reasonable to expect that additive ions are usually of sufficient concentration to readily form salts with ionic analytes. As a general practice in method development, the mobile

Table 4.2 Peak areas for 0.5 μg of the semi-volatiles folic acid, oxalic acid, and guanine, with and without salt formation, based on data from Menz et al. [34].

	Peak area (pA*min)		
Compounds	**Without TEA**	**With TEA**	**With TEA, after salt correction**
Folic acid	0.319	2.430	1.977
Oxalic acid	0.626	5.583	1.719
Guanine	0.089	3.389	2.029

phase pH is often chosen to be ideally 2 units away from the pK_a of the analytes so that they are either predominantly ionized or predominantly neutral. This can simplify the approach toward single calibrant quantitation but is not always possible. In the common case of low pH reversed phase (RP) chromatography, (e.g., water–ACN eluents containing 0.1% FA) ionization of weakly acidic analytes is suppressed while most basic analytes are fully protonated and carry a charge. Under these conditions, higher response would therefore be obtained only for basic analytes. With FA as the only modifier, the corrected response would be obtained by multiplying the measured response by $M_{\text{analyte}}/(M_{\text{analyte}} + M_{\text{formate}})$. Since M_{formate} is only 45 g/mol, the correction may be relatively small (e.g., <10% for a 500 g/mol analyte). Another example is the analysis of anions and cations, a common application area for CAD (Chapter 6). These analyses most often use eluents with ammonium formate or ammonium acetate buffers, and the analytes can be assumed to be detected as their ammonium (M = 17), formate (M = 45), or acetate (M = 59) salts. It should be emphasized that the need to correct for salt formation is only a consideration when quantitation in the absence of individual standards is necessary. Depending on the application, correcting for salt formation may be deemed unnecessary, especially when using low M additives and higher M analytes. However, when using higher M additives the error introduced for ionizable analytes can become significant and such corrections may be needed to obtain the required level of accuracy.

Menz et al. [34] studied CAD response by FIA of 58 chemically diverse analytes whose M spanned the range of 59.1–604.5 g/mol (Mean 241, SD 116.7). Several analytes that behaved as volatiles or semivolatiles were excluded. Response for the remaining 36 analytes, after accounting for salt formation, is shown in Figure 4.2. The variability (%RSD) in response for these analytes was 5.8, which is comparable to that of other studies. It should be noted that additional sources of variability from this study may include purity of the starting material, changes

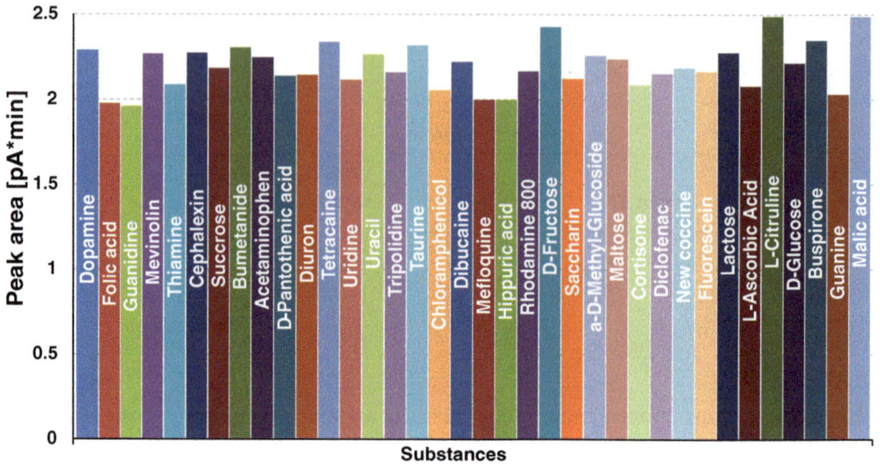

Figure 4.2 CAD response to 0.5 μg of each of 36 compounds by flow injection.

during preparation (e.g., water absorption), weighing and dilution errors, analyte degradation and solubility. Also, given the number of low M analytes, use of a T_e lower than the described 35 °C may further improve response uniformity.

4.4.3 Composition of Mobile Phase

Solvent composition of the mobile phase has the second greatest impact amongst the method parameters, and it affects the response of any solute to a relevant extent, regardless of its chemical properties. It can cause up to fivefold response variation in a full range gradient from purely aqueous to 100% ACN, with a nonlinear dependence. Response curves as a function of organic content can somewhat vary between solutes and often exhibit a response maximum between 60% and 80% ACN content. An empirical response model of Hutchinson et al. [6] is based on the RFs of four chemically diverse substances from 0% to 100% ACN in 10% steps. Second-order polynomial equations correlated detector response to %ACN with correlation coefficients of >0.999. The model was initially verified in FIA experiments of 19 chemically diverse solutes at 50% ACN. The solutes were characterized by principal component analysis using log$P$ value, log$D$ value (pH 2.6), pK_a, molecular weight, boiling point, enthalpy of evaporation, polar surface area, refractive index, surface tension, charge at pH 2.6, density and polarizability. The calculated RFs were compared to the experimentally measured factors with FIA and showed relative errors of 0–10% for eight solutes from the set and of <40% for another 10 compounds. Five compounds, all with semi-volatile character, showed 40–380% deviations. In the last step, the correction function was extended for consideration of peak dispersion in a linear gradient from 0% to 100% ACN over 5 minutes at a flow rate of 1 mL/min, applied to an Acclaim™ Polar Advantage II column (75 mm × 3.0 mm, 3 μm). The correction factor to consider the decrease of solute concentration in the peak apex is not needed when the response is linearized using the power value function available on more recent CAD models (see Chapter 1, Section 1.5.1). Application of a PV (power value, formerly power function value, PFV) makes this approach more attractive, because each new column format no longer requires determination of a specific correction factor. To test the model over 0–100% ACN, responses for 16 nonvolatile samples were calculated, followed by real testing on the HPLC system. The relative errors for 12 compounds were <24%. Errors for 4 of those 12 were even <10%. The worst four samples showed much higher deviation, but the authors presented plausible root causes why the specific analytes did not follow the same response pattern as the compounds used to construct their model or when practical peak area evaluation problems occurred [7].

Technical approaches to address the solvent effect without the need to build a response model, but to keep the true detector response as uniform as possible with gradient methods will be presented in Section 4.5.

4.4.4 Influence of Solute Density

The impact of particle bulk density on CAD response is typically considered minor compared to other factors. As described in Chapter 1, this influences response weakly because of the cube root relationship between particle diameter and mass (e.g., Equation 1.5) and can be compensated by multiplying the response by the cube root of the analyte's density. This compensation may not be fully accurate, however, since the packing density that results from spray-drying may differ from analyte density and the residue particles are likely to include impurities. Also, density values, whether determined empirically or calculated, may not always be accurate and are not available for unknowns. For ionizable analytes, density of the salt form may be more relevant.

Literature commonly reports that response of ELSD and other nebulization-based detectors only has a small dependence on density [18]. Russell et al. [28] could not identify an effect of solute density on detector response of 29 probe compounds, even though their applied theoretical model predicted it. Liu et al. reported a FIA-based response comparison of 11 mostly nonvolatile compounds covering a density range from 1.1 to 1.7 g/cm^3 (at 25 °C) [37]. They found that the density showed a negative correlation with response intensities. Thus, higher compound density resulted in lower response as expected. Based on their results with relatively small density range amongst the investigated compounds, they considered this influence minor and not adverse to response uniformity.

The above reports of negligible influence of density on response may be attributed to the weak (cube root) influence that may be further dampened by the presence of impurities. Some studies, however, have shown more significant effects. Matsuyama et al. [8] investigated a total of 16 (mostly) brominated flame retardants dissolved in chloroform or toluene applying FIA or direct infusion experiments. Densities were taken from literature if available or pycnometrically measured and ranged from 1.21 to 3.54 g/cm^3. While the expected response difference, based on a cube-root relationship, is expected to be a factor of 1.5, the reported difference in response was as high as a factor of 2.5. This discrepancy is most likely related to the reported nonlinear response. This illustrates the importance of using an optimal power value to linearize response as described in Section 4.5.4. Additional examples of significant effects on particle density, include the higher sensitivity obtained for Li ion compared to other ions (Chapter 6) and the differences in response found for large linear and branched polyethylene glycols (see Section 4.5.3). From these examples, it is apparent that opportunities exist to refine current models to gain an increased understanding and to develop more robust calibration models that account for density. Efforts, for example, to further understand control parameters in spray-drying applications for pharmaceutical manufacturing may be relevant to this end. Regardless of the correctness of the model or the tightness of the correlation, it is a valid conclusion that uniform response with RSDs in the single-digit percentage range as was reported in many publications and can be expected in most studies.

4.5 Practical Strategies for Obtaining Uniform Response to Realize Global Quantification

In Section 4.4 of this chapter, we reviewed the influence of physicochemical solute parameters on the CAD response. We also examined the specific chromatography method conditions, which can be controlled to a certain extent by the analyst in the laboratory. We have seen that nonvolatile solutes that elute neither as anions nor cations and that exhibit a narrow density range in solid state show very uniform response with CAD. A uniform response requires substances to be introduced into the detector in a well-defined, constant mobile phase, which is not always possible in HPLC. In this section, we will describe methods for improving uniform response by ensuring a constant mobile phase to the detector, including isocratic methods and IGC. We will also detail practical approaches to optimizing the IGC.

4.5.1 Uniform Response Using Isocratic Methods

The most obvious approach is to keep the composition of the mobile phase constant (aka isocratic), at least for the part of the separation where the analytes of interest are eluted. Such methods can still include wash and conditioning steps performed by gradient pumps outside of the elution windows. Poplowska et al. [38] demonstrated the unified quantification of synthetic cannabinoids with LC-CAD, even though reference standards were unavailable for most of the analytes. They applied LC–ESI-QTOF-MS/MS for identification and quantitative NMR to verify correctness of quantification. Isocratic separation was achieved in RP mode on a C18 column with mobile phase of water/ACN/FA (40 : 60 : 0.1, v/v/v) for elution of 21 relevant cannabinoids. After each separation, the column was cleaned for 14 minutes with ACN/FA (100 : 0.1, v/v), (2 minutes linear gradient transition), and then re-equilibrated. Two available reference standards were selected as universal reference calibrators. The LC-CAD method was validated for unified quantification of synthetic cannabinoids without the need for corresponding reference standards and 19 different herbal mixtures could be successfully analyzed with this method.

4.5.2 Uniform Response with Inverse Gradient Compensation

The example given above takes advantage of uniform response with CAD and isocratic methods. There are many analytical challenges, where the application of constant eluent composition is not possible, because of a too wide polarity range of the solutes. Such separations require programmed increase of mobile phase solvent strength, aka solvent gradients. Since a full range RP gradient alters analyte response up to fivefold, a countermeasure is needed to achieve uniform response across the solvent gradient. In the attempt to resolve the general elution problem without applying a solvent gradient in LC-CAD analysis, Khandagale et al. applied a temperature and flow gradient at constant mobile phase composition and thus obtained normalized response with such a different gradient method [39]. The study involved eight sulfonamide drug compounds spanning a range from the quite hydrophilic solute sulfamerazine up to the strongly hydrophobic indapamide, a mixture impossible to separate on a C18 column with isocratic RP chromatography. An Acclaim C18, (2.1 × 100 mm, 3 μm) was used. At a constant column temperature of 30 °C, a linear gradient from 20% ACN to 80% ACN resolved the elution challenge successfully but caused a more than threefold response variation. Alternatively, at a constant mobile phase composition of 20% ACN, all compounds could be eluted with a temperature gradient running from 30 to 80 °C, albeit with a 75% increase of analysis time. Another iteration was a combination of the 20 to 80 °C temperature gradient with a flow gradient from 0.2 to 0.6 mL/min on the 2.1 mm i.d. column format, which only started after the elution of the first six compounds. The additional flow gradient in the later phase of the elution reduced the analysis time only by 15%, decreased the peak width by driving the zones at threefold speed through the detector and resulted in almost twofold peak height increase for the latest eluting compound. Peak areas of all sulfonamides were quite close with this combination of gradients. While this study introduced an intriguing alternative to composition gradients for uniform response with LC-CAD, its practical relevance is questionable. None of the major vendors of HPLC instruments can perform temperature gradients in the default configuration, hence a customization with special column thermostats is required.

The concept of IGC was first introduced by Gorecki et al. in 2006 [5] and, with close to 300 citations, is considered as a milestone for implementation of uniform response in CAD applications. The authors used a C18 column and separated six sulfonamides with a multi-segment gradient from 10% ACN to 90% ACN, the aqueous component included 0.1% FA. To implement the IGC concept, the binary gradient system was equipped with a second binary gradient pump of the same type and a second flow path with a column equivalent to the separation column as well as a fluidic connection to mimic the volume of the autosampler. As detailed in Section 4.5.3., the addition of a second column is not recommended. The outlet capillaries of both columns were connected via a T-piece with the CAD. The second pump was programmed to deliver the mirrored (i.e., inverse) gradient to the elution gradient, starting with 90% ACN and ending at 10% ACN, thus keeping the eluent composition in the nebulizer of the CAD constant at 50% ACN. The flowrate on both pumps was 1 mL/min, hence the flowrate entering the detector was 2 mL/min. With the application of the IGC, the RFs of all six sulfonamides became almost identical. At 800 ppm concentration the peak area of sulfaguanidine was only 20% higher than that of sulfamethizole (representing the two extremes). However, when no IGC was applied, the last eluting compound sulfadimethoxine exhibited a 200% (factor 3) higher peak area than the first eluting compound sulfaguanidine. The authors positioned their solution to be the "door opener" for quantitation of unidentified species with single-compound calibration. Gorecki et al. already predicted the applicability of IGC to the ELSD in their paper. The successful transfer to ELSD was published 1 year later by De Villiers together with the authors of the IGC for CAD paper and feasibility was demonstrated with the same set of sulfonamides as solutes [40].

An application note from Menz et al. on the quantification of paclitaxel degradation products demonstrated the implementation of IGC using a Thermo Scientific Vanquish Dual Gradient instrumental set-up as illustrated in Figure 4.3 [41]. This system can include an additional UV detector. The UV detector can provide additional analytical information, especially if it is a diode array detector type that can also record UV spectra. If a UV detector is included, it should be installed between the separation column outlet and the T-piece that combines the two flow paths. Unlike CAD with its mass-dependent response, UV-detector response is concentration-dependent so its response would thus suffer from the twofold dilution behind the T-piece with the IGC setup.

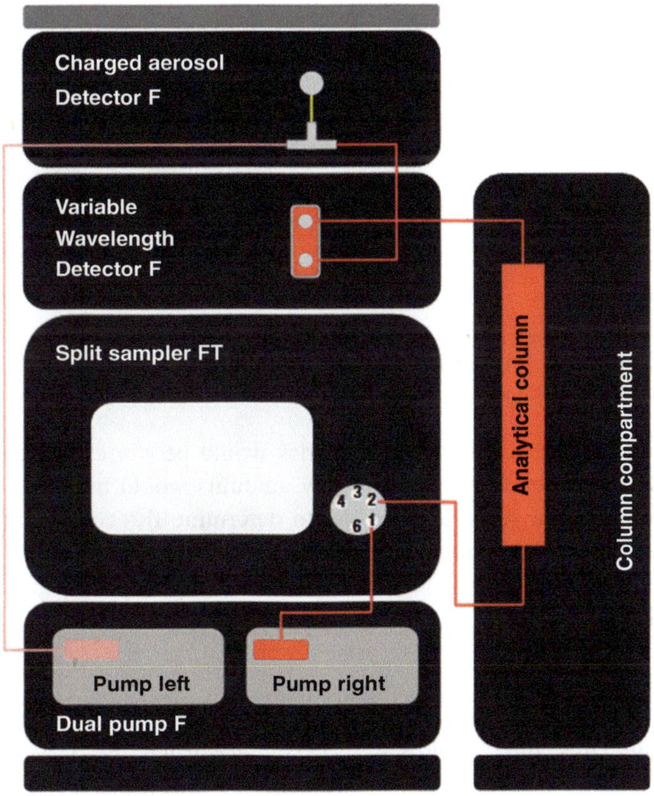

Figure 4.3 Fluidic scheme and instrumental set-up of the Thermo Scientific Vanquish inverse gradient solution for CAD.

In scope of the study with degradation products from paclitaxel, the IGC concept and instrumental setup was used to evaluate the benefit of IGC over the non-compensated LC-CAD analysis, as well as the benefit of the additional UV detector for comprehensive characterization of a stressed solution. First, the RFs with IGC-CAD and UV detection were compared for two known impurities of paclitaxel with available standards. The respective results are shown in Figure 4.4 as overlaid IGC CAD calibration curves and as normalized bar charts to compare IGC CAD and $UV_{227\,nm}$ for response uniformity.

While the CAD RFs were perfectly consistent, UV response varied markedly. Second, the IGC solution was then applied to analyze thermally stressed paclitaxel samples. The total peak areas for all impurities without IGC summed up to 63.9% of the API peak area, but only to 53.8% of the API when using IGC. This large (>10%) difference in determined impurity content is attributed to the influence of solvent composition on response with CAD and it is also an obvious

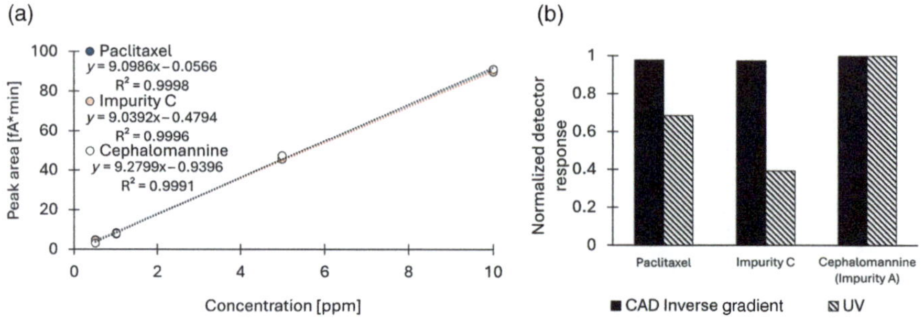

Figure 4.4 IGC-based CAD calibration curves obtained for paclitaxel and its related compounds (left side) [41]. Comparison of UV response at 227 nm and IGC CAD response at 10 μg/mL for the three analytes normalized to cephalomannine (right side).

conclusion that the content of late-eluting impurities would be overestimated without applying IGC, while early-eluting impurity amounts would be underestimated. IGC with CAD is a powerful solution to determine the content of unknown impurities in pharma samples, while the common application of relative peak areas with UV detection to determine what falls below the disregard limit is a questionable practice.

4.5.3 Troubleshooting Inverse Gradient Compensation

If a particular application does not provide a uniform response when the IGC is in use, the following checks should be implemented. Any trend in response that is associated with changes in retention time is likely due to a problem with the IGC implementation. For effective IGC implementation, the eluent from both pumps must reach the detector simultaneously. The gradient delay volume of both pumps plus the void volume of the column, the volumes of the capillaries on both sides of the system and any other pre-detector contributions to system dead volume must be summed up. Even with a precise calculation, the volumes should be verified. Eluents similar to those used in the application but with a void volume marker in eluent B can be used to test the system. For example, a RP application can be evaluated using a C18 column, with 80/20 ACN/water (v/v) as eluent A and 0.1% uracil marker spiked into 80/20 ACN/water (v/v) for eluent B. Uracil is a good choice for a void volume marker because it has minimal interaction with RP columns. Incorrect validation of the volumes in the inverse and analytical gradient quickly becomes evident due to signal deviations during testing [42]. A deviation of just 96 μL leads to the differences in gradient composition shown in Figure 4.5.

The addition of a second column, identical to that used in the analytical flow path, to the inverse gradient flow path is not recommended. The uniformity of the

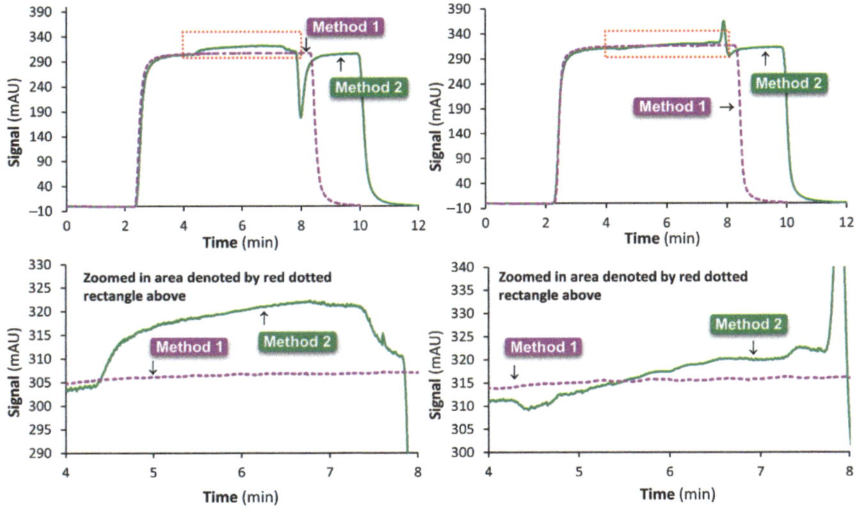

Figure 4.5 Left, top and bottom: Demonstration of the effect of improper timing of the inverse gradient by 96 µL at a flow rate of 0.6 mL/min, a mismatch of 9.6 seconds. Right, top and bottom: Effect of proper timing of the inverse gradient after reducing the flow path by a volume of 96 µL. *Source:* [42] / with permission of Elsevier.

combined IGC with one or two columns in the flow path was tested for an acidified water–ACN gradient by measuring flow rate, mobile phase density and assessing mixing effectiveness of analytical and inverse gradient flows (Figure 4.6). The second column adds no benefits to the uniformity of the mixed flow. The mobile phase composition is expected to be constant with IGC, hence the stability of the mobile phase density is a sensitive measure for the H_2O/ACN proportioning. The figure shows no difference in the stability of the mobile phase composition when identical columns are used on both flow paths compared with when only an analytical column in the analytical flow path. Additionally, the addition of a second column does not eliminate the need to compensate the volume mismatch between the two flow paths. The analytical gradient flow path includes the autosampler and the inverse flow path does not, so the inverse gradient still needs to be delayed, even if a second column is inline.

If the peak areas of early eluting peaks are smaller than expected and the late eluting peaks are greater than expected, there may be an irregularity in the supply of organic solvent from the inverse gradient pump [42].

A literature example of response associated with retention time can be found in Lisa et al. [43]. This report investigated the potential for CAD to

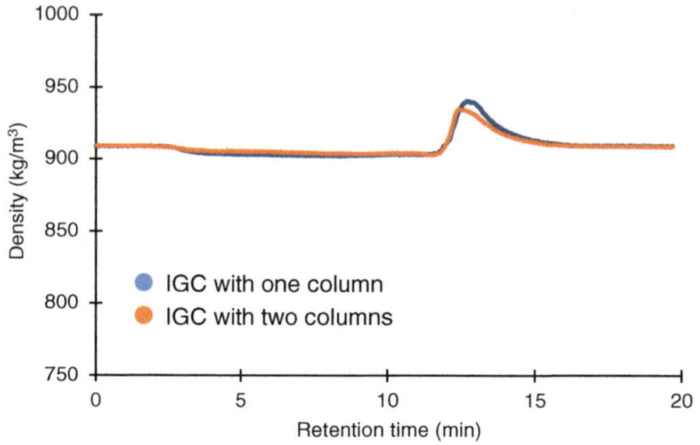

Figure 4.6 Comparison of mobile phase density progression for the combined flow with (orange) and without (blue) a second column in the inverse flow path. Experiments were done with an Accucore™ aQ C18 column and a gradient over 10 minutes from 100% water with 0.1% FA to 100% ACN with 0.1% FA. A flow/density meter mini CORI-FLOW™ ML120V00 (Bronkhorst®, Newmarket, UK) was used to measure flow rate, density and assess mixing effectiveness of analytical and inverse gradient flows.

provide simple quantitative analysis of triacylglycerol in plant oils without using analyte-dependent RFs. The team reported excellent uniformity for their purpose since most of the triacylglycerides (TAGs) of interest fell within the flat region of the gradient. The analytes were expected to behave as nonvolatiles and deviations in the response for triacylglycerols with both shorter and longer chain fatty acids was surprising. The deviations correlate with retention order, as shown in Figure 4.7, part a. The problem cannot just be a mismatched timing of the inverse gradient, as the linear gradual gradient would only be mismatched at the beginning and then the mixed gradient composition would be constant over the entire run, although with less %B than expected. The unusual RP mobile phase, consisting of a mixture of isopropanol and hexane in A and ACN in B, and the columns containing high levels of silanols installed on both the inverse and analytical gradient flows may lead to unexpected partitioning effects between the columns and the mobile phases or unexpected behavior from the pump due to the large pressure difference between isopropanol/hexane and ACN, resulting in an uneven application of the IGC. A rough calculation of mean droplet diameter using the Nukiyama-Tanasawa model of Chapter 1 indicates that ACN would give a better response than 50 : 50 IPA:hexane. The compensation did improve the response uniformity. The C9 : 0 (trinonanoin) response factor was 0.69 without compensation and 0.83 with compensation. Another literature example of response associated with retention time is published by

Peng et al. [44]. The extensive saponin analysis was used to predict quantities of saponins in notoginseng. Complicated structure-response relationships were developed with few data points based on molecular structure. As is well established in Chapter 1, CAD response does not depend on molecular structure. Two more likely explanations for the diverse RFs are a timing mismatch of the inverse gradient and disparate purities of the standards. The relationship between retention time and response factor is plotted in Figure 4.7, part b. The difference between the analytical and inverse gradients was calculated as 0.3 minutes (300 μL), but the Chromeleon Inverse Gradient Wizard calculates a delay of roughly 3 minutes. The lower-than-expected response of early-eluting components and higher-than-expected response of late-eluting components suggests an incorrectly applied IGC. Another possible source of uncertainty is the purity of the saponin standards used. A quality control purity of ≥98% by HPLC often indicates a quality check by UV detector, which cannot detect salts or impurities that lack chromophores. Standards should be checked for chromophore-free impurities by injection into the CAD.

Another report [45] argues for chain-length dependent response for homologs of polyethylene glycol hexadecyl ether (Brij C10, alcohol ethoxylates), obtained from self-prepared standards. The fraction collector required 65 runs of a HILIC method to obtain the homologous standards. The longer-chain samples with later elution times and lower RFs would have been isolated along with as much as 3 mg of ammonium formate (~0.01–0.04 mg per run, depending on gradient delay time). The mild drying step at room temperature may not have removed all the ammonium formate from the later samples and there were no control fractions to prove a clean evaporation of the additive in the eluent. The generally inverse relationship between ammonium formate in the fractions and the response (slope) shows a more promising correlation (Figure 4.7, part c) than the relationship between homolog length and response shown in the paper. The purity of standards plays a large role in detector response, and a surrogate standard of the highest possible purity should be obtained for quantification of unknowns.

Pawellek et al. [46] performed machine learning on CAD data from a fatty acid panel of seven fatty acids from C12 to C18 and found that the mobile phase flow rate, the evaporation temperature and the content of organic solvent in the mobile phase all affected the CAD response. They also found that the distance of the carboxyl groups from the geometrical center of the molecule (carbon chain length) and the topological distribution of molecular charge and mass played a role in the CAD response. Practically speaking, these factors affect the packing of the particle, and their influence can be somewhat offset by correcting for density using the model introduced in Chapter 1 (multiplication of response by the cube root of density). Taking the nonvolatile fatty acids and correcting for density yields the result shown in Figure 4.7, part d and reduces the %RSD among the five substances

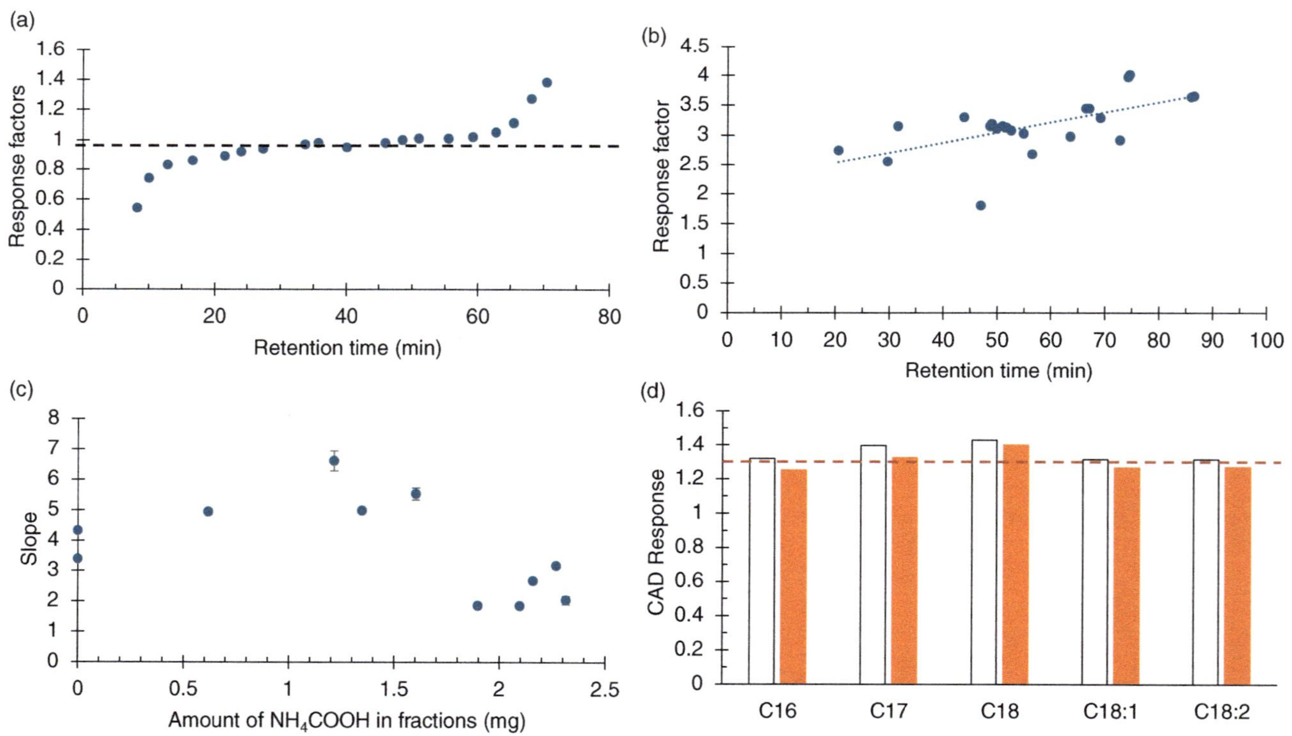

Figure 4.7 Exploration of the effect on real data of some of the implementation tips for IGC. (a) Association of the retention time of triacylglycerols and the response factor found in Lisa et al. [43]. (b) Association of the retention time of saponins and the response factor found in Peng et al. [44]. (c) Dependence of the slope (response) on the amount of ammonium formate in the fractions collected to produce the alcohol ethoxylate standards [45]. Error bars reflect the reported uncertainty. (d) Correction of the original CAD response to fatty acids (black outline bars) for density (orange bars) from data in Pawellek et al. [46]. *Source:* Figure produced by the authors for this work. (a) Adapted from [43]. (b) Adapted from [44]. (c) Adapted from [45]. (d) Adapted from [46].

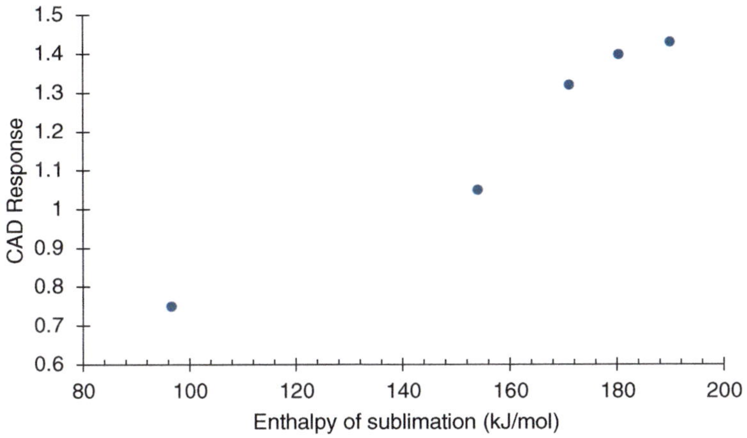

Figure 4.8 Inverse dependence of CAD response on the enthalpy of sublimation for the C12 (96.7 kJ/mol), C14 (154.1 kJ/mol), C16 (171.1 kJ/mol), C17 (180.4 kJ/mol), and C18 (190.0 kJ/mol) saturated fatty acids (data from right to left) [47, 48].

from 10% to 9%. Beyond the density, which affects both CAD and ELSD particle, much of the 9% variability may be due to the fact that the saturated fatty acids sublime, as shown in Figure 4.8. Therefore, even after the formation of a particle, additional mass may be lost in the detector. The unsaturated fatty acids do not sublime because they do not form solids at CAD-relevant temperatures. As discussed in Chapter 5, even the C18 and longer chain fatty acids show some volatility at ≤35 °C, further contributing to RSD.

An important comparative study of CAD versus differential refractive index detection (dRI) and multi-angle light scattering detection (MALS) for analysis of large linear and branched polyethylene glycol compounds showed 10% RSD for the response to linear PEGs between 1.400 and 21 kDa molecular weight [49]. Similarly, the result was 11% RSD for 4-arm PEGs of 5, 10, 20, or 40 kDa molecular weight. These %RSDs were worse than those for dRI and, comparing only PEGs with the same molecular weights but different structures, worse than those for MALS. An explanation in the article ascribes the difference in response to structure. But any attempt to correlate CAD response with structure omits the fact that, even for these larger molecules, it is very clear from Equations 1.1 and 1.2 in Chapter 1 that the particles measured by CAD consist of many molecules. For example, a 10 nm particle of the largest PEG in the study (40 kDa) would be comprised of, very roughly, about 10 molecules, while a 50 nm particle would have more than 1000 molecules. At the mass levels injected (>500 ng) and with a 100% aqueous eluent, most of the signal should come from particles ≫10 nm (see Figure 1.8 in [50]). The smaller PEGs would have many more molecules per particle. The fact that higher MW PEG or less branched PEG has a larger diameter than other types of PEG does not affect response in CAD because each particle is comprised of so many molecules. Additionally, the molecular density of a polymer in an ideally packed state does not equal the particle density after spray drying, which is relevant for CAD response, as discussed in Chapter 1, so it is not clear that, as stated in the text, PEGs with more arm branches should have a higher density compared with PEGs with less arm branches. Measurement of density after spray-drying is possible but not commonly performed. From Equations 1.1 and 1.2 in Chapter 1, one might speculate that the bulk particle density after spray-drying would decrease with increasing number of arms leading to an increased response.

Comprehensive analysis of oils and fats requires separation of all contained lipid species that can encompass free fatty acids (FFAs), monoacylglycerides (MAGs), diacylglycerides (DACs), and TAGs, the latter being the main ingredients. The elution of such highly hydrophobic compounds from a C18 column requires solvents with greater elution strength than ACN, including solvents that are immiscible with water. This approach is a subset of RP LC and is typically named nonaqueous RP LC (NARP). Causevic et al. [51] report the development of a NARP method with CAD detection for analysis of vegetable oils and fats, applying the IGC instrumental set-up as depicted in Figure 4.3, but without a variable-wavelength detector. The method was tuned to separate FFAs, fatty acid esters (FAEs), MAGs, DAGs, and TAGs using a total of 33 standard compounds with representatives from all these groups. To generate sufficient peak capacity, they coupled two long UHPLC columns (Nucleodur® C18, 250 mm × 2.1 mm, 1.8 μm), used a flow rate of 0.2 mL/min and applied a shallow ternary gradient program over 80 minutes. The composition started with 100% ACN held for 16 minutes, then increased elution strength by adding isopropyl alcohol (IPA) in segments up to 35/65 (v/v) ACN/IPA at 45 minutes. The last segment added heptane as a third component successively replacing IPA to further increase elution strength and ended with a ternary composition of 35/48/17 v/v/v ACN/IPA/heptane at 80 minutes. To achieve sufficient retention and selectivity for unsaturated FFAs, FAEs, and MAGs, a very low column temperature of 9 °C was kept for the first 18 minutes of the gradient and then switched to (a still sub-ambient) temperature of 15 °C for the remainder of the gradient where the saturated FFAs, FAEs, and MAGs, as well as all DAGs and TAGs were eluted. The authors report substantial variation in RFs between different groups of lipid species and attribute this to solute volatility. FAEs showed by far the poorest RFs, up to 30-fold lower than the highest TAG values. The RFs of unsaturated MAGs and FFAs clustered closely together at levels about fivefold higher than those of FAEs. Saturated MAGs also formed a close group, with RFs about twice those of FFAs. DAG and TAG RFs varied within a 2.5-fold range and overlapped with those of MAGs and FFAs. Surrogate standards for specific lipid species classes were defined that also allowed the quantification of substances where no calibrants were available. The authors referred to the possibility of compound identification with MS to ensure correct assignment of peaks to a lipid class. Interestingly, their attempt to apply an IGC with the dual ternary gradient pump applying a 0.2 mL/min flow rate on both pumps did not provide any visible improvement at response uniformity. The authors attributed this to the purely organic mobile phase gradient with the NARP method. Several factors may have contributed to the lack of improvement in response uniformity when using IGC. For example, the ability to accurately compensate a ternary analytical gradient with an inverse gradient may be especially challenging. Also, lower RFs were obtained for nearly 50% of the analytes including FAEs, FAs and some MAGs. This may be attributed to volatility. Even with IGC, uniform response cannot be obtained with semivolatiles. Also, peak shapes for TAGs, which are assumed to behave as nonvolatiles, varied from very sharp (ca. 4 seconds at base) to significantly tailing (>120 seconds at base). With a data sampling rate of 2 Hz the sharpest peaks might have been under-sampled, effectively reducing their RFs. On the other hand, RFs for severely tailing peaks may have been too high because response was found to be sub-linear (Chapter 1). Peak shape that deviates from Gaussian shape can be improved with an optimal PFV, as discussed in Chapter 1 and Figure 1.9. It would be interesting to see if further optimization of IGC, a lower T_e, a higher data acquisition rate and an optimal PV could improve the response uniformity of this method.

The use of CAD with packed column supercritical fluid chromatography (pSFC) was first reported by Brunelli et al. in 2006 [52]. The paper also describes the application of a post-column addition to account for a constant mobile phase composition in the detector. As commonly known, SFC applies carbon dioxide (CO_2) compressed and thermostatted above its critical parameters (p = 74 bar, T = 31 °C) as mobile phase and a polar stationary phase, hence building a normal phase (NP) chromatography separation system. A backpressure regulator (BPR) behind the column is required to control the pressure inside the column independent of the flow rate and it thus controls the CO_2 density which is important for its elution strength. Typically, an organic modifier is added by an additional pump to generate an elution gradient. MeOH is mostly used as modifier. It increases the mobile phase polarity and thus its elution strength, following the NP chromatography principle. To maintain a constant MeOH-to-CO_2 ratio at the CAD inlet, a second pump delivers an inverse gradient mixed with a T-piece to the column-effluent in front of the BPR. The instrumental setup is shown in Figure 4.9.

A UV-detector was also included upstream of the mixing-T, like the IGC setup for HPLC (see Figure 4.3). The CAD was connected via a stainless steel transfer

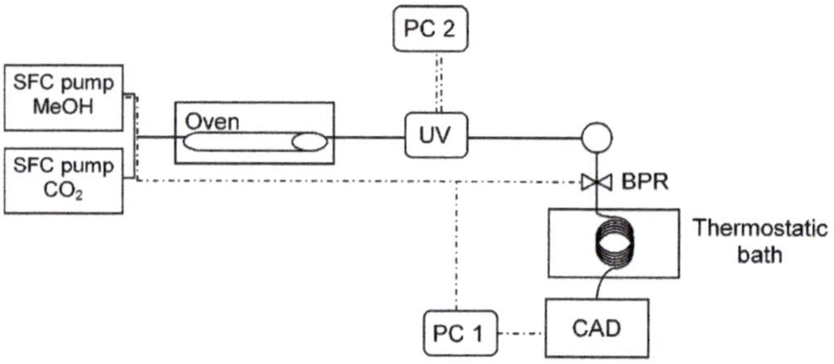

Figure 4.9 Configuration of the pSFC-UV-CAD system with MeOH gradient compensation. *Source:* [53] / with permission of the American Chemical Society.

line (1 m × 0.25 mm i.d.) placed in a thermostatic bath at 45 °C. The authors performed a proof of principle study to demonstrate uniform response employing nine nonvolatile organic compounds. The separation was run on an ethylpyridine column (250 mm × 4.6 mm i.d., 3 μm), thermostatted at 40 °C, operated at a flow rate of 2.0 mL/min and applying a gradient from 10% MeOH to 35% MeOH. Two methods for compensating the MeOH gradient to achieve constant CAD response were evaluated. The first method involved adjusting the flow rate of the additional MeOH to keep the amount of MeOH reaching the detector constant. The second method involved adjusting the amount of additional MeOH to maintain a nearly constant CO_2/MeOH ratio in the detector. This second method provided a more uniform response, in fact almost perfectly uniform for eight compounds (theophylline, testosterone, cortisone, and five sulfonamides). The only exception was naproxen, which exhibited a response of only 70% relative to the other compounds, a deviation that was not further discussed by the authors. The developed method was then applied to a mixture of seven steroids and proved its uniform response characteristics. A comparison with UV-detection for both analyses demonstrated the clear uniform response advantage of CAD with gradient compensation. Depending on the applied wavelength, UV detection produced major RF variations of up to 30-fold. Even at the optimal wavelength, the RFs varied by more than threefold.

Scholz et al. more recently demonstrated the ability of this approach to provide uniform response for biosurfactants [54]. Four exophilin and four liamocin species with mannitol and arabitol as head groups, were separated by means of SFC on a charge modulated hydroxyethyl amide functionalized silica-based column. With this method, it was possible to separate the species either based on polyol head group or different polyester tails, a clear advantage over RP HPLC methods. A quantification method for the individual species by SFC-CAD was developed using an external calibration curve of the standard compound n-octyl-β-d-glycopyranoside and validity confirmation by mass balancing, i.e., the total amount of all individual measured concentrations of the liamocin and exophilin species was reported to be in good accordance with the dry weight of the heavy oil of an aliquot of the sample material (less than 4% difference).

4.5.4 The Benefits of Linearizing Response to Universal Quantitation

In some laboratories, a highly desirable feature of uniform response is that the response be both uniform and linear. For those laboratories, particularly those dealing with quantification of polymer distributions, who cannot, may not or do not have the software to transform data using a quadratic or log–log fit, a linear concentration dependency must be ensured for CAD data by obtaining data using the PV. The mathematical basis for the PV is detailed in Chapter 1 and its practical implementation is described in Chapter 3. It is used to linearize response when the analyte is present at concentrations that are outside of the quasi-linear response range of CAD.

The benefits of optimizing PV may be considered for all methods, even those that use nonlinear calibration. Common pitfalls include calculating metrics such as chromatographic resolution, column efficiency or limits of detection without use of an optimal PV. These metrics are calculated by using the linear-scaled analyte concentration profiles and thus yield distorted results when a nonlinearly translated signal profile is used. The peak broadening and deviation from Gaussian shape discussed in Chapter 1 and Figure 1.9 can also be minimized with an optimal PFV.

For methods that rely on relative response between peaks for quantitation, optimization of the PV is highly recommended. Figure 4.10 is an example where

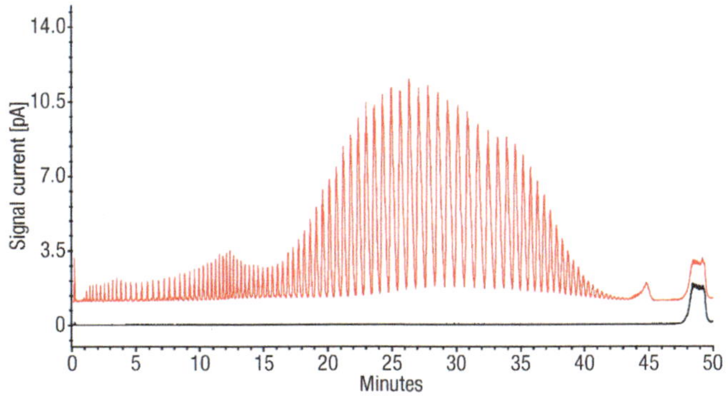

Figure 4.10 Chromatogram of polyethylene glycol 3000 (red) and water blank (black) using reversed-phase HPLC and CAD. [55] Since relative peak areas are used to evaluate the mass distribution of subspecies, it is recommended to optimize the PV to obtain the most accurate results.

relative peak areas are used to determine the mass distribution of polymers [55]. With sublinear response, the smaller peaks at the edge of the distribution would likely be overestimated. Optimizing the PV would therefore provide more accurate results.

A paper by Karlsen et al. [56] describes the use of UV as a reference detector to derive an optimal PV but notes that this approach has significant limitations if the polymer doesn't have a chromophore. A solution without the need for UV would be to analyze a dilution series of the polymer at PV 1.5 (PFV 1.0) and determine the optimal PV based on the highest or most central peak.

4.6 Conclusions

LC would greatly benefit from a true universal detector capable of detecting all analytes with uniform response factors, regardless of their physicochemical properties. Although aerosol-based detectors are not ideal universal detectors, they offer a powerful detection alternative, providing relatively uniform response factors for a large number of nonvolatile analytes, and achieve response deviations as low as single-digit RSDs over a wide range of chemical properties. CAD proved to be greatly superior to ELSD and CNLSD at response uniformity and also achieves lower LOD and LOQ for the vast majority of analytes. It provides a complementary detection technique to MS and UV and is also combined with those in multi-detector set-ups. The factors affecting uniform response are well understood and encompass for strict nonvolatiles the aerosol particle density, salt formation of acidic or basic solutes with potential counterions from the mobile phase, as well as the viscosity and surface tension of the mobile phase. The latter is mostly determined by its composition ratio between water and organic solvents. Intentional salt formation can be used to increase response of semivolatile compounds by reducing the volatility of bases or acids and by the resulting aerosol mass increase, but the resulting signal increase is often compromised by a severe increase in baseline noise. Salt formation furthermore compromises the benefit of uniform response. To make up for the impact of the solvent composition change with gradient methods, the concept of IGC by adding a flow from a second pump with a mirrored gradient between column and CAD is successfully used for numerous RP gradient applications. This is also described for HILIC methods and SFC with organic modifier gradients. It offers several advantages, such as being compatible with separation methods consistent with the ideals of green chemistry and offering the ability to quantify unknowns due to the universal response factors. IGC typically doubles the consumption of organic solvents and should thus be used with small-bore separation columns to minimize the resulting environmental footprint.

References

1 Zhang, K., Kurita, K.L., Venkatramani, C., and Russell, D. (2019). Seeking universal detectors for analytical characterizations. *Journal of Pharmaceutical and Biomedical Analysis* 162: 192–204.

2 Lane, S., Boughtflower, B., Mutton, I. et al. (2005). Toward single-calibrant quantification in HPLC. A comparison of three detection strategies: evaporative light scattering, chemiluminescent nitrogen, and proton NMR. *Analytical Chemistry* 77 (14): 4354–4365.

3 Activated Research Company (ARC) Model Solvere-Carbon Selective Detection for LC. Available at: https://www.energy-xprt.com/products/model-solvere-carbon-selective-detection-for-lc-953769.

4 Gamache, P.H., McCarthy, R.S., Freeto, S.M. et al. (2005). HPLC analysis of nonvolatile analytes using charged aerosol detection. *LCGC North America* 23 (2): 150–161.

5 Górecki, T., Lynen, F., Szucs, R., and Sandra, P. (2006). Universal response in liquid chromatography using charged aerosol detection. *Analytical Chemistry* 78 (9): 3186–3192.

6 Hutchinson, J.P., Li, J., Farrell, W. et al. (2010). Universal response model for a corona charged aerosol detector. *Journal of Chromatography. A* 1217 (47): 7418–7427.

7 Hutchinson, J.P., Remenyi, T., Nesterenko, P. et al. (2012). Investigation of polar organic solvents compatible with corona charged aerosol detection and their use for the determination of sugars by hydrophilic interaction liquid chromatography. *Analytica Chimica Acta* 750: 199–206.

8 Matsuyama, S., Orihara, Y., Kinugasa, S., and Ohtani, H. (2015). Effects of densities of brominated flame retardants on the detection response for HPLC analysis with a corona-charged aerosol detector. *Analytical Sciences* 31: 61–65.

9 Hutchinson, J.P., Li, J., Farrell, W. et al. (2011). Comparison of the response of four aerosol detectors used with ultra high pressure liquid chromatography. *Journal of Chromatography. A* 1218 (12): 1646–1655.

10 Cohen, R.D. and Liu, Y. (2014). Advances in aerosol-based detectors. In: *Advances in Chromatography*, vol. 52 (ed. E. Grushka and N. Grinberg), 1–53. Boca Raton, FL: CRC Press.

11 Kou, D., Manius, G., Zhan, S., and Chokshi, H.P. (2009). Size exclusion chromatography with Corona charged aerosol detector for the analysis of polyethylene glycol polymer. *Journal of Chromatography. A* 1216 (28): 5424–5428.

12 Merle, C., Laugel, C., Chaminade, P., and Baillet-Guffroy, A. (2010). Quantitative study of the stratum corneum lipid classes by normal phase liquid chromatography: comparison between two universal detectors. *Journal of Liquid Chromatography and Related Technologies* 33 (5): 629–644.

13 Shaodong, J., Lee, W.J., Ee, J.W. et al. (2010). Comparison of ultraviolet detection, evaporative light scattering detection and charged aerosol detection methods for liquid-chromatographic determination of anti-diabetic drugs. *Journal of Pharmaceutical and Biomedical Analysis* 51 (4): 973–978.

14 Takahashi, K., Kinugasa, S., Senda, M. et al. (2008). Quantitative comparison of a corona-charged aerosol detector and an evaporative light-scattering detector for the analysis of a synthetic polymer by supercritical fluid chromatography. *Journal of Chromatography. A* 1193 (1–2): 151–155.

15 Vervoort, N., Daemen, D., and Török, G. (2008). Performance evaluation of evaporative light scattering detection and charged aerosol detection in reversed phase liquid chromatography. *Journal of Chromatography. A* 1189 (1–2): 92–100.

16 Wipf, P., Werner, S., Twining, L.A., and Kendall, C. (2007). HPLC determinations of enantiomeric ratios. *Chirality* 19 (1): 5–9.

17 Mourey, T.H. and Oppenheimer, L.E. (1984). Principles of operation of an evaporative light-scattering detector for liquid chromatography. *Analytical Chemistry* 56 (13): 2427–2434.

18 Oppenheimer, L.E. and Mourey, T.H. (1985). Examination of the concentration response of evaporative light-scattering mass detectors. *Journal of Chromatography* 323: 297–304.

19 Righezza, M. and Guiochon, G. (1988). Effect of the wavelength of the laser beam on the response of an evaporative light scattering detector. *Journal of Liquid Chromatography* 11 (11): 2709–2729.

20 Mathews, B.T., Higginson, P.D., Lyons, R. et al. (2004). Improving quantitative measurements for the evaporative light scattering detector. *Chromatographia* 60 (11–12): 625–633.

21 Koropchak, J.A., Sadain, S., Yang, X. et al. (1999). Nanoparticle detection technology. *Analytical Chemistry* 71 (11): 386A–394A.

22 Hering, S.V., Stolzenburg, M.R., Quant, F.R. et al. (2005). A laminar-flow, water-based condensation particle counter (WCPC). *Aerosol Science and Technology* 39 (7): 659–672.

23 Kupc, A., Bischof, O., Tritscher, T. et al. (2013). Laboratory characterization of a new nano-water-based CPC 3788 and performance comparison to an ultrafine butanol-based CPC 3776. *Aerosol Science and Technology* 47 (2): 183–191.

24 Liu, W., Kaufman, S.L., Osmondson, B.L. et al. (2006). Water-based condensation particle counters for environmental monitoring of ultrafine particles. *Journal of the Air & Waste Management Association* 56 (4): 444–455.

25 Eckardt, M., Kubicova, M., and Simat, T.J. (2018). Universal response quantification approach using a corona charged aerosol detector (CAD) – application on linear and cyclic oligomers extractable from polycondensate plastics polyesters, polyamides and polyarylsulfones. *Journal of Chromatography. A* 1572: 187–202.

26 Squibb, A.W., Taylor, M.R., Parnas, B.L. et al. (2008). Application of parallel gradient high performance liquid chromatography with ultra-violet, evaporative light scattering and electrospray mass spectrometric detection for the quantitative quality control of the compound file to support pharmaceutical discovery. *Journal of Chromatography. A* 1189 (1–2): 101–108.

27 Cohen, R.D., Liu, Y., and Gong, X. (2012). Analysis of volatile bases by high performance liquid chromatography with aerosol-based detection. *Journal of Chromatography. A* 1229: 172–179.

28 Russell, J.J., Heaton, J.C., Underwood, T. et al. (2015). Performance of charged aerosol detection with hydrophilic interaction chromatography. *Journal of Chromatography. A* 1405: 72–84.

29 Sinclair, I. and Gallagher, R. (2008). Charged aerosol detection: factors for consideration in its use as a generic quantitative detector. *Chromatography Today* 1 (3): 5–9.

30 McConville, J.A., Bullock, S., Warner, F., and O'Donohue, S. (2006) Sub-ambient ELSD for enhanced detection of semi-volatile compounds. Poster available at https://www.researchgate.net/publication/268254139_Sub-Ambient_ELSD_for_Enhanced_Detection_of_Semi-Volatile_Compounds.

31 Megoulas, N.C. and Koupparis, M.A. (2004). Enhancement of evaporative light scattering detection in high-performance liquid chromatographic determination of neomycin based on highly volatile mobile phase, high-molecular-mass ion-pairing reagents and controlled peak shape. *Journal of Chromatography. A* 1057 (1–2): 125–131.

32 Petritis, K., Dessans, H., Elfakir, C., and Dreux, M. (2002). Volatility evaluation of mobile phase/electrolyte additives for mass spectrometry. *LC GC Europe* 15 (2): 98–103.

33 Deschamps, F.S., Baillet, A., and Chaminade, P. (2002). Mechanism of response enhancement in evaporative light scattering detection with the addition of triethylamine and formic acid. *Analyst* 127 (1): 35–41.

34 Menz, M., Eggart, B., Lovejoy, K. et al. (2018) Charged aerosol detection - factors affecting uniform analyte response. Thermo Fisher Scientific Technical Note 72806.

35 Wu, J., Crist, R.M., McNeil, S.E., and Clogston, J.D. (2019). Ion quantification in liposomal drug products using high performance liquid chromatography. *Journal of Pharmaceutical and Biomedical Analysis* 165: 41–46.

36 Bailey, B., Plante, M., Thomas, D. et al. (2017). Practical use of CAD. In: *Charged Aerosol Detection for Liquid Chromatography and Related Separation Techniques*, 1e (ed. P.H. Gamache), 163–189. Hoboken, NJ: John Wiley & Sons, Inc.

37 Liu, S., Lu, B., Peng, Z. et al. (2023). HPLC-CAD as a supplementary method for the quantification of related structure impurities for the purity assessment of organic CRMs. *Analytical and Bioanalytical Chemistry* 415 (17): 3375–3384.

38 Popławska, M., Błażewicz, A., Kamiński, K. et al. (2018). Application of high-performance liquid chromatography with charged aerosol detection (LC–CAD) for unified quantification of synthetic cannabinoids in herbal blends and comparison with quantitative NMR results. *Forensic Toxicology* 36 (1): 122–140.

39 Khandagale, M.M., Hilder, E.F., Shellie, R.A., and Haddad, P.R. (2014). Assessment of the complementarity of temperature and flow-rate for response normalisation of aerosol-based detectors. *Journal of Chromatography. A* 1356: 180–187.

40 de Villiers, A., Górecki, T., Lynen, F. et al. (2007). Improving the universal response of evaporative light scattering detection by mobile phase compensation. *Journal of Chromatography. A* 1161 (1–2): 183–191.

41 Menz, M., Acworth, I., and Steiner, F. (2018) Quantification of paclitaxel, its degradants, and related substances using UHPLC with charged aerosol detection, Application Note 72594.

42 Barnhart, W.W., Farooq, M.Q., and Ahmad, I.A.H. (2024). A simplified tutorial on charged aerosol detection: understanding the basics, optimization, and troubleshooting. *Journal of Chromatography Open* 6: 100181.

43 Lísa, M., Lynen, F., Holčapek, M., and Sandra, P. (2007). Quantitation of triacylglycerols from plant oils using charged aerosol detection with gradient compensation. *Journal of Chromatography. A* 1176 (1–2, 142): 135.

44 Peng, M., Zhang, T., Ding, Y. et al. (2016). Structure-based prediction of CAD response factors of dammarane-type tetracyclic triterpenoid saponins and its application to the analysis of saponin contents in raw and processed: panax notoginseng. *RSC Advances* 6 (43): 36987–37005.

45 Dufour, A., Thiébaut, D., Loriau, M. et al. (2020). Corona charged aerosol detector non-uniform response factors of purified alcohol ethoxylated homologues using liquid chromatography. *Journal of Chromatography. A* 1627: 461402.

46 Pawellek, R., Krmar, J., Leistner, A. et al. (2021). Charged aerosol detector response modeling for fatty acids based on experimental settings and molecular features: a machine learning approach. *Journal of Cheminformatics* 13: 53.

47 Chickos, J.S. Heat of Sublimation Data. In: *NIST Chemistry WebBook, NIST Standard Reference Database Number 69* (ed. P.J. Linstrom and W.G. Mallard). Gaithersburg, MD: National Institute of Standards and Technology. https://doi.org/10.18434/T4D303 (accessed 21 May 2026).

48 Jones, A.H. (1960). Sublimation pressure data for organic compounds. *Journal of Chemical & Engineering Data* 5 (2): 196–200.

49 Wang, J., Tang, S., and Zhang, K. (2024). Quantitation of polyethylene glycol by size exclusion chromatography with charged aerosol, differential refractive index, and multi-angle light scattering detectors. *Journal of Pharmaceutical and Biomedical Analysis* 238: 115854.

50 Gamache, P.H. and Kaufman, S.L. (2017). Principles of charged aerosol detection. In: *Charged Aerosol Detection for Liquid Chromatography and Related Separation Techniques*, 1e (ed. P.H. Gamache), 1–65. John Wiley & Sons, Inc.

51 Causevic, A., Olofsson, K., Adlercreutz, P., and Grey, C. (2021). Non-aqueous reversed phase liquid chromatography with charged aerosol detection for quantitative lipid analysis with improved accuracy. *Journal of Chromatography. A* 1652: 462374.

52 Brunelli, C., Górecki, T., Zhao, Y., and Sandra, P. (2007). Corona-charged aerosol detection in supercritical fluid chromatography for pharmaceutical analysis. *Analytical Chemistry* 79 (6): 2472–2482.

53 Brunelli, C., Górecki, T., Zhao, Y., and Sandra, P. (2007). Corona-charged aerosol detection in supercritical fluid chromatography for pharmaceutical analysis. *Analytical Chemistry* 79 (6): 2472–2482.

54 Scholz, K., Lipphardt, A., Wienken, C.M. et al. (2020). Hyphenation of supercritical fluid chromatography with different detection methods for identification and quantification of liamocin biosurfactants. *Journal of Chromatography. A* 1631: 461584.

55 Gamache, P., Muellner, T., Eggart, B. et al. (2019) Charged aerosol detection-use of the power function and robust calibration practices to achieve the best quantitative results TN73299.

56 Karlsen, C.E., Cravino, J.A., Li, F. et al. (2025). Quantitative analysis of oligostyrenes using a charged aerosol detection: linearisation of quasi-linear calibration curves using power functions. *Microchemical Journal* 218: 115105.

Section 2

Charged Aerosol Detection of Specific Analyte Classes

5

Analysis of Lipids with Charged Aerosol Detection

Paul H. Gamache

Consultant, Nashua, NH, USA

CHAPTER MENU

5.1 Summary

The use of CAD for quantitative analysis of lipids is now quite common and includes its coupling with both liquid chromatography (LC) and supercritical fluid chromatography (SFC). Several papers have described method development and validation for a range of applications including for diverse lipid classes, within-class molecular species and individual molecules. This chapter provides an overview of these methods and discusses advantages and limitations of the technique and best practices to address challenges often encountered. This includes those related to analyte volatility, wide concentration ranges, the requirement for complex separation methods and difficulties in using individual calibration standards. Key takeaways: When optimizing the evaporation temperature (T_e) and power function value (PFV) and when assessing a calibration model's goodness of fit it is especially important to evaluate low analyte levels; best detection performance is typically obtained when using low-molar-mass additives at low concentrations in the mobile phase and conditions and columns that produce minimal bleed.

5.2 Introduction

Lipids are a diverse class of organic compounds that are soluble in nonpolar solvents and poorly soluble in water. Some lipids are amphiphilic and can form bilayers in aqueous matrices, which is fundamental to their role as main components of biological membranes and important to their use in drug delivery. These and many other functions such as cellular signaling, energy storage and surfactant activity afford lipids with great biological significance and high importance in cosmetic, food, medical, pharmaceutical, and other disciplines. A large demand exists for reliable techniques that can provide accurate analysis. This can be quite challenging since sample extracts are often complex mixtures of both closely related and diverse chemical structures, many of which are unknown and are present over wide and variable concentration ranges. Acylglycerols (AGs), for example, include monoacylglycerols (MAGs), diacylglycerols (DAGs), or triacylglycerols (TAGs) depending on the number of fatty acid (FA) groups and there is significant diversity in the FAs themselves based on chain length, degree of saturation, stereochemistry, and other modifications. Representative structures are shown in Figure 5.1.

In a typical sample, AGs are often present with FAs, phospholipids (PLs), sterols (STs), and other classes, each containing many subspecies. The main approaches to their analysis include chromatographic separation of lipid classes

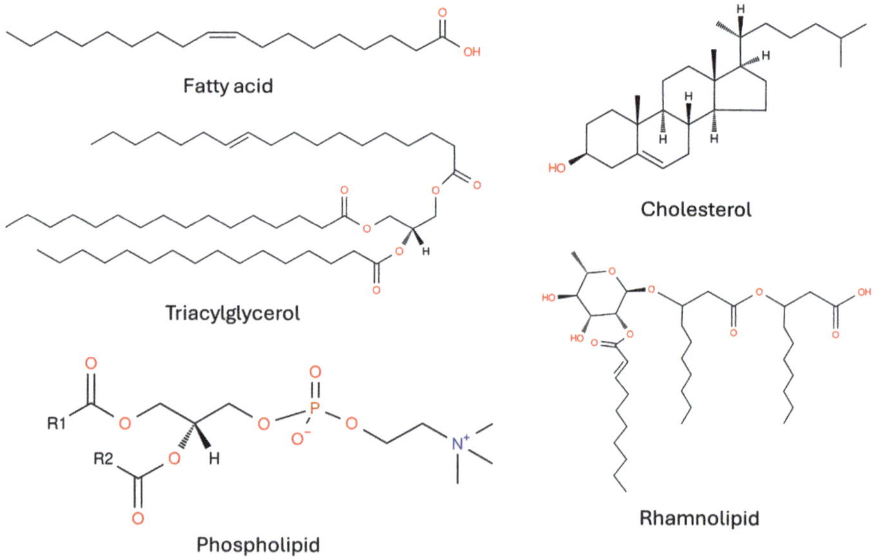

Figure 5.1 Representative lipid chemical structures. R1, R2 = fatty acid groups.

over a wide polarity range, within-class separation of individual species and everything in between. Gas chromatography (GC) is useful for many lipid studies but usually requires derivatization, which is well established for FAs but has significant limitations for other lipids. LC is a primary technique for lipid analysis and, more recently, SFC has shown great utility for challenging separations. Most lipids are weak chromophores and there is high variability in molar absorptivity among those with stronger chromophores. These factors and the common lack of individual calibration standards severely limit the use of absorbance-based detection. Mass spectrometry (MS) and evaporative aerosol detection (EAD) techniques including evaporative light scattering detection (ELSD) and charged aerosol detection (CAD) are quite often used for these analyses. MS is a primary tool for lipid analysis as it can provide high sensitivity, high specificity and critical qualitative analysis but, especially in the absence of proper calibration standards, it has limited quantitative capabilities. EAD techniques are most often chosen since refractive index detection (RID) lacks sensitivity and is incompatible with gradient elution. ELSD has been used extensively for lipid analysis beginning in the 1980s. As with other analyte classes, detector comparison studies (summarized in Section 5.3) have consistently shown significant advantages of CAD over ELSD for a range of lipid applications and chromatographic methods. Many laboratories have more recently opted to use CAD, which is reflected by the growth in publications since its commercial introduction (Chapter 2). This chapter will focus on the use of CAD for analysis of lipids primarily in biological (animal, plant, microbial) sources. Analyses related to lipid-based pharmaceutical formulations (e.g., emulsions, liposomes, lipid nanoparticles) represent a significant area for CAD and are separately discussed in Chapter 11.

Basic considerations for use of CAD for lipid analysis are similar to those of other analyte classes. This includes the choice of mobile and stationary phases; detector settings such as T_e and PFV; and calibration approaches. These topics are detailed in Chapter 3 where the following aspects may be particularly relevant when using CAD for lipid analysis:

- Lipid separations most often use eluents with high proportions of organic solvents with low viscosity and surface tension. This leads to high mass transport efficiency within EADs and therefore high signal to both analytes and impurities. This may result in higher baseline noise and drift than typically observed with predominantly aqueous eluents. It is therefore important to use pure solvents, i.e., those specified with low evaporation residue. Solvent purity is also a key consideration with lipid extraction techniques that involve evaporation to dryness of larger volumes of organic solvents. In these cases, impurities can be pre-concentrated, which can produce problematic chromatographic artifacts.

- Lipid separations commonly use unbonded silica (Si) or polar-ended alkyl-bonded Si stationary phases some of which have been shown to exhibit significant column bleed. Care should therefore be taken to use conditions that minimize bleed and choose "low bleed" columns.
- While lipids are generally not pH-sensitive, notable exceptions include FAs, PLs, some sphingolipids (SLs), and aminolipids that are often used in pharmaceutical formulations. In these cases, volatile acids or buffers are used as mobile phase additives to control ionization or serve as ion-pairing (IP) reagents. In general, it is recommended to use low-molar-mass additives at low concentrations whenever possible. A key consideration is to avoid using higher molar mass acidic additives [e.g., trifluoroacetic acid (TFA)] together with any basic additive (ammonium). Likewise, caution should be exercised when using higher molar mass basic additives [e.g., triethylamine, (TEA)] together with some acidic additives [e.g., acetic acid (HAc)]. These oppositely charged additives can, through spray-drying, combine with each other and/or separately with ionic eluent impurities to produce high background currents, baseline noise, drift, and ghost peaks.
- Many lipid analyses involve quantitation of species, either chromatographically separated or eluting as a group, for which specific calibration standards are unavailable. Quantitative accuracy may therefore depend on factors that affect response uniformity (see Chapter 4). This includes consideration of T_e, choice of mobile phase additives, solvent gradient effects, and use of an optimal PFV.

The following sections will highlight literature examples that describe the use of CAD for various applications with a focus on best practices, challenges and limitations. This will include the analysis of individual lipid species, lipid class separations, and other illustrative applications.

5.3 Direct Comparisons of CAD and ELSD

Hazotte et al. [1] compared detectors for lipid analysis using micro high-temperature LC. They used both isocratic and gradient methods with a porous carbon column and mobile phase components of butan-1-ol and ethanol. Cholesterol, squalene, and ceramide type III (a mixture of ceramides and FAs) were used as models for their investigation. They found that, for all analytes and conditions, CAD response was linear over two orders of magnitude while ELSD response was nonlinear with varying degrees of curvature depending on the analyte, and chromatographic conditions including column T. CAD provided lower LODs than ELSD for all analytes and more universal detection than full scan APCI-MS. Merle et al. [2] developed a method to quantify three major lipid classes of the stratum Corneum, the outermost layer of the epidermis. They used a polyvinyl alcohol (PVA)-Si column and a binary gradient with mobile phase components of heptane, chloroform and acetone to separate FAs, ceramides, and cholesterol. They reported that CAD provided better accuracy, precision, repeatability and sensitivity. CAD and ELSD response factors (RFs) significantly differed between analytes, which was mainly attributed to the use of uncompensated gradient elution and, for FAs, volatility. Similar response factors were, however, obtained with CAD for several ceramides, which enabled the use of a single external standard for quantitation of multiple ceramide species. This was not the case with ELSD.

Ramos et al. [3] analyzed 11 main PL classes from membranes of the parasitic organism *Leishmania*. They used a PVA-Si column as an alternative to pure Si and a ternary gradient with mobile phase components of heptane, isopropanol (IPA), $CHCl_3$, methanol (MeOH), HAc, and TEA to separate phosphatidic acid (PA), phosphatidylglycerol (PG), cardiolipin (CL), phosphatidylinositol (PI), phosphatidylethanolamine (PE), phosphatidylserine (PS), lysophosphatidylethanolamine, phosphatidylcholine (PC), sphingomyelin (SM), and the lyso forms of PE (LPE) and PC (LPC) in the same analysis. They reported that CAD was more precise and sensitive. From a detailed analysis of the response curves, they found that although the response of both detection methods can be fitted to a power function, CAD response can also be described by a linear model with determination coefficients (r^2) ranging from 0.993 to 0.998 for an injected mass of 30 ng to 20.00 μg. It should be noted that this range of ca. four orders of magnitude should be more reflective of CAD's dynamic range than its linear range. As emphasized in Chapter 3, a curve fit with an r^2 of 0.9990 may still be poor, with the largest error typically near the low end. Also, reliance only on r^2 can be especially problematic when using a log–log curve fitting model. The authors described a robust assessment of linearity and found that linear calibration can be used with CAD over a smaller range (typically two orders of magnitude). By contrast, linear calibration could not be used with ELSD. They also noted that log transformation distorts the experimental error and that estimated parameters such as correlation coefficients are in fact inaccurate [4] as further described in Chapter 3.

Damnjanović et al. [5] studied six PL classes (PA, PE, PI, PC, lyso-PI, and lyso-PC) in phospholipase D-modified soy lecithin. They used a Si column and a binary gradient with mobile phase components of chloroform, MeOH, formic acid and TEA. They reported that CAD provided better precision, a broader dynamic range, a 10-fold broader linear range and approximately three to five-fold better detection limits for all analytes. Olsson et al. [6] analyzed lipid classes and their molecular species. They used a phenyl-coated Si stationary phase and a MeOH: water binary gradient to elute all tested lipid classes including monoacylglycerols, acidic PLs, bulky cardiolipins, and long-chain triacylglycerols (TAGs) within 30 minutes. They used a post-column flow splitter to quickly compare ELSD and CAD for simultaneous analysis of six levels of representative standards in the range of 0–5.0 μg. For all tested analytes, they observed linear response with CAD and significant curvature with ELSD. Nair and Werling [7] analyzed free FAs in a pharmaceutical suspension formulated with PLs as stabilizing agents. They used a C18 column with isocratic separation with mobile phase components of acetonitrile (ACN), water and HAc. They found that CAD was more accurate, linear, precise, and sensitive. Abreu et al. [8] described an optimized normal-phase method for multiple lipid classes applicable to a range of biological matrices (Section 5.4.4). They observed that CAD provided a response that was more representative of analyte quantity than ELSD whose response for minor components appeared to be less than their actual quantity while that of major components dominated the chromatograms.

The advantages described above are consistent with theory as discussed in Chapter 1 where the downstream mechanism of diffusional charging of aerosols employed in CAD underlies its more uniform response, lower detection limits, wider dynamic range, and less complex response curves than ELSD.

5.4 Case Studies

Lipid analysis typically requires measurement of both nonvolatile and semivolatile species whose concentrations vary over several orders of magnitude and, in many cases, with individual chromatographic peaks containing multiple species. Also, the ability to use individual calibration standards for quantitation is often limited. Several approaches to address these challenges have been described in literature and are used below as case studies to help inform the use of optimal T_e settings, nonlinear calibration models, optimal PFV settings to linearize CAD's signal output, and inverse gradient compensation (IGC) to normalize the response among analytes in gradient methods.

5.4.1 Volatility and Evaporation Temperature (T_e)

CAD and ELSD use the same basic principle of spray- drying a chromatographic eluent to produce an aerosol residue. While their downstream detection mechanisms differ, the signal in both cases is proportional to the number concentration and size of the dried aerosol particles. Not surprisingly, they have an identical detection scope. Several groups have reported that most lipids behave as nonvolatiles under default conditions (e.g., VCAD T_e of 35 °C). Exceptions include FA esters (FAEs), many of which behave as volatiles producing little or no response and short-chain FAs, whose behavior at a given T_e mostly depends on chain length [9–11]. T_e is a main factor that influences the extent to which an analyte behaves as a nonvolatile or semivolatile and this has strong implications for method development and performance. As described in Chapter 1, particle evaporation rate is proportional to its surface area. Consequently, an analyte that behaves as a semivolatile produces both an overall lower response and a characteristic exponential signal loss at lower mass levels. As seen inherently with ELSD, an exponential (supralinear) signal loss compromises detection limits, precision, and quantitative accuracy. When optimizing T_e, it is therefore important to evaluate low analyte levels since volatility may go unnoticed when only analyzing higher-level standards. Described below are some case studies describing the use of CAD for lipids that may exhibit some volatility with a focus on T_e and its interplay with other variables that may affect performance.

The analysis of FAs with CAD represents an interesting case study where the analytes of interest clearly exhibit varying degrees of volatile behavior according to their chain length. An examination of the strategies used for FA analysis with CAD, and its limitations can provide useful insight into any method that involves semivolatile analytes. Ilko et al. [12] described a reversed-phase (RPLC) method for analysis of FAs in surfactants as a complement to GC-based pharmacopeial methods that require derivatization. The method was applied to measurement of free FAs (FFAs) in polysorbate (PS) and to FA compositional analysis of polysorbate (PS) after saponification. They reported limits of quantitation (LOQs) between 2.1 and 5.4 ng on column for FAs with a ≥ C16 carbon chain length, 6.1 ng for a C14 FA, and poor sensitivity for more volatile FAs with ≤ C12. The method was therefore deemed unsuitable for FA analysis in PS20, which contains significant amounts of short-chain FAs, but suitable for PS40, PS60 and PS80. While the LC-CAD method could not adequately detect shorter chain FAs, it was capable of directly detecting related substances including a degradation product previously not found with current pharmacopeial methods. As further detailed in Chapters 8 and 10, this more comprehensive detection capability of CAD and its frequent combination with UV detection can provide significant advantages for impurity control compared to existing compendial and other manufacturing QC methods.

Schilling et al. [13] adapted the above-described HPLC method of Ilko to UHPLC conditions and achieved a > 75% faster analysis time and > 40% reduction in solvent usage. They also studied the effects of systematic variation of T_e, filter constant and PFV on performance for a series of homologous FAs ranging from C14 to C18. Their approach is an example of using best practices for optimization of CAD parameters as detailed in Chapter 3. This includes a focus on analyte levels near the targeted detection limits (10 ng in this case) where the effects of T_e are more pronounced and the potential to be misled by nonlinear response is minimized.

Figure 8.4 (Chapter 8) plots peak height versus T_e for 10 ng levels of representative FAs from C12 through C18. This shows a much lower overall response, attributed to volatility, for C12 and C14 FAs compared to C18 FAs. At this 10 ng level, all C18 FAs also show evidence of volatile behavior even at lower T_e. Although the effects are more pronounced at higher T_e, the results suggest that the lowest practical setting is preferable. Other studies have similarly described a significant loss of CAD response for the C18 FA oleic acid at $T_e = 50\,°C$ compared to 35 °C and observed that other lipids, including monoacylglycerols (MAGs), were not significantly affected [14, 15]. However, there is evidence that additional analytes including MAGs may, in some cases (Section 5.4.3), exhibit a significant level of volatility at lower mass levels even with a T_e of 35 °C.

As described in Chapter 3, T_e is a useful instrument parameter to optimize SNR by altering the extent of vaporization of solutes. This is often a compromise between maximizing the vaporization of eluent impurities to reduce baseline noise and minimizing vaporization of target analytes. As noted, it is recommended to use the lowest T_e setting that consistently produces the required sensitivity limits and where the ideal goal is for all analytes to behave as nonvolatiles. This generally provides more universal detection, more uniform response and enables more accurate quantitation.

Figure 5.2 plots the corresponding SNR versus T_e for three of the analytes from Schilling's study. This shows that SNRs were improved at lower temperatures, but even at the lowest T_e, myristic acid (C14) showed a significantly lower response. In this study, the described improvements in SNR at low T_e were likely achieved by the practice of ensuring and maintaining clean column eluents as outlined in Chapter 3. It should be noted, however, that many lipid analyses require the use of more complex separation methods than those described for FAs and other lipid species. This includes the use of complex ternary or quaternary gradients, less

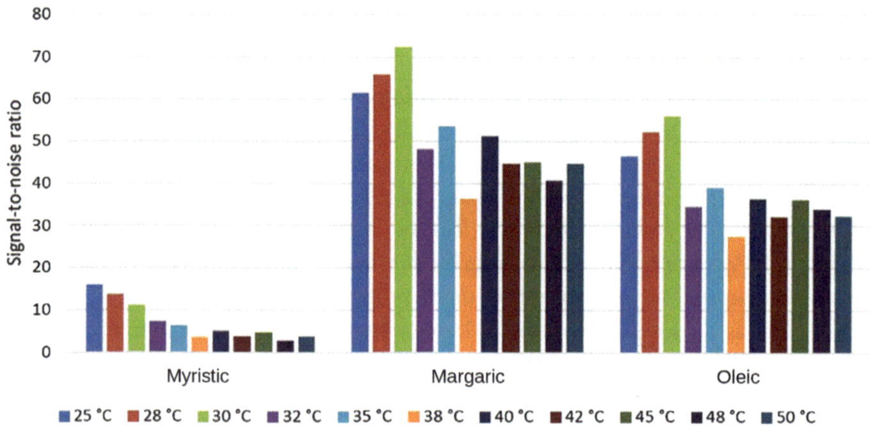

Figure 5.2 SNR versus T_e for 10 ng injections of three different fatty acids. *Source:* [13] / with permission of Elsevier.

stable columns and both solvents and additives that may lead to higher baseline fluctuations, noise and drift. While certainly necessary in most cases, this can pose challenges for detection with CAD. For example, Takeda et al. [16] described an SFC method employing a diethylamine-bonded Si stationary phase. While they achieved impressive separations for multiple lipid classes, high column bleed led to the choice of a very high T_e (95 °C). At this setting, while some lipid classes may not have been significantly affected, cholesterol response was nearly eliminated and the response for FAs and other species would most likely be reduced with the most significant effects at lower analyte levels. The topic of chromatographic variables that may affect CAD performance is further discussed in Section 5.4.5.

5.4.2 Calibration and Power Function Value (PFV)

The following case studies describe various approaches to address the shape of CAD's response curve. CAD response for nonvolatiles is sufficiently linear over approximately 1.5–2.0 orders of magnitude but is nonlinear over wider ranges. Various approaches are employed after data acquisition and peak integration. This includes the use of nonlinear calibration models (e.g., quadratic) and transformation (e.g., log–log or power), of peak area or height data to facilitate the use of a linear regression. Also, the Power Law Tool in Chromeleon™ software (Thermo Fisher Scientific) allows post-acquisition power transformation of all data points in the chromatogram. This is used to better linearize response prior to peak integration. This and other tools described in Chapter 3 provide a means to determine an optimal PFV with minimal experimentation, requiring only the generation of response curve data at one PFV for any chromatography platform. This optimal PFV can then be implemented as an instrument setting, which is the recommended approach as discussed below. The PFV setting is a data acquisition parameter like data collection rate or filter time constant that directly affects the signal output and is used to optimize the response with respect to linearity for a given method and range of quantitation.

Roslon et al. [17] separated 4 PL classes using a diol stationary phase and a binary gradient with mobile phase components of n-hexane, 2-propanol, and water containing both TEA and HAc. Peak areas were normalized to an internal standard to correct for extraction efficiency and then transformed using a power function (Equation (5.1)) where A = normalized peak area, X = power function value (PFV), scale = 500 pA full-scale range of the detector and $A_{(\mathrm{lin})}$ is the transformed peak area. Note: This study used an earlier model CAD, which did not include a PFV instrument setting, and the transformation was therefore performed after acquisition and peak integration.

$$A_{\mathrm{lim}} = \frac{A^X}{\mathrm{scale}^{(X-1)}} \tag{5.1}$$

Calibration curves were generated over the range of 8.0–188 (PE), 4.0–200 (PC), 0.8–34 (PS), and 1.4–35 (PI) μg on column. From these data, an optimal PFV of 1.85 was determined according to the method of Ahmad et al. [18], which is mainly based on minimizing the %RSD of RFs for the required concentration range and is further described in Chapter 3. Equation (5.1) is mathematically similar to the power function feature of newer CAD instruments (e.g., VCAD), which is used for data linearization. These newer models include a user-defined PFV setting that directly affects the signal output and therefore applies to all individual data points in a chromatogram. While the power transformation in this study allowed linearization of peak area results, it did not provide the additional benefits often described when applying an optimal PFV to entire chromatograms. These benefits include improved signal-to-noise ratios (SNRs), precision and peak shapes, higher throughput and simpler quantitation and have been observed when PFV > 1.0 is used. These improvements are to be expected when sublinear response is obtained at PFV = 1.0 as is often the case for nonvolatiles. When a higher PFV is used, by definition peaks become sharper and SNRs increase. It should be emphasized, however, that use of a PFV that is too high would create an ELSD-like supralinear response where the signal for lower levels may be lost. As with T_e, it is therefore important to analyze standard levels near the targeted LOQ and to optimize PFV based on linearity and not SNR. The optimal PFV of 1.85 described in this case [17] is toward the high end of theoretically useful values for this parameter. This may be attributed to the relatively high overall (0.8–200 μg on column) mass range of quantitation described in this study and possibly higher background signal due to the combined use of TEA and HAc.

Several LC-CAD methods have been described for the study of dietary lipids whose main components are TAGs and main digestion products are DAGs, MAGs and FAs. This may therefore include a mixture of analytes that behave as nonvolatiles and semivolatiles. A key requirement in kinetic studies of digestion is to obtain accurate quantitation over a wide dynamic range with sufficient sensitivity for products formed at low levels early in the digestion timeline. Infantes-Garcia et al. [14] developed a normal-phase quaternary gradient method using a monolithic Si column with mobile phase components of isooctane, acetone, ethyl acetate, IPA, HAc and ethanolamine. The method was applied to synthetic triolein emulsions and described as suitable for the study of single triacylglycerol oils. A T_e of 35 °C was used since the response of oleic acid was observed to be significantly reduced at 50 °C. Note: the effect of T_e was shown using a 500 ng injection where significant effects for low levels of additional analytes might go unnoticed. Individual standards of triolein, sn-1,2/2,3-diolein, sn-1,3-diolein, sn-2-monoolein, sn-1/3-monoolein, and oleic acid were analyzed over the range of 25–2500 ng on column and tested using PFV settings of 1.0, 1.2, 1.4, 1.6 and 1.8. The optimal PFV was determined as described in Ahmad et al. [18]

and further detailed in Chapter 3. From these results, a PFV of 1.6 was chosen since it presented the lowest %RSD for the majority of analytes and was equal to or lower than that obtained with PFV 1.0 for the remaining analytes. The authors reported that a PFV of 1.6 was found to not only provide the best overall linear dynamic range for all analytes but also resulted in lower baseline noise and drift compared to the default PFV of 1.0.

A very similar approach was described in Guevara-Zambrano et al. [15]. An RP method was developed using a C18 column and ternary gradient with mobile phase components of dichloromethane, ACN, and water containing 0.02% HAc. The method was applied to commonly consumed plant-based oils, which are more complex than the single triacylglycerol oils described above. Individual standards of oleic acid, linoleic acid, 1-monolinolein, sn-1/3-monoolein, 1-linolein, 2-olein, sn-1/3-diolein, triolein, and sn-1,2-diolein-3-linolein were analyzed over the same mass range and at the same PFV settings as described above and the optimal PFV was determined by the same approach. Figure 5.3 plots the RFs versus. ng on column for each analyte at different PFVs. Note that with PFV of 1.0, the RFs for both TAGs increase toward the lower mass levels indicating sublinear response as expected for non-volatiles. By contrast, both MAGs and both FAs clearly exhibit supralinear response indicative of semivolatile behavior. Also, with any PFV above 1.2, it appears that no response was obtained for either FA at the 25 ng level, a consequence of supralinear response. In this study, a PFV of 1.2 was determined as the best overall choice among the analytes of interest. As with the Infantes-Garcia study, use of the optimal PFV increased the linear dynamic range of all analytes and resulted in lower baseline noise and drift compared to the default PFV of 1.0. It is interesting to note that for similar analytes and an identical mass range, the chosen optimal PFV was quite different between the two methods (1.2 versus 1.6). For the same analytes, the plots of RF versus ng, see reference [14], are also different where only oleic acid exhibited clear semivolatile behavior. A possible explanation for the more volatile behavior in Guevara's study may be related to a lower liquid flow rate and different eluent conditions where an overall higher proportion of smaller, less stable, particles may have been transported to the downstream aerosol detector. In a study described by Causevic et al. (see Section 5.4.3) a similarly higher level of volatility is evident under similar eluent conditions. In these cases, a lower optimal PFV is expected since it is a compromise between multiple analytes where a majority behave as semivolatiles. This is also consistent with the results for the FA method of Schilling [13] where a PFV of 1.1 was found to be an optimal setting for a series of FAs, all exhibiting some volatility. For these methods, use of a lower T_e might be beneficial to reduce overall analyte volatility.

de Waal et al. [19] described a gradient method with IGC for the study of lipid digestion products in intestinal fluid. Neutral lipid classes were separated using

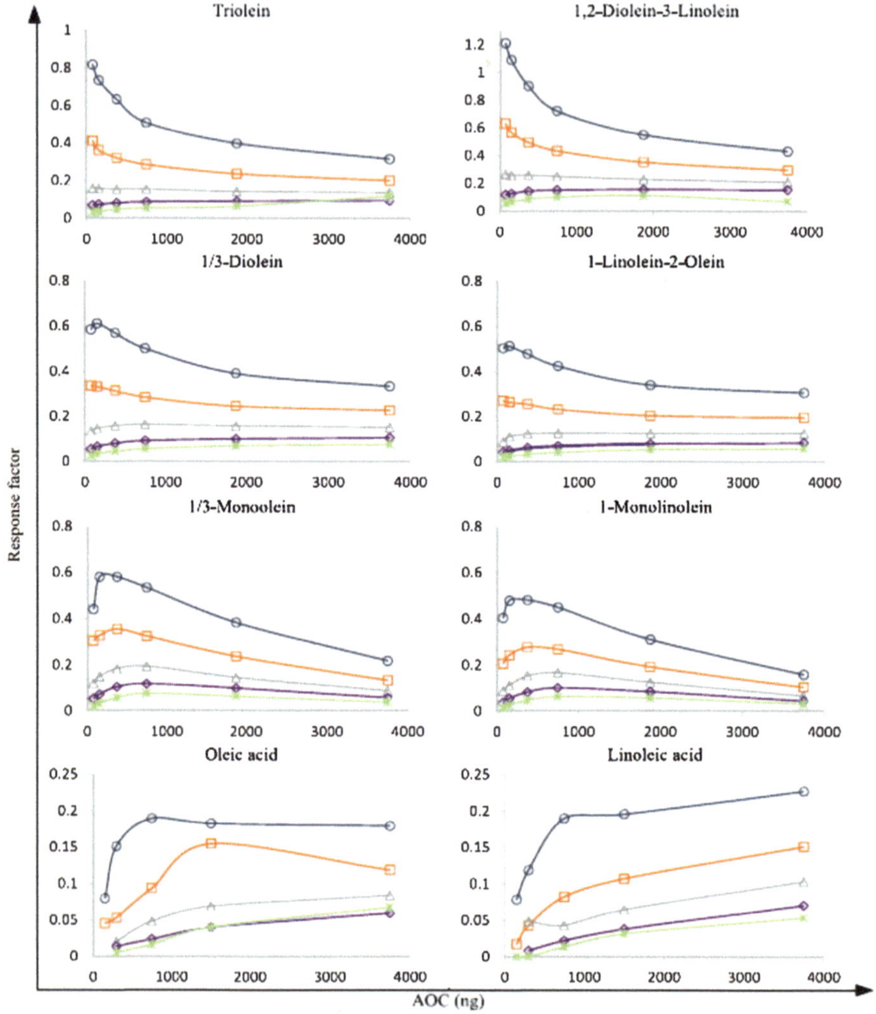

Figure 5.3 Response factor (peak area/AOC) versus AOC (ng) for eight lipid species with data collection at different PFV settings ○ PFV 1, □ PFV 1.2, △ PFV 1.3, ◇ PFV 1.4, * PFV 1.5. *Source:* [15] / with permission of Elsevier.

a diol column and binary gradient with mobile phase components of isooctane, acetone, ethyl acetate, and HAc. Individual surrogate reference standards from six neutral lipid classes (TAGs, DAGs, MAGs, FFAs, Chol, and CholE) were used for calibration over the range of 19–250 ng on column for cholesterol, cholesterol palmitate, tripalmitin, dilinolein, and mono-oleate. The calibrated range for oleic acid was 7.8–1000 on column. They used a CAD T_e of 35 °C and a different

PFV between 1.2 and 1.6, depending on the lipid class. This was accomplished using the "power law" feature in Thermo Scientific Chromeleon software [20] as further discussed in Chapter 3. Information on the specific PFV used for individual analytes and on response uniformity was not provided.

5.4.3 Response Uniformity

The following case studies describe various approaches to quantify sample components without individual standards. This topic, which includes the use of IGC is detailed in Chapter 4.

Lísa et al. [21] investigated the potential for CAD to provide simple quantitative analysis of TAGs in plant oils without the need to use analyte-dependent RFs. They used nonaqueous reversed-phase HPLC (NARP-HPLC) with two 25 cm Hypersil ODS columns, connected in series, and a binary gradient with mobile phase components of ACN, IPA and hexane. IGC was applied to suppress the response dependency of the analytes on the eluent composition (see Chapter 4). Calibration curves of 16 saturated (from C7 : 0 to C22 : 0) and 3 unsaturated (C18 : 1, C18 : 2, C18 : 3) single-acid TG standards were generated, and their RFs were compared with a previously described method using atmospheric pressure chemical ionization-mass spectrometry (APCI-MS). The variation in response of the most common TGs (containing fatty acid chains from 12 to 19 carbons) could be reduced to less than 5% making the combination of NARP-HPLC with CAD and mobile phase compensation an adequate tool for fast quantitative analysis of TGs in common plant oils. The FA content calculated from this method is in good agreement with the results obtained using a quantitative APCI-MS method, which requires knowledge of the RFs.

de la Mata-Espinosa et al. [22] described a method for analysis of TAGs in olive oil. They used a C18 column and a binary gradient with mobile phase components of ACN, hexane, and IPA. Representative standards of LLL, OOL, OOP, and SSS were selected to cover the ECN range of (42–54) found in vegetable oils. Response uniformity among these standards was found to be acceptable without the use of IGC (<10% RSD of the RFs). The proposed method was described as having advantages of faster analysis time and better sensitivity compared to an RID method adopted by European Commission for the identification of TAGs in olive oil: lower analysis time and better sensitivity. The same authors used this method for an efficient differentiation of olive oil and several types of vegetable oils where the chromatographic data were directly imported to chemometric software [23].

Lucci et al. [24] described a method for analysis of TAGs in vegetable oils and its potential use for authentication of extra virgin olive oil (EVOO). They used a C18 column and binary gradient with IGC and mobile phase components of acetone, IPA and ACN. The CAD PFV was set to 1.65 and the evaporation T was 50 °C.

They used LLL as a calibration standard and reported a correlation coefficient of >0.9999 over the mass range of 50–10,000 ng. (LOD 2.5 LOQ 8.0 for LLL). Response uniformity was assessed by analyzing 1 μg each of 5 representative TAG standards (LLL, OOO, OPO, POP, and OOP). Peak area differences among these TAGs were found to be <5% RSD, indicating a lack of dependency of CAD response on the length and the number of double bonds in the fatty acid chains. Ismaiel et al. [25] adapted the method of Lucci et al. [24] with modifications for analysis of TAGs in coffee. They used a different C18 column and a slightly modified gradient profile with the same eluent strengths and CAD detector settings. The use of IGC was not specifically mentioned but similar response uniformity (<6%) was reported for the same five TAG standards.

Green et al. [26] described a chemometric approach for detection of olive oil adulteration based on TAG profiles obtained with NARP-CAD. Samples were simply diluted 100-fold before analysis using a C18 column and a binary gradient with mobile phase components of ACN and IPA. They used a CAD evaporation T of 50 °C and a PFV of 1.00. Peak area ratios were calculated for 11 TAGs in each sample by comparing each peak area to the total TAG area. The 11 TAGs were known to be either primary components of EVOO or important in distinguishing other adulterants. This approach was reported to quickly and easily detect adulteration of olive oil at levels of 10% or greater and was described as capable of replacing both fatty acid and sterol traditional methods.

Behrens et al. [27] described a method for the quantification of rhamnolipid biosurfactants present in microorganisms. They used a Si-based column with mixed C18 and phenyl ligands and a binary gradient with mobile phase components of ACN and ammonium formate. IGC was applied and its ability to provide absolute quantification of rhamnolipid congeners was investigated. They showed that the response curves for four rhamnolipid standards were nearly identical and that the percentage composition obtained for other congeners in sample material with either the corresponding calibration equation, a calibration equation from a neighboring congener or an averaged calibration equation resulted in no significant differences. They reported that the method produces a uniform response not only for the analytes but also for structurally different (nonvolatile) compounds. These results were also confirmed by comparison to a laborious HPLC-UV derivatization method.

Causevic et al. [9] described the analysis of lipid components and digestion products in edible oils. The approach was similar to those discussed in Section 5.4.2, but with a primary goal to separate additional acylglycerol (AG) regioisomers. This was mainly accomplished with a low and changing column T (9 °C first 18 minutes, 15 °C for 62 minutes), which allowed separation of multiple species within the five main groups of FAs, FAEs, MAGs, DAGs, and TAGs along with specific AG regioisomers. Response curves for 33 standards were generated to study linearity and response uniformity. IGC, using a separate column and a capillary of a determined volume was used to match the eluent composition. The authors reported that peak shape highly affected response and response uniformity, even when IGC was employed. They also reported sublinear response over the desired 10–1000 ng range and used a quadratic curve fit. A correlation between RFs and peak width was found and used as the basis for a simplified calibration approach. It was shown, in a complex oil sample, that use of only 6 calibrated components resulted in virtually the same composition as when using 33 calibrated components. This greatly simplified the approach and method validation showed good reproducibility and accuracy, making it an excellent tool for extensive analysis of complex lipid mixtures. In this study, several factors may have contributed to the lack of improvement in response uniformity when using IGC. First, as discussed in Chapter 4, the ability to accurately compensate a ternary analytical gradient with an inverse gradient may be especially challenging. Also, lower RFs were obtained for nearly 50% of the analytes including FAEs, FAs and some MAGs. This may be attributed to volatility as described above in a similar study (Figure 5.3 above). Even with IGC, uniform response cannot be obtained with semivolatiles. Also, peak shapes for TAGs (not shown), which are assumed to behave as nonvolatiles, varied from very sharp (ca. 4 seconds at base) to significantly tailing (>120 seconds at base). With a data sampling rate of 2 Hz the sharpest peaks might have been under-sampled, effectively reducing their RFs. On the other hand, RFs for severely tailing peaks may have been artifactually increased since response was found to be sublinear (Chapter 1). It would be interesting to see if further optimization of IGC, use of a lower T_e and a higher data acquisition rate and use of an optimal PFV could provide significant improvements in response uniformity of this method.

Takahashi et al. [28] described a NP-RP 2D-LC for analysis of phospholipid molecular species. In the first dimension, main lipid classes were separated using ternary gradient NP conditions with a PVA-Si column and mobile phase components of n-hexane, MTBE, and MeOH. The targeted PLs, PE, and PC. were trapped on a C18 column before separation by isocratic RP-LC using a C18 column with mobile phase components of ACN, MeOH, AmAc with flow split to CAD and MS. Quantification was performed using external calibration curves from known PE and PC standards (16 : 0–16 : 0PE, 18 : 0–18 : 0PE, and 18 : 1–18 : 1PE; 16 : 0–16 : 0PC, 18 : 0–18 : 0PC, 18 : 1–18 : 1PC, and 18 : 2–18 : 2PC) within the mass on column range of 12.5 ng to 5.0 μg. A power model was fitted to the resulting calibration curve. The authors reported that the RSD of the peak area among various compounds was approximately 8%.

Scholz et al. [29] described a method for quantification of liamocins, which are a class of polyol lipid biosurfactants from yeast-like fungus, and are of growing interest for a variety of industrial uses. They used SFC on a charge modulated hydroxyethyl amide functionalized Si-based column for separation of liamocins and structurally related compounds. The authors noted that to implement IGC in the same way as described for HPLC, a second complete SFC would be needed. In this project, the compensation was achieved by adding a make-up flow, which is the same solvent as the modifier, via the splitter kit before the backpressure regulator as first described by Brunelli et al. [30]. The flow rate of the make-up flow was adjusted so that the ratio of CO_2 and the organic solvent was nearly constant when entering the detector. With this approach, the concentrations of the individual liamocin species were determined using a single external calibration standard and the total amount of these concentrations agreed with the dry weight of an aliquot of the heavy oil.

5.4.4 Analysis of Diverse Lipid Classes

Abreu et al. [8] described an improved NPLC method for analyzing the lipid classes present in mammalian, vegetable as well as microorganism (yeast and bacteria) lipid samples. The method effectively separated 30 lipid classes or subclasses with a special focus on medium-polarity lipids. They used an unbonded Si stationary phase and the influence of mobile phase composition and addition of ionic modifiers on the chromatographic behavior of particular lipid classes was studied. HAc and TEA were added at the end of the gradient to enhance PL separation and the addition of ethyl acetate in isooctane appeared to improve the separation of early eluting lipids. The presence of impurities in isooctane led the investigators to re-purify this solvent using a semi-preparative Si column.

Takeda et al. [16] described an SFC-CAD method for analysis of diverse lipid classes. They evaluated a range of columns and reported that a DEA column provided the best separation but produced high column bleed and unstable RTs. They implemented a column-washing procedure and used the DEA column with a CAD T_e of 95 °C (Figure 5.4). They showed a relatively small decrease in the intensity of a representative DAG with increasing T_e (from 65 to 95 °C) and a ca. 16-fold decrease in cholesterol. Note: This was shown for 100 ng levels, and the possible effect on lower quantities was not mentioned. Response curves over the range of 1.0–1000 ng were generated at different PFV settings. As the PFV increased, the SNR improved and a PFV of 1.80 was used. Linear ranges were reported to be similar except for acidic lipid classes (LPA, LPS, PA, and PS) whose peak shapes were tailing. The LOQs of each lipid class differed due to the changing background levels in response to the mobile phase composition. Linear calibration was used for most analytes except LPA, lysophosphatidylinositol (LPI), LPS, and PI. The authors noted that an optimal PFV is specific for individual analytes but the PFV could not be changed during sample analysis. While uniform

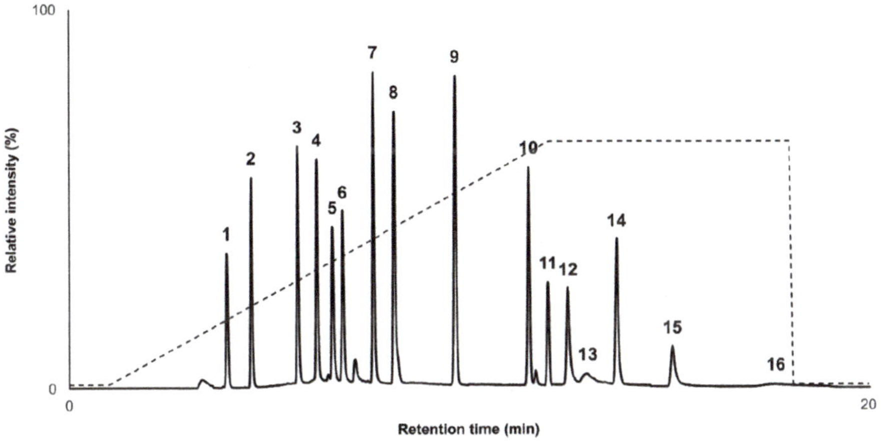

Figure 5.4 Separation of lipid standards using SFC-CAD. Standard solutions (100 μg/mL) were prepared in 65 : 35 : 8 MeOH/$CHCl_3$/water, v/v/v. Dotted line shows the gradient condition of SFC (B pump %). 1, DG 18 : 0–20 : 4; 2, Cer d18 : 1–17 : 0; 3, PC 17 : 0–17 : 0; 4, SM d18 : 1–17 : 0; 5, LPC 18 : 0; 6, PE 17 : 0–17 : 0; 7, CPE, d17 : 1–12 : 0 (internal standard); 8, LPE 17 : 1; 9, PG 17 : 0–17 : 0; 10, LPG 17 : 1; 11, PI 18 : 0–20 : 4; 12, PS 17 : 0–17 : 0; 13, PA 17 : 0–1 7 : 0; 14, LPI 17 : 1; 15, LPS 17 : 1 and 16, LPA 17 : 0. *Source:* [16] / with permission of Elsevier.

response was not observed among analyte classes similar response curves were obtained for PC standards that eluted at similar times.

5.4.5 Chromatographic Variables Affecting CAD Performance

Table 5.1 lists the analytes or analyte classes, stationary phase, elution method (i.e., isocratic, gradient) and mobile phase components with reported precision and LOQs for the case studies described above. This shows a broad range of chromatographic variables used with CAD. The LOQs are expressed in ng on column to facilitate a comparison between methods and since CAD behaves primarily as a mass-flow-sensitive device. Since LOQs are calculated by a variety of methods and response near these limits is often nonlinear, comparisons may not always be accurate. Also, while specific root causes for poorer LOQs are sometimes difficult to assign, their likely sources are discussed below.

For analytes that behave as semivolatiles, poorer detection limits, linearity, precision, and response uniformity is expected and has been consistently reported. This may be addressed by reducing T_e albeit with limitations related to increased noise from reduced evaporation of eluent impurities. Here we focus on nonvolatiles whose signal is relatively independent of analyte properties but significantly depends on eluent viscosity and surface tension, which affects

Table 5.1 Summary of LC-CAD methods for lipids.

First author (references)	SP	Elution[a)]	MP components	Analytes	*n*	LOQ range (ng)	Precision (%RSD)
Ilko et al. [12]	C18	B	ACN, Water, Formic Acid	FA (≥C16)	7	3.2–4.0	0.01–1.7
				FA (C14)	1	6.1	
Schilling et al. [13]	C18	B	ACN, Water, Formic Acid	FA (≥C16)	7	1.3–2.2	0.1–2.5
				FA (C14)	1	8.1	
Kim et al. [31]	C8	B	ACN, Water, Formic Acid	FA (≥C16)	5	0.7–2.1	0.1–1.6
Zhao et al. [32]	C18	B	ACN, Water, Formic Acid	FA (≥C16)	8	0.6–3.6	<5
				FA (C14)	1	4.4	
Wu et al. [33]	C18	I	ACN, Water, Formic Acid	FA (≥C16)	5	11–68	<3
Kielbowicz et al. [34]	C18	B	Water, ACN, Formic Acid, TEA	FA (≥C16)	9	4.9–50	<0.5
Lísa et al. [21]	C18	B	IPA, Hexane, ACN	TAG	19	1.3–6.6[b)]	0.4–1.3
Lucci et al. [24]	C18	B	ACE, IPA, ACN	TAG	1	8.0	4.2
Green et al. [26]	C18	B	ACN and IPA	TAG	2	1.2–1.5	0.3–5.1
Infantes-Garcia et al. [14]	Si	Q	ISO, ACE, Ethyl Acetate, IPA, Water, HAc, Ethanolamine	TAG	1	2.7	0.4–2.3
				DAG	2	19–24	
				MAG	2	5.2–5.5	
				FA (≥C16)	1	13.0	
Guevara et al. [15]	C18	T	CH_2Cl_2, ACN, Water, HAc	TAG	1	7.8	0.7–2.5
				DAG	3	5.9–9.5	
				MAG	2	4.8–5.8	
				FA (≥C16)	2	6.9–31	
Causevic et al. [9]	C18	T	ACN, IPA, Heptane, HAc	TAG	13	2.0–6.0	1.1–3.3
				DAG	6	1.6–12	
				MAG	5	4.5–9.6	
				FA (≥C16)	5	6.1–18	
				FAE	4	20–113	
Ramos et al. [3]	PVA-SiT		Heptane, IPA, $CHCl_3$, MeOH, HAc, TEA	PL Class	11	45–707	3.2–13

Table 5.1 (Continued)

First author (references)	SP	Elution[a)]	MP components	Analytes	*n*	LOQ range (ng)	Precision (%RSD)
Damnjanović et al. [5]	Si	B	$CHCl_3$, MeOH, Formic Acid, TEA	PL Class	6	19–136	1.0–7.9
Takeda et al. [16]	DEA	B	supercritical CO_2, Water, MeOH	Lipid Class	15	1.9–72	< 0.9
Rosłon et al. [17]	Diol	B	Hexane, IPA, Water, HAc, TEA	PL Class	4	60–400	3.5–9.0
Zhao [35]	Si	T	Hexane, IPA, MeOH, Water, HAc, TEA	Lipid Class	10	4.0–80	2.33–7.1
Khoomrung et al. [36]	Diol	T	Hexane, HAc, ACE, IPA, HAc, TEA	Neutral and polar lipids	12	2.7–17	0.6–6.0
Barry et al. [37]	Si	B	CH_2Cl_2, MeOH, TEA, HAc	PL Class	6	6.3–183	n/a
Kiełbowicz et al. [38]	Diol	T	Formic Acid, Hexane, IPA	PL Class	6	38–175	< 0.80
Behrens et al. [27]	C18 Phenyl	B	ACN AmF	Bio-surfactants	4	6.0–10	n/a

a) I, isocratic; B, binary gradient; T, ternary gradient; Q, quaternary gradient.
b) Estimated from the reported LOD.

detector transport efficiency. Since transport efficiency applies equally to analytes and impurities, there should be a significant positive correlation between analyte signal and baseline current and noise. The SNRs obtained for nonvolatile analytes between methods or within a solvent gradient method should therefore be relatively similar as long as peak widths are similar and there is an equivalent concentration of eluent impurities. This argues that differences in eluent impurity concentrations are a main source of differences in reported LOQs. As described in Chapter 3, eluent impurities originating from column bleed, mobile phase components and other sources can have a profound effect on baseline noise and drift affecting both nonvolatile and semivolatile analytes. Peak broadening and tailing is another main cause of poorer LOQs, which are typically based on measurement of signal height at a peak's apex. Lower peak heights are expected for broader and tailing peaks.

FA methods listed in Table 5.1 typically use binary gradient RPLC with simple eluents and Si-based C18 columns and conditions (pH, column T) that are well known to produce minimal bleed. For these methods, LOQs of ca. 5 ng or better were reported for FAs with a ≥ C16 carbon chain length with higher LOQs for the more volatile, shorter chain, analytes. Within a single gradient method where peak widths are similar, the LOQs among different longer chain FAs are also quite similar. This is consistent with expectations as discussed above. Higher LOQs for ≥ C16 FAs were, however, reported by Wu et al. [33]. Here baseline noise may have been higher with a time-constant filter setting of "none" and the signal height was lower due to the relatively broad peaks obtained with isocratic conditions. Higher LOQs were also reported by Kielbowicz et al. [34]. A main difference between this and other FA methods is the use of both TEA and formic acid mobile phase additives. As previously discussed, this is a common source of higher background currents and baseline noise where TEA and formate can combine, both together and separately with ionic impurities, to form detectable salt particles.

For the described TAG methods, analytes are observed to behave as nonvolatiles and the separation methods also commonly use simple mobile phases with C18 columns and "minimal bleed" conditions [21, 22, 24–26]. Here the LOQs are quite similar to those reported for the longer chain FAs and quite similar among the different TAGs. The similarity in LODs ranging from 0.4 to 2.0 ng among 19 TAG species reported by Lisa et al. is particularly notable and may, in part, be due to their use of IGC where eluent impurity concentrations remain constant. While FAs and TAGs are commonly analyzed with simple eluents, simple gradients and low-bleed columns/conditions, this is certainly not the case for many other separations including those used for lipid classes (Section 5.4.4), polar lipids and other studies. These separation methods, while necessary, can pose challenges for detection with CAD, which can adversely affect detection limits, precision and other figures of merit. As evidenced in Table 5.1, detection limits obtained with more complex methods are often higher and more variable within and among lipid classes. For example, methods described above for analysis of edible oils and their digestion products (e.g., TAGs, DAGs, MAGs, and FAs) typically use normal phase or RP with more complex ternary or quaternary gradients. This includes isocratic segments, step changes and changes in gradient slope, which produce a variety of peak shapes. The method of Causevic et al. [9] is one example where peaks varied from very sharp to broad and symmetrical to significantly tailing. As reported, the LOQs correlated with peak shape where higher LOQs were obtained for nonvolatile analytes whose peaks were broader or tailing. It should be noted that analytes that behave as semivolatiles typically produce artifactually sharper peaks due to their supralinear response toward lower mass concentrations. This so-called peak shaving is commonly observed even for nonvolatiles with ELSD due to its inherently supralinear response toward lower mass levels.

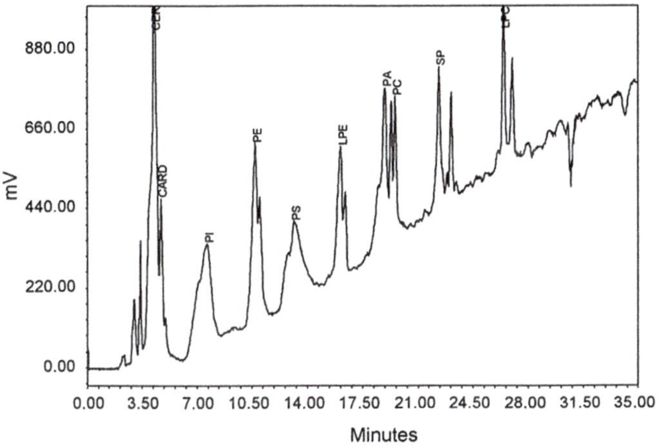

Figure 5.5 Chromatographic separation of phospholipid standards by HPLC-CAD. Cerebrosides (CER); cardiolipin (CARD); l-α-phosphatidylinositol (PI); l-α-phosphatidylethanolamine (PE); l-α-phosphatidylserine (PS); l-α-lisophosphatidylethanolamine (LPE); phosphatidic acid (PA); l-α- phosphatidylcholine (PC); sphingomyelin (SP); l-α-lisophosphatidylcholine (LPC) using a binary gradient from 0% to 100% B. Eluant A: CH_2Cl_2–MeOH–NH_4OH (75 : 24 : 0.5, v/v/v); Eluant B: CH_2Cl_2–MeOH–NH_4OH–water (43 : 43 : 1 : 13, v/v/v/v). *Source:* [39] / with permission of Elsevier.

Besides peak shape, differences and changes in eluent-impurity concentration are likely to be main causes of poorer LOQs for some or all analytes. Since ionic additives can increase baseline signal and noise, changes in their concentration, both gradual and stepwise, might differentially affect analytes depending on their retention time.

In Table 5.1, higher LOQs are quite frequently reported for methods that employ NP and/or HILIC based separations. This includes the use of either unbonded Si or polar alkyl-bonded stationary phases, which are often described as exhibiting column bleed (Chapters 3, 6 and 7). Several of these lipid papers specifically describe the effects on baseline noise and drift from column bleed. Cascone et al. [39] reported a high level of column bleed in a PL method with a Si column. The observed baseline drift and increased baseline noise was attributed to an increase in the water content during the gradient (Figure 5.5). This was also associated with an increase in ammonium hydroxide concentration, where a higher pH may increase Si dissolution.

Kielbowicz et al. [38] compared the baseline level and drift for two columns (unbonded Si and diol) at different column T and with gradient changes in eluent composition. With the unbonded Si column, baseline signal, noise and drift significantly increased with column temperature over the range of 20–40 °C while the diol column was relatively stable. Takeda et al. [16] compared baseline noise and drift between seven different columns, some of which showed significant evidence of bleed even with a CAD T_e of 65 °C. As with other analyte classes (e.g., ions, carbohydrates) that use NP and HILIC columns with both CAD and MS, column bleed can be a significant issue. Hopefully, further technological improvements to reduce column bleed will soon become available. When choosing a column for use with CAD, a good practice is to examine baseline currents, noise and drift and to compare the results to those obtained without a column. This should be done over the expected range of T_c and mobile phase variables such as pH, solvent, and additive composition and include gradient elution.

References

1 Hazotte, A., Libong, D., Matoga, M., and Chaminade, P. (2007). Comparison of universal detectors for high-temperature micro liquid chromatography. *Journal of Chromatography. A* 1170 (1): 52–61.

2 Merle, C., Laugel, C., Chaminade, P., and Baillet-Guffroy, A. (2010). Quantitative study of the stratum corneum lipid classes by normal phase liquid chromatography: comparison between two universal detectors. *Journal of Liquid Chromatography and Related Technologies* 33 (5): 629–644.

3 Ramos, R.G., Libong, D., Rakotomanga, M. et al. (2008). Comparison between charged aerosol detection and light scattering detection for the analysis of *Leishmania* membrane phospholipids. *Journal of Chromatography. A* 1209 (1): 88–94.

4 Motulsky, H.J. and Ransnas, L.A. (1987). Fitting curves to data using nonlinear regression: a practical and nonmathematical review. *The FASEB Journal* 1 (5): 365–374.

5 Damnjanović, J., Nakano, H., and Iwasaki, Y. (2013). Simple and efficient profiling of phospholipids in phospholipase D-modified soy lecithin by HPLC with charged aerosol detection. *Journal of the American Oil Chemists' Society* 90 (7): 951–957.

6 Olsson, P., Holmbäck, J., and Herslöf, B. (2014). A single step reversed-phase high performance liquid chromatography separation of polar and non-polar lipids. *Journal of Chromatography. A* 1369: 105–115.

7 Nair, L.M. and Werling, J.O. (2009). Aerosol based detectors for the investigation of phospholipid hydrolysis in a pharmaceutical suspension formulation. *Journal of Pharmaceutical and Biomedical Analysis* 49 (1): 95–99.

8 Abreu, S., Solgadi, A., and Chaminade, P. (2017). Optimization of normal phase chromatographic conditions for lipid analysis and comparison of associated detection techniques. *Journal of Chromatography. A* 1514: 54–71.

9 Causevic, A., Olofsson, K., Adlercreutz, P., and Grey, C. (2021). Non-aqueous reversed phase liquid chromatography with charged aerosol detection for quantitative lipid analysis with improved accuracy. *Journal of Chromatography. A* 1652: 462374.

10 Libong, D., Héron, S., Tchapla, A., and Chaminade, P. (2017). Lipid analysis with the corona CAD. In: *Charged Aerosol Detection for Liquid Chromatography and Related Separation Techniques* (ed. P.H. Gamache), 221–287. John Wiley and Sons.

11 Moreau, R.A. (2006). The analysis of lipids via HPLC with a charged aerosol detector. *Lipids* 41 (7): 727–734.

12 Ilko, D., Braun, A., Germershaus, O. et al. (2015). Fatty acid composition analysis in polysorbate 80 with high performance liquid chromatography coupled to charged aerosol detection. *European Journal of Pharmaceutics and Biopharmaceutics* 94: 569–574.

13 Schilling, K., Pawellek, R., Lovejoy, K. et al. (2018). Influence of charged aerosol detector instrument settings on the ultra-high-performance liquid chromatography analysis of fatty acids in polysorbate 80. *Journal of Chromatography. A* 1576: 58–66.

14 Infantes-Garcia, M.R., Verkempinck, S.H.E., Guevara-Zambrano, J.M. et al. (2021). Development and validation of a rapid method to quantify neutral lipids by NP-HPLC-charged aerosol detector. *Journal of Food Composition and Analysis* 102: 104022.

15 Guevara-Zambrano, J.M., Michels, D., Verkempinck, S. et al. (2023). HPLC-CAD method to quantify lipolysis products from plant-based oils rich in unsaturated fatty acids. *Journal of Food Composition and Analysis* 121: 105400.

16 Takeda, H., Takahashi, M., Hara, T. et al. (2019). Improved quantitation of lipid classes using supercritical fluid chromatography with a charged aerosol detector. *Journal of Lipid Research* 60 (8): 1465–1474.

17 Rosłon, M., Jaworska, M., and Anuszewska, E.L. (2022). Determination of glycerophospholipids in biological material using high-performance liquid chromatography with charged aerosol detector HPLC-CAD—a new approach for isolation and quantification. *Molecules* 27 (10): 3356.

18 Ahmad, I.A.H., Blasko, A., Tam, J. et al. (2019). Revealing the inner workings of the power function algorithm in charged aerosol detection: a simple and effective approach to optimizing power function value for quantitative analysis. *Journal of Chromatography. A* 1603: 1–7.

19 de Waal, T., Brouwers, J., Berben, P. et al. (2023). Characterization of aspirated duodenal fluids from parkinson's disease patients. *Pharmaceutics* 15 (4): 1243.

20 Gamache, P., Muellner, T., Eggart, B. et al. (n.d.). Charged aerosol detection – use of the power function and robust calibration practices to achieve the best quantitative results, Thermo Fisher Scientific Technical Note.;73299.

21 Lísa, M., Lynen, F., Holčapek, M., and Sandra, P. (2007). Quantitation of triacylglycerols from plant oils using charged aerosol detection with gradient compensation. *Journal of Chromatography. A* 1176 (1): 135–142.

22 de la Mata-Espinosa, P., Bosque-Sendra, J.M., and Cuadros-Rodríguez, L. (2011). Quantification of triacylglycerols in olive oils using HPLC-CAD. *Food Analytical Methods* 4 (4): 574–581.

23 de la Mata-Espinosa, P., Bosque-Sendra, J.M., Bro, R., and Cuadros-Rodríguez, L. (2011). Olive oil quantification of edible vegetable oil blends using triacylglycerols chromatographic fingerprints and chemometric tools. *Talanta* 85 (1): 177–182.

24 Lucci, P., Moret, S., Buchini, F. et al. (2018). Improved analysis of olive oils triacylglycerols by UHPLC-charged aerosol detection. *Journal of Food Composition and Analysis* 66: 230–236.

25 Ismaiel, L., Fanesi, B., Kuhalskaya, A. et al. (2023). The determination of triacylglycerols and tocopherols using UHPLC–CAD/FLD methods for assessing the authenticity of coffee beans. *Food* 12 (23): 4197.

26 Green, H.S., Li, X., De Pra, M. et al. (2020). A rapid method for the detection of extra virgin olive oil adulteration using UHPLC-CAD profiling of triacylglycerols and PCA. *Food Control* 107: 106773.

27 Behrens, B., Baune, M., Jungkeit, J. et al. (2016). High performance liquid chromatography-charged aerosol detection applying an inverse gradient for quantification of rhamnolipid biosurfactants. *Journal of Chromatography. A* 1455: 125–132.

28 Takahashi, R., Nakaya, M., Kotaniguchi, M. et al. (2018). Analysis of phosphatidylethanolamine, phosphatidylcholine, and plasmalogen molecular species in food lipids using an improved 2D high-performance liquid chromatography system. *Journal of Chromatography B* 1077-1078: 35–43.

29 Scholz, K., Lipphardt, A., Wienken, C.M. et al. (2020). Hyphenation of supercritical fluid chromatography with different detection methods for identification and quantification of liamocin biosurfactants. *Journal of Chromatography. A* 1631: 461584.

30 Brunelli, C., Gorecki, T., Zhao, Y., and Sandra, P. (2007). Corona-charged aerosol detection in supercritical fluid chromatography for pharmaceutical analysis. *Analytical Chemistry* 79 (6): 2472–2482.

31 Kim, H.S., Moon, B.C., Yang, S. et al. (2019). Determination of fatty acids in the seeds of Lepidium apetalum Willdenow, Descurainia sophia (L.) Webb ex Prantl, and Draba nemorosa L. by ultra-high-performance liquid chromatography equipped with a charged aerosol detector. *Journal of Liquid Chromatography and Related Technologies* 42 (5–6): 128–136.

32 Zhao, X., He, Y., Chen, J. et al. (2021). Identification and direct determination of fatty acids profile in oleic acid by HPLC-CAD and MS-IT-TOF. *Journal of Pharmaceutical and Biomedical Analysis* 204: 114238.

33 Wu, Z., Li, L., Li, N. et al. (2017). Optimization of ultrasonic-assisted extraction of fatty acids in seeds of *Brucea javanica* (L.) Merr. from different sources and simultaneous analysis using high-performance liquid chromatography with charged aerosol detection. *Molecules* 22 (6): 931.

34 Kiełbowicz, G., Chojnacka, A., Gliszczyńska, A. et al. (2015). Positional analysis of phosphatidylcholine and phosphatidylethanolamine via LC with a charged aerosol detector. *Talanta* 141: 137–142.

35 Zhao, G.-H. (2022). A direct and facile simultaneous quantification of non-polar and polar lipids in different species of marine samples using normal-phase HPLC–CAD. *Journal of Food Composition and Analysis* 114: 104813. --2022 v.114.

36 Khoomrung, S., Chumnanpuen, P., Jansa-Ard, S. et al. (2013). Rapid quantification of yeast lipid using microwave-assisted total lipid extraction and HPLC-CAD. *Analytical Chemistry* 85 (10): 4912–4919.

37 Barry, K.M., Dinan, T.G., Murray, B.A., and Kelly, P.M. (2016). Comparison of dairy phospholipid preparative extraction protocols in combination with analysis by high performance liquid chromatography coupled to a charged aerosol detector. *International Dairy Journal* 56: 179–185.

38 Kiełbowicz, G., Micek, P., and Wawrzeńczyk, C. (2013). A new liquid chromatography method with charge aerosol detector (CAD) for the determination of phospholipid classes. Application to milk phospholipids. *Talanta* 105: 28–33.

39 Cascone, A., Eerola, S., Ritieni, A., and Rizzo, A. (2006). Development of analytical procedures to study changes in the composition of meat phospholipids caused by induced oxidation. *Journal of Chromatography. A* 1120 (1): 211–220.

6

Analysis of Inorganic and Organic Ions with Charged Aerosol Detection

Paul H. Gamache

Consultant, Nashua, NH, USA

CHAPTER MENU

6.1 Summary

The use of HPLC-CAD for analysis of inorganic ions and small organic ionizable molecules has now become commonplace. This has been greatly facilitated by advances in mixed mode and hydrophilic interaction column technologies along with a greater understanding of the variables affecting their selectivity. Several papers have described the implementation of this approach in biopharmaceutical and pharmaceutical process development and quality control. This chapter provides an overview of these methods and discusses advantages and limitations of the technique and best practices to address the challenges. Main advantages of HPLC-CAD over traditional methods such as ion chromatography (IC) include the ability to reliably measure inorganic and organic anions and cations along with neutral analytes in a single analysis. This can often be achieved with simplified sample preparation due to greater sample matrix compatibility and with sufficient sensitivity and dynamic range to quantify main components and impurities. Also, the use of CAD doesn't require a dedicated IC system and can be connected to a conventional HPLC system that is usually equipped with a UV detector and using methods that are readily compatible with MS. The overall approach can therefore provide a more comprehensive assessment of what is in a sample. Technical considerations specific to analysis of ions with CAD include those related to the choice of stationary phase; the choice of mobile phase buffers and their method of preparation; the added complexity of method development with mixed mode columns; column bleed and possible sources of contamination.

6.2 Introduction

Ions are ubiquitous in everyday life. Inorganic ions such as calcium, chloride, sodium, and phosphate play important biological roles including those related to bone structure, fluid and pH balance, nerve and muscle function, and as enzyme co-factors. These ions along with small organic ionizable molecules including maleate, tromethamine, and choline are common pharmaceutical and biopharmaceutical excipients (e.g., counterions, stabilizers, buffer components) and, in some cases, serve as active pharmaceutical or nutritional ingredients. Ions are common and useful components in the environment, food and beverages, industrial processes, and pharmaceutical formulations but in many cases may exist as impurities and contaminants with potentially adverse effects. The identification and quantitative analysis of ions is therefore important in many areas including environmental and industrial process monitoring, food quality control and for discovery, development, and quality control of biopharmaceuticals and pharmaceuticals.

Ions can be measured directly by potentiometric titration and inductively coupled plasma-atomic emission spectrometry (ICP-AES) and after separation by high-performance liquid chromatography (HPLC), capillary electrophoresis (CE), and ion chromatography (IC). IC is a primary tool for ion analysis and can be viewed as a subdiscipline of HPLC with specialized instrumentation, consumables, and detectors. The topic of IC has been well covered in books and reviews [1, 2]. IC with suppressed conductivity can provide very high sensitivity for most ionic analytes. However, IC is generally limited to the separate detection of either anionic or cationic species and is compatible with only a limited range of mobile phases and detectors compared to conventional HPLC. Recent developments in stationary phase technology for hydrophilic interaction chromatography (HILIC) and mixed mode chromatography (MMC) and their more facile use with evaporative aerosol detectors (EADs) and mass spectrometry (MS) have led to a significant expansion in the number of methods reported for HPLC analysis of ions [3]. While UV/vis absorption is the most common HPLC detection technique, it is often unsuitable for ion analysis where many analytes lack a sufficient chromophore. MS provides a specific and sensitive technique for HPLC, but detection is highly dependent on analytical conditions, and the instrumentation can be costly and operationally complex. Refractive index detectors (RID) have considerable limitations in sensitivity and stability and are incompatible with gradient elution. Evaporative light scattering detection (ELSD) exhibits limitations in precision, sensitivity, dynamic range, and the nature of calibration curves. Condensation nucleation light scattering detection (CNLSD) is appropriate for detecting semivolatile and nonvolatile compounds with better sensitivity than ELSD, but its response substantially depends on analyte properties.

Charged aerosol detection (CAD) represents a significant technological improvement over other EADs (ELSD, CNLSD) as a universal detector for LC and SFC. The main benefits of CAD include universal detection of nonvolatile and many semivolatile analytes, response for nonvolatile analytes that is largely independent of chemical properties, broad dynamic response range with detection limits spanning from low ng to high μg, good precision for a diverse range of analytes, and simple and reliable operation. The combination of MMC with CAD and other EADs provides the unique ability to simultaneously measure anions, cations, and other non-ionic species on a standard LC system. With CAD, like other EADs: response is not perfectly linear and its magnitude varies with mobile phase composition; it lacks analyte-specific information such as with diode array or MS; it requires the use of volatile mobile phases and is limited to detection of analytes that behave as nonvolatiles or semivolatiles with the latter having poorer sensitivity. Various approaches to address these limitations are discussed throughout this book, especially in Chapters 3 and 4. This chapter describes the main practical considerations to obtain the best performance with CAD specifically for ion analysis and provides an overview of key applications.

6.3 Practical Considerations

Analytes described in this chapter include inorganic ions and smaller, moderately strong organic acids and bases. Within the typical pH range used for separation, these analytes should elute as mostly ionized solutes. CAD response is then based on the salt formed between the analyte and oppositely charged co-eluting solutes (Chapter 1). The most likely co-eluting ions are mobile phase additives, but impurities and sample matrix components may also come into play. The choice of mobile phase additive type and concentration is especially important for ion analysis with CAD as is the choice of stationary phase. While ion-pair reversed-phase liquid chromatography (IP-RPLC) is commonly used for separation of ions, several issues have been described with this approach. With newer developments in column technology for HILIC and MMC, several options are now available to avoid many of the issues with IP-RPLC. Columns designed for MMC and

HILIC that exhibit low-bleed and provide good separation with minimal additive concentration can provide notable advantages over IP-RPLC for use with CAD (e.g., see Chapter 10). Practical considerations related specifically to ion analysis with CAD are detailed below. Additional best practices, applicable to any analyte class, are detailed in Chapter 3.

6.3.1 Column Selectivity

Column selectivity is a main factor to consider when choosing a column for ion analysis with CAD. While IP reagents can be used to effect ion retention with RPLC, their use is often associated with chromatographic artifacts, incompatibility with MS, and limited ability to tailor selectivity [4]. Also, the use of IP reagents with CAD often leads to several issues including increased baseline currents, drift, and noise (Chapter 3). Columns used for HILIC can retain ions and small polar molecules that are not retained by RPLC without the use of IP reagents. Russell et al. [5] studied the separation of ions under HILIC conditions using an ethylene bridged hybrid (BEH™, Waters, Corp.) amide stationary phase. They obtained good separation of 6 inorganic ions with a charge-neutral amide stationary phase and noted that cations were more strongly retained than anions. This was attributed to the retention of cations and repulsion of anions by ionized silanol groups, which exist but are typically unwanted on all silica-based columns. Ion separation was therefore attributed to a combination of intended HILIC interactions and unintended secondary ionic interactions. As described by Zhang and Liu [3], mixed mode stationary phases have been developed with the specific intent to provide desirable and repeatable secondary or tertiary interactions through the use of carefully designed functional groups. These stationary phases intentionally employ a combination of two or more interaction modes [e.g., hydrophilic, hydrophobic and ionic (anionic, cationic, zwitterionic)]. Reported advantages of mixed mode columns with ion-exchange functionalities compared to conventional ion-exchange columns include: a greater ability to develop suitable separations for sample mixtures containing both ionic and neutral analytes; better sample matrix compatibility often facilitating simplified sample preparation; the ability to use conventional HPLC without the need to purchase and maintain a dedicated IC system; the ability to separate anions and cations in a single method without the need for separate methods with different columns and IC instrument configurations; and greater compatibility with MS [3, 4].

Lantz et al. [6] described an early use of MMC with ELSD for simultaneous analysis of active pharmaceutical ingredients (APIs), impurities, and counterions. They evaluated a series of silica-based columns bonded with either anionic (amine) or cationic (carboxylate) functionalities in addition to phenyl, C8 or C18 functionalities. This provided an initial demonstration of the potential utility of combining MMC with universal detectors for the simultaneous analysis of APIs and counterions. Huang et al., described the use of a polymer-based zwitterionic column (ZIC-pHILIC, Merck) operated in HILIC mode with CAD as a generic method for the determination of inorganic pharmaceutical counterions in drug substances [7]. In this study, three silica-based HILIC columns were also evaluated. The experimental data suggested that the presence of ion-exchange functionality on the stationary phase was necessary to obtain desirable selectivity and decent peak shape for ions. Pack and Risley [8] used a silica-based zwitterionic column in HILIC mode with ELSD to separate and measure 33 commonly used counterions. They performed a thorough evaluation of column retention characteristics by varying key mobile phase parameters including pH, buffer strength and organic modifier concentration. Crafts et al. [9] used this information to develop a method for simultaneous analysis of anionic and cationic counterions on a polymer-based zwitterionic column in HILIC mode with CAD. They reported low ng LODs for several ions and demonstrated the ability to measure ionic impurity levels at the 0.1% level relative to the API.

Liu et al. reported a novel anion-exchange/cation-exchange/reversed-phase trimodal column that separates APIs and corresponding counterions (including anions and cations) using CAD [10, 11]. Its column chemistry is shown in Figure 6.1, and an exemplary application is shown in Figure 6.2. Zhang et al. used this trimodal column (Acclaim™ Trinity™ P1, Thermo Fisher Scientific) to develop a generic screening LC-CAD method for simultaneous determination of positive and negative pharmaceutical counterions [12]. In this study, several mixed mode columns, including Primesep AB (SIELC), ZIC-pHILIC (Merck),

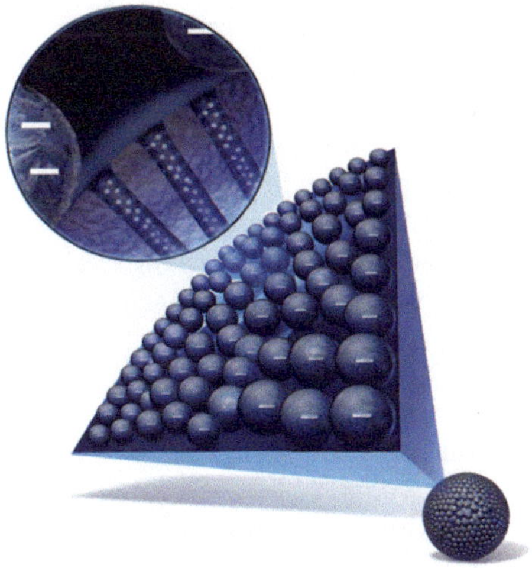

Figure 6.1 Column chemistry of Acclaim Trinity P1—a reversed-phase/anion-exchange/cation-exchange trimodal mixed mode column based on Nanopolymer Silica Hybrid (NSH) technology. (a) Overview of a silica particle coated with nanopolymer beads; (b) enlarged view of the silica surface coated with negatively charged nanopolymer beads; (c) inner-pore area consists of reversed-phase and weak anion-exchange functionalities and outer surface provides strong cation-exchange interaction. *Source:* Reproduced with permission of Thermo Fisher Scientific Inc.

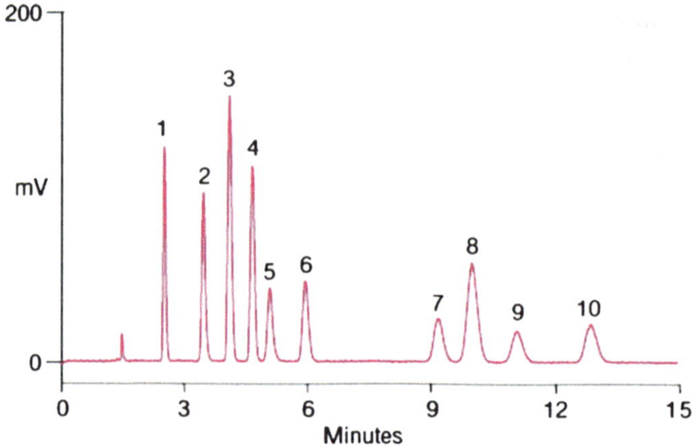

Figure 6.2 Separation of pharmaceutical counterions using a reversed-phase/anion-exchange /cation-exchange trimodal mixed mode column. Column, Acclaim Trinity P1, 3.0 × 100-mm format; mobile phase, acetonitrile/ammonium acetate buffer, pH 5 (20 mM total concentration) (60 : 40, v/v); flow rate, 0.5 mL/min; injection volume, 2 μL; temperature, 30 °C; and detection, corona ultra (gain = 100 pA; filter = med; nebulizer temperature = 30 °C). Sample: 0.05–0.1 mg/mL. Peaks: (1) Choline or N,N,N-trimethylethanolammonium; (2) tromethamine or tris(hydroxymethyl)aminomethane; (3) sodium; (4) potassium; (5) meglumine or N-methyl glucamine; (6) mesylate or methanesulfonate; (7) nitrate; (8) chloride; (9) bromide; and (10) iodide. *Source:* [10] / with permission of Elsevier.

Obelisc N (SIELC), and Acclaim Trinity P1 (Thermo Fisher Scientific) were compared, and the Acclaim Trinity P1 was found to provide the best selectivity for separating a total of 25 pharmaceutical-related counterions.

A later study compared the Acclaim Trinity P1 to a silica-based zwitterionic HILIC (ZIC-HILIC, Merck) column for determination of APIs and their corresponding counterions. The results indicated that the trimodal column provided several advantages over the zwitterionic column, such as desirable selectivity for pharmaceutical counterions, better versatility for simultaneous separation of APIs and corresponding counterions (Figure 6.3), greater flexibility in method development, and less usage of organic solvent [13]. As discussed in Section 6.4, the approach described in Zhang et al. [12] has served as a basis for several key applications of CAD for ion analysis.

6.3.2 Column Bleed

Given its behavior as a mass-dependent universal detector with substantial sensitivity, column bleed is more readily detected by CAD compared to other techniques. In the context of CAD, column bleed refers to the addition of less volatile impurities into the eluent by the column itself. This can lead to high

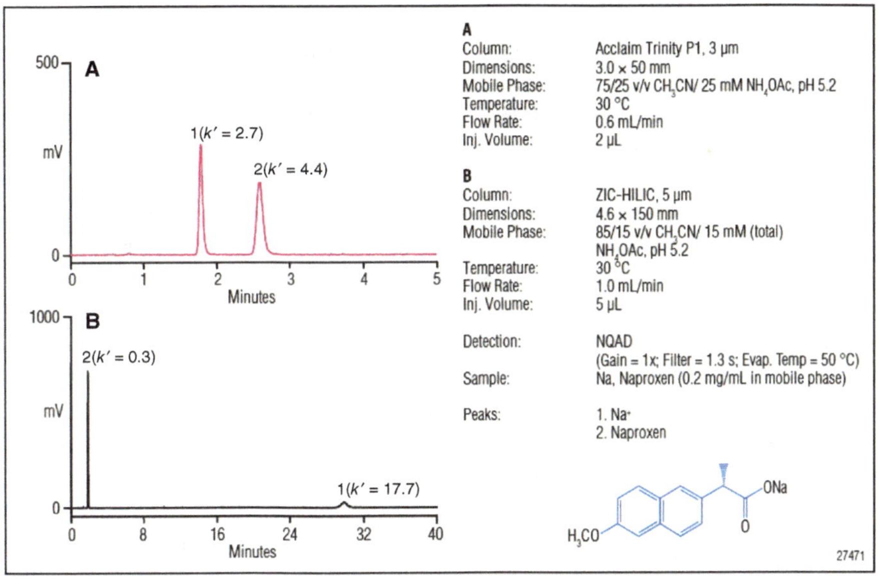

Figure 6.3 Simultaneous separation of a hydrophobic acidic drug and its counterion—naproxen sodium salt. *Source:* Reproduced with permission of Thermo Fisher Scientific Inc.

baseline currents, drift, noise, and other adverse effects (Chapter 3). This is a particular concern with ion analysis since most of the separation methods use HILIC and mixed mode columns. While most modern silica-based reversed-phase columns exhibit low bleed when used within their recommended pH and temperature range, there have been reports of significant bleed observed with CAD or MS when using silica-based HILIC columns even when operated within their recommended ranges [14]. Dissolution of the silica support at higher pH is a main source of bleed with any silica-based column, but the issue appears to be more pronounced with columns used for normal phase (NP) and HILIC separations (see also Chapters 5 and 7). Possible factors contributing to higher column bleed with HILIC include: less protection of the support by an organic bonded phase layer (e.g., C18); more facile hydrolysis of attached functional groups and/or dissolution of the base silica substrate by the water-enriched layer on the stationary phase surface; greater exposure of the support to higher organic eluents, which shifts the pH higher relative to the aqueous pH [14–16]. Studies that have examined the condition-dependent stability of stationary phases usually monitor changes in chromatographic efficiency, peak tailing and retention times. However, as noted in several studies, significant column bleed can be observed with CAD in the absence of chromatographic effects. An example of this is provided by Huang et al., who evaluated the bleed of several HILIC columns with CAD for ion analysis [7]. They studied three silica-based columns (including ZIC-HILIC, Waters Atlantis HILIC, and Phenomenex Luna HILIC) and one polymer-based column (ZIC-*p*HILIC) and found that baseline currents varied significantly among the columns (see Figure 3.2, Chapter 3). When operating on silica-based columns (ZIC-HILIC and Atlantis HILIC), very high baseline currents (>50 pA) were observed with CAD. Under the same condition, both ZIC-*p*HILIC and Luna HILIC exhibited lower baseline currents, suggesting minimal column bleed. This was attributed to the relative stability of the polymeric support and to protection of the silica by a cross-linked bonded phase, respectively. In a separate study, the baseline currents of different mixed mode columns including Primesep AB, ZIC-pHILIC, Obelisc N, and Acclaim Trinity P1 were compared [12]. Among these columns, the Primesep AB column exhibited significant baseline currents (>50 pA). The remaining three columns gave baseline currents lower than 5 pA. As noted in Chapter 3, a good practice is to examine not only baseline currents, but also noise and drift and to compare the results to those obtained without a column. This allows one to more directly evaluate the source of impurities. For gradient methods, it may also be useful to examine the trend in baseline currents over several cycles. While baseline currents of <5 pA as noted above have been associated with very successful applications, this may still be high relative to RP methods.

6.3.3 Mobile Phase

The concentration and type of both the buffer and the organic modifier along with pH are critical for the retention of ionic analytes. Pack and Risley described a comprehensive investigation on how these variables impact retention of ionic analytes using a zwitterionic stationary phase in HILIC mode [8]. Liu and Pohl likewise showed the significant influence of these variables on the selectivity and retention characteristics when using an Acclaim Trinity P1 trimodal column [10]. These studies provide great insight and demonstrate the flexibility to optimize selectivity for specific requirements. This finding was substantiated by Zhang et al. for pharmaceutical counterion ion analysis who provided further insight into the retention mechanisms by studying the behavior of analytes grouped into four categories based on their physicochemical properties [12]. With this information in mind, the following discussion focuses on the influence of mobile phase composition on CAD response.

6.3.4 Solvent Composition

For CAD and other EADs, mass transport from the nebulizer to the downstream aerosol detector is higher for eluents with lower viscosity and surface tension (e.g., CH_3CN). As a result, signal is higher when compared to eluents with higher viscosity and surface tension (e.g., H_2O). Mitchell et al. [17] reported that sensitivity (signal/ng) with CAD was approximately 5–10 times higher in HILIC mode [90% acetonitrile (ACN)/10% buffer] than in RP mode (5% ACN/95% buffer) for a group of 12 hydrophilic compounds. It should be noted however, that higher sensitivity as defined in this case does not necessarily mean better detection limits since solvent-dependent differences in transport apply equally to both analytes and eluent impurities. Improved detection limits are therefore mainly achieved when eluent impurities (e.g., from column bleed, solvent, additive) are minimized.

In solvent gradient methods, response factors differ between analytes that elute at different points, also due to changes in mass transport. This may pose a problem when the goal is to achieve uniform response, which would allow the use of single calibration standards to quantify multiple analytes. Gorecki et al. [18] described an elegant approach called inverse gradient compensation (IGC) where a secondary stream of the exact inverse gradient is added post-column to achieve a constant composition at the detector inlet. As exemplified throughout this book, IGC is used in many laboratories for quantitation of analytes for which individual standards are not available. This topic is detailed in Chapter 4.

6.3.5 Additives

The type and concentration of mobile phase additives is of particular importance for ion analysis with CAD since its response is based on the salt formed between the analyte and, primarily, mobile phase additives. Most LC-CAD methods for ion analysis involve the use of either ammonium acetate (AmAc) or ammonium formate (AmF) buffers (Table 6.1) where anionic analytes should be detected as their ammonium salts and cationic analytes as either acetate or formate salts. The additives, in the absence of analyte, should behave as volatile species with minimal contribution to particle formation. Under these conditions, ammonium, formate, and acetate would not be detectable as analytes. It is interesting to note, however, that Wu et al. [26] showed the capability to easily detect ammonium ions with CAD by judicious choice of trifluoroacetic acid (TFA) as the only mobile phase additive. Here ammonium was detected as its less volatile TFA salt. When using TFA under these conditions, any basic additive would likely result in prohibitively high baseline current and noise since this would form a detectable salt. This exemplifies the fundamental trade-off between universal detection versus selective detection and how the choice of mobile phase additives can be either beneficial or detrimental. As described in Chapters 1 and 3, volatile acidic and/or basic mobile phase additives can be used to increase detection scope and influence sensitivity for ionizable analytes. This can be advantageous by enabling

Table 6.1 Summary of HPLC-CAD methods for quantitation of ions.

First author year	Analytes/application	Stationary phase	Mobile phase additive	LOD (ng)
Huang et al. [7]	Counterions	Zic pHILIC	AmAc pH 7.0, AmF pH 3.5	N/A
Zhang et al. [12]	Counterions	Trinity P1	AmF pH 4.0	4 ± 3.1, 21 analytes[a)]
Ilko et al. [19]	Counterions	Trinity P1	AmF pH 4.0	5.3 ± 5.9, 21 analytes[b)]
Wigman et al. [20]	Raw material ID	Trinity P1	AmF pH 4.0	N/A
Dai et al. [21]	Li impurities in drugs	Trinity P1	AmF pH 4.0	0.1
Winterberg and Kirk [22]	Intracellular Na, K	Trinity P1	AmAc pH 5.0	0.13 and 0.29
Li et al. [23]	Ions and amino acid, IV soln	Trinity P1	AmF pH 4.5	N/A
Huang et al. [24]	Excipients, biopharmaceuticals	ZIC HILIC and ZIC pHILIC	AmAc pH 6.9, AmF pH 4.5	N/A
Li et al. [25]	Excipients, oligonucleotide formulations	Trinity P2	AmF pH 3.5	N/A
Wu et al. [26]	Am, sulfate in liposomal formulations	Sielc Primesep 100 for Am, Trinity P2 for sulfate	TFA for Am, AmF pH 3.65 for sulfate	90 for Am and 19.2 for sulfate
Ahirraho et al. [27]	Sulfate Impurities, zidebactam	Obelisc N	AmF pH 3.3	15.2 and 30.8
Liu et al. [28]	Bisphosphonates	Primesep SB	TFA	50
Stojanovic et al. [29]	Ionic Liquids	C18	IP RPLC	0.8–2.9
Rmandić et al. [30]	Mg, pyridoxine, thiamine in dietary supplements	ZIC HILIC	AmF pH 4.4	N/A
Lovejoy et al. [31]	Choline, counterions, impurities	Zic pHILIC	AmAc pH 4.7	N/A

a) For the same four analytes (citrate, gluconate, lactate and malate) in both Zhang et al. and Ilko et al., LODs were found to be significantly higher (see discussion).

b) LODs listed for Ilko et al. were estimated from the reported LOQs based on LOD = LOQ/3.3.

better detection of volatile analytes (e.g., volatile basic analyte + volatile acidic modifier) or disadvantageous by leading to higher background signal and noise (e.g., volatile basic impurity + volatile acidic modifier) or decreased response uniformity among analytes [e.g., differential interaction between analytes and a given volatile modifier (Chapter 4)].

Cohen et al. [32] provided useful insight into the main factors influencing response for ionizable solutes. They observed that the degree of ionization of the oppositely charged species is a main factor influencing detectability. This can be inferred from the eluent pH and the pK_a's of the acidic species and conjugate acid of the basic species. The most common buffers for ion analysis with CAD (Table 6.1) encompass pH ranges of 4.7–7.0 for AmAc and 3.3–4.5 for AmF, respectively, with pK_a's of 4.8 (acetic acid), 3.8 (formic acid) and 9.25 (ammonium ion). Within the described pH ranges, ammonium should be fully ionized, acetate should be at least 44% ionized (pH 4.7), while formate should be at least 24% ionized (pH 3.3). Considering that additive concentrations, in most cases, should be several orders of magnitude higher than typical analyte concentrations within the column eluent [33], it is reasonable to expect that additive ions are of sufficient concentration to readily form salts with ionic analytes.

6.3.6 Impurities, Contaminants, and Other Potential Issues

As detailed in Chapter 3, the best performance with CAD is obtained when the concentration of less volatile impurities in the eluent is minimized by using solvents with low evaporation residue, columns and conditions that produce minimal bleed, LC systems, and operational practices that minimize contamination, and low concentrations of low molar mass additives. Since buffers are the most used additives for ion analysis with HPLC-CAD, their method of preparation may be a considerable source of problems. Here, it is highly recommended to prepare any buffers gravimetrically rather than to titrate to the desired pH with a pH electrode to ensure greater reproducibility and avoid contamination from pH calibration buffers [34]. As previously mentioned, ions are ubiquitous, which also applies to the laboratory environment. Specific sources of ion contaminants that can contribute to the background signal or present as chromatographic peaks or artifacts include glassware, solvents and sample matrix. Ilko et al. [19] observed the presence of Mg, Ca, and sulfate ions in blank injections along with occasional contaminant peaks of Na, Cl, and possibly K ions, which were traced to the use of glass sample vials. They recommended the use of polypropylene equipment throughout sample preparation. A further example is shown in Figure 3.13 of a Thermo Fisher Scientific technical guide [35] where borate, silicate, sodium and unidentified contaminant were observed in blank injections and attributed to leaching from glass sample vials. Also relevant, are contaminants such as Li,

Na, K, and Fe that are frequently observed as adducts with LC-MS, where ACN and formic acid have been cited as likely sources [36]. Toussaint et al. studied the effects of chloride ions from high NaCl sample matrices on the analysis of polar compounds at low concentrations by HILIC with CAD [37]. Their experiments indicate that large amounts of ions introduced from sample matrices may interact with an amide stationary phase through complex mechanisms, which can in turn interact with the analyte in a second step. Besides contaminants, additional considerations for ion analysis include interactions with metal surfaces within the HPLC system or column such as those described between surface metal oxides and carboxylate and phosphate-containing analytes [38] and the formation of chelation complexes between metal ions and amino, hydroxyl, carbonyl, and phosphate-containing analytes [39]. These issues can lead to peak broadening and tailing, analyte loss on column, carryover, and peak artifacts. For example, Ilko et al. observed that the first injection for malate and the first two injections for citrate had to be disregarded as they gave a lower response than the following injections. The need for these priming injections was found to be more or less pronounced on different HPLC systems and attributed to surface effects [19]. The development of more inert surfaces using techniques such as chemical vapor deposition may be quite helpful in that regard.

6.3.7 Sensitivity and Linearity

Section 6.4 provides an overview of HPLC-CAD methods for ion analysis described in literature. These were selected with priority given to those that include details of method development and validation results. Table 6.1 summarizes the analytes, stationary phases, mobile phase additives, and limits of detection (LODs).

The LODs in Table 6.1 are expressed in ng on column to facilitate a comparison between methods and since CAD behaves primarily as a mass-flow-sensitive detector. In general, the results for most analytes are in the low ng range, which is consistent with that obtained for nonvolatile analytes in other classes. Figure 6.4 compares the sensitivity of CAD and ELSD for analysis of Na and Cl ions under the same conditions. For CAD, LODs as low as 0.1–0.3 ng have been reported in some studies. These lower values are likely associated with higher mass transport when using ACN-containing eluents, low baseline noise obtained with pure eluents and signal enhancement that is obtained by salt formation with mobile phase additives. These factors should similarly apply to ELSD. However, the downstream light scattering measurement technique of ELSD is fundamentally limited in its ability to measure the small particles produced at these analyte levels (Chapter 1).

CAD behaves as a mass-flow sensitive detector with response for nonvolatiles that, except for a minor dependency on particle bulk density, is independent of analyte physicochemical properties. Dai et al., however, reported higher sensitivity

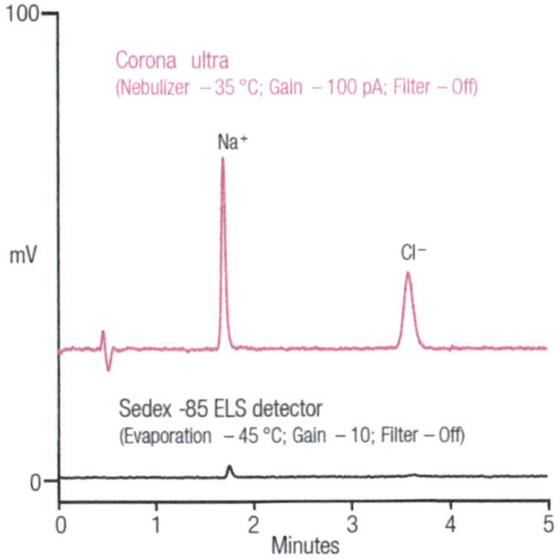

Column: Acclaim Trinity P1, 3 µm
Dimensions: 3.0 x 50 mm
Mobile Phase: 60/15/25 v/v/v CH_3CN/0.1 M NH_4OAc, pH5.2/D.I. H_2O
Temperature: 30 °C
Flow Rate: 0.6 mL/min
Inj. Volume: 5 µL
Inj. Amount: 5 ng
Detection: Corona ultra or Sedex -85 ELS detector
Sample: NaCl (1ppm based on Na^+) in D.I. H_2O

Detector	Na^+ (S/N)	Na^+ (LOD)	Cl^- (S/N)	Cl^- (LOD)
Corona ultra	71	0.2 ng	27	0.9 ng
Sedex -85	13	1.2 ng	2	> 11 ng

Figure 6.4 Sensitivity comparison of CAD and ELSD for Na^+ and Cl^-. *Source:* Reproduced with permission of Thermo Fisher Scientific Inc.

(signal/ng) and a lower LOD for Li compared to Na [21] despite having similar peak widths and retention times. Toussaint et al. also observed a trend in sensitivity of Li > Na > K in flow injection analysis [37]. This is somewhat surprising given CAD's expected response uniformity. With CAD and other EADs, dried particle diameter and therefore response is inversely proportional to the cube root of particle bulk density (Chapter 1). For ions, density of the salt form (e.g., Li-formate) is most relevant. While Li is the least dense solid element and its salts tend to have lower density than Na salts, this does not completely account for the observed differences. Additional factors involved in the process of spray drying may also influence residue particle size. Toussaint et al. [37] provided useful insight where the trend in CAD sensitivity follows the hydration strength order and the hydration radii where $Li^+ > Na^+ > K^+$. In this case, it seems that a higher relative hydration of Li^+ may lead to production of larger residue particles. This is further supported by Toussaint's study where the response for Li ions decreased more so than Na and K ions at higher evaporation temperatures. Both Zhang et al. [12] and Ilko et al. [19] reported higher LODs for the same four analytes. For citrate, malate and gluconate this was attributed to extra-detector surface interactions as discussed above; while a lower response for lactate was attributed to its volatility. The higher LODs reported for Am by Wu et al. [26] are likely due to both higher noise with TFA as mobile phase additive and to the volatility of the resulting Am-TFA salt.

In terms of linearity and calibration, most methods used linear regression over relatively small mass ranges. However, some used linear regression after log transformation of peak response and analyte amount. As detailed in Chapter 3, it is highly recommended to optimize the power function value to optimize the detector output with respect to linearity. An optimal PFV extends the linear range and enables straightforward quantitation while often improving accuracy, signal-to-noise ratios, precision, and peak shapes. The use of log transformation is generally not recommended. Log–log plots can be visually deceptive as small deviations on the plot can represent large absolute differences. For example, during method development it is important to know whether an analyte behaves as a nonvolatile or semivolatile. This is readily seen as a supralinear peak response versus analyte amount plots but may go unnoticed in log–log plots. Also, linear regression calculations such as correlation coefficients, are valid only when the experimental uncertainty of replicate y values is not related to the values of x or y. This assumption may not be fully valid after data have been transformed. Chapter 3 describes several options to determine an optimal PFV with minimal experimentation. These tools can be used with any chromatography platform and require only the generation of response curve data at one PFV. This optimal PFV can then be implemented as an instrument setting, which is the recommended approach as discussed below. The PFV setting is a data acquisition parameter like data collection rate or filter time constant that directly affects the signal output and is used to optimize the response with respect to linearity for a given method and range of quantitation.

6.4 Applications

This section provides an overview of HPLC-CAD methods for ion analysis described in literature. These were selected with priority given to those that include details of method development and validation results.

6.4.1 Pharmaceutical and Biopharmaceutical Applications

Ions often serve as counterions for many drug substances, as excipients in parenteral and biopharmaceutical formulations and as key components of upstream processes such as cell culture, fermentation and drug loading of liposomes. Salts can also be present as undesirable impurities originating from raw materials, residual salts from upstream processes, side reactions and degradation.

6.4.1.1 Counterions

Drug substances have been routinely developed as salts to improve solubility, stability and processability. About half of the active drug substances approved by the US FDA are in salt form [12]. The final concentration and purity of counterions affect drug potency, safety, stability and bioavailability. Thus, the control of counterions and their impurities is a key element of a drug substance's specification. Because most counterions have no UV chromophores, CAD has gained great popularity in counterion analysis [7, 9, 12].

Huang et al. [7] described a simple generic approach to quantitation of inorganic counterions in drug substances. CAD was used with Sequant ZIC-*p*HILIC columns for two different methods based on the valence of the ions: (i) a 150-mm column with pH 7.0 AmAc buffer: ACN (25 : 75) as mobile phase for monovalent ions such as NO_3^-, Cl^-, Br^-, Na^+, and K^+ (ii) a 50-mm column with pH 3.5 AmF: ACN (30 : 70) as mobile phase for multivalent ions such as Ca^{2+}, Mg^{2+}, SO_4^{2-}, and PO_4^{3-}. Excellent system precision was obtained for all tested ions with RSD less than 2.0%. Accuracy was within 2% of the theoretical ion content of the drug. Better accuracy was observed for Cl^- in several drug substances as compared to IC. This was attributed to poor compatibility between the sample matrix of more hydrophobic drug substances and the mainly aqueous mobile phase requirements of IC with suppressed conductivity. With IC, it is likely that more hydrophobic drug substances would partially precipitate in the eluent along with a small portion of Cl^- resulting in low recovery due to a mass balance deficit. With HILIC, hydrophobic drug substances are more soluble in the higher organic mobile phase and were clearly resolved, eluting much earlier than their counterions. This indicates complete dissociation between counterions and their parent drug substances. With IC, additional sample preparation and several different detection and separation modes would be needed to address the range of counterions and drug substances encountered. By contrast, the HILIC-CAD technique permitted use of "generic" sample diluents and generic methods for quantitation.

As mentioned in Section 6.3.1, Zhang et al. [12] developed a generic HPLC-CAD method for simultaneous analysis of 25 commonly used anionic and cationic pharmaceutical counterions using the above-described Acclaim Trinity P1 trimodal column. Analytes were resolved in 20 minutes as shown in Figure 6.5. As part of method development and optimization, they performed a detailed study to understand the retention mechanisms of the trimodal column. This was accomplished by varying mobile phase organic modifier concentration, buffer type, ionic strength, pH, and column temperature and examining the effects on four categories of retention mechanism probes among the analytes. They reported that the column provides reversed-phase, strong cation-exchange and weak anion-exchange properties at the same time and HILIC properties at high organic modifier concentrations. While method development was not straightforward, the trimodal column provided great flexibility for analyte retention and selectivity. Figure 6.6 shows the separation of APIs and their respective counterions for naproxen sodium, adenine hydrochloride, and compound X in hemi-fumarate salt. As also observed by Huang et al. (see above), counterions and APIs were generally well separated with no sample matrix issues. The method showed good

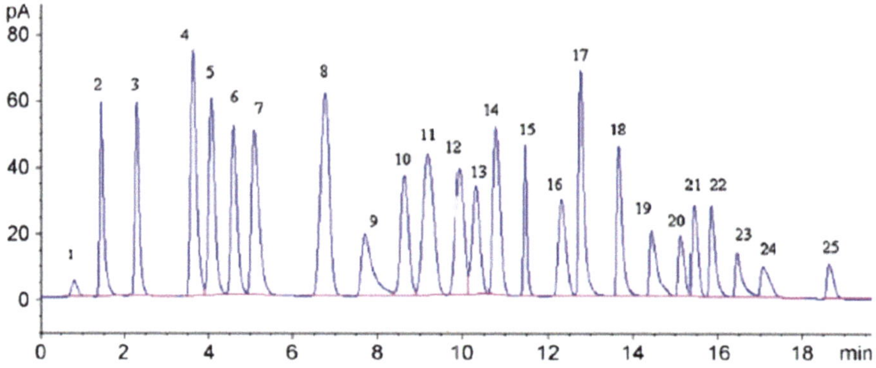

Figure 6.5 Chromatogram of the separation of 25 common pharmaceutical ions. Column, Acclaim Trinity P1, 3.0 × 50-mm format; mobile phase, A, acetonitrile; B, 200 mM ammonium formate (pH 4.0); C, DI water; gradient elution; flow rate, 0.5 mL/min; injection volume, 5 μL; temperature, 35 °C; and detection, Corona CAD. Peaks: (1) lactate, (2) procaine, (3) choline, (4) tromethamine, (5) sodium, (6) potassium, (7) meglumine, (8) mesylate, (9) gluconate, (10) maleate, (11) nitrate, (12) chloride, (13) bromide, (14) besylate, (15) succinate, (16) tosylate, (17) phosphate, (18) malate, (19) zinc, (20) magnesium, (21) fumarate, (22) tartrate, (23) citrate, (24) calcium, and (25) sulfate. *Source:* [12] / with permission of Elsevier.

linearity in the defined range, and the accuracy versus theoretical amount was in the range of 99.0–101.0% with %RSD less than 2.0%. This method is an example of a generic, or platform, method for use in development and manufacturing that provides large analyte coverage and broad sample compatibility with the required accuracy to control stoichiometry and the sensitivity for purity control and to establish mass balance.

Several papers have described methods based on that of Zhang et al. [12]. The method was directly adopted and further validated by the laboratory department of the European Pharmacopoeia (Ph. Eur.) and applied to counterion identification and quantification in drug substances and for control of inorganic ions as impurities [19]. Method performance was demonstrated by analysis of Ph. Eur. reference standards and pharmaceutical substances and was readily coupled to MS without any modification for identification of API as well as organic ions. For all but four of the ions, the limit of quantification (LOQ) was in the range of 2–40 ng on column, which indicates more than sufficient sensitivity for quantification of impurity ions at a <0.1 level when analyzing a 10 mg/mL test solution. Higher LOQs obtained for malate and citrate were attributed to surface adsorption in the HPLC system. Gluconate produced a very broad peak and lower lactate response was attributed to its volatility. Analysis of 12 Ph. Eur. drug and impurity reference substances demonstrated, in all cases, that unambiguous identification of the counterion was feasible with clear chromatographic separation from the

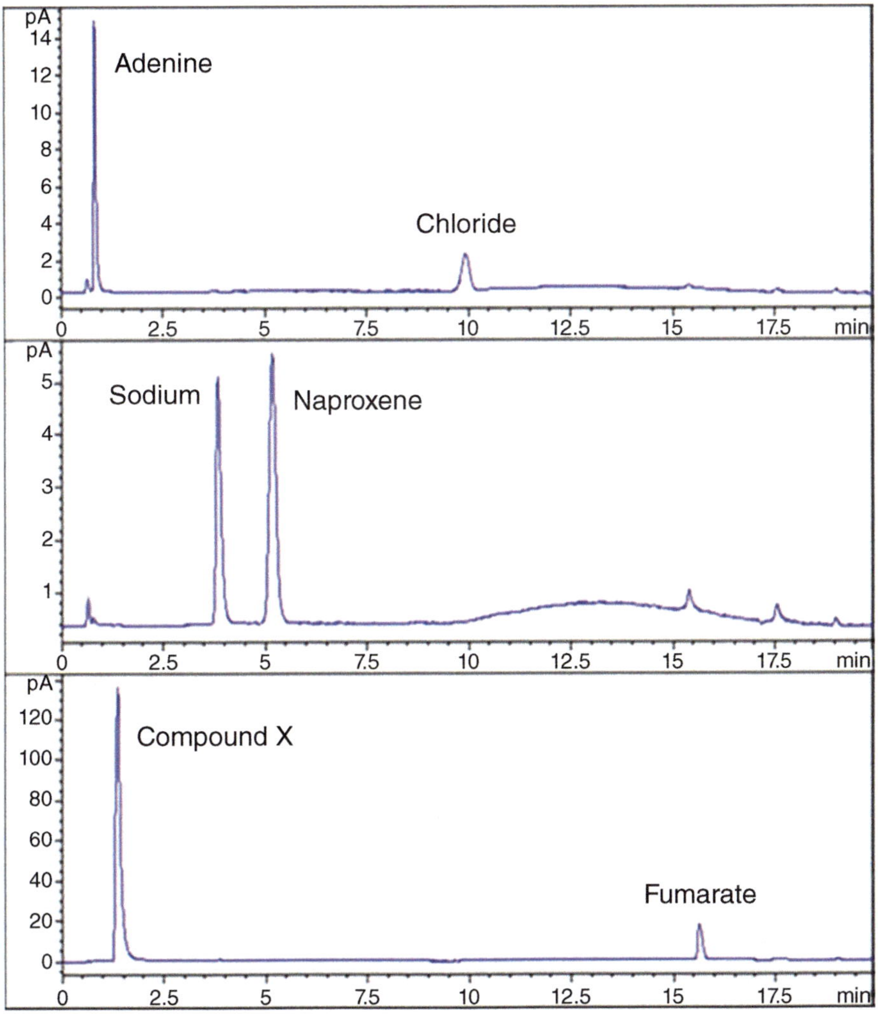

Figure 6.6 Chromatograms of the separation of active pharmaceutical ingredients and their counterions. Column, Acclaim Trinity P1, 3.0 × 50 mm format; mobile phase, A, acetonitrile; B, 200 mM ammonium format (pH 4.0); C, DI water; gradient elution; flow rate, 0.5 mL/min; injection volume, 5 μL; temperature, 35 °C; and detection, Corona CAD. Samples: adenine hydrochloride (top), naproxen sodium (middle), and compound X fumarate salt (bottom). *Source:* [12] / with permission of Elsevier.

drug substances. The determined stoichiometric values were in good agreement with theoretical values and the ability to determine trace level impurities was demonstrated. The authors reported that the method can provide a very useful contribution to a thorough characterization of Ph.Eur. reference standards and can be used as a universal method for the identification and control of the most pharmaceutically relevant ions in Ph.Eur. monographs.

6.4.1.2 Raw Material Identification

Wigman et al. [20] adapted the method of Zhang for raw material identification for pharmaceutical manufacturing. Bromide and lithium were added to 11 of the ions described in the original method to provide additional coverage of common raw materials described in USP <191>. Modifications to the gradient method were made to decrease analysis time. The ability to separate all 13 substances for identification purposes and the capability for simultaneous quantitation and impurity detection in the same analytical run was demonstrated. This was exemplified by the identification and, with proper calibration, quantification of potassium phosphate raw material and its sulfate and sodium impurities. The technique was reported to provide a rapid and flexible alternative to USP <191>, and other analytical identification techniques. The authors noted that, while the universal detection capabilities of CAD increases the probability of detecting impurities, it lacks the specificity and qualitative information necessary to identify impurities in more complex samples. In these cases, specific techniques such as mass spectrometry can be used. The authors also reported that the inability of CAD to detect volatile compounds, such as ammonium, acetate and formate is a limitation for raw material identification but stated "this is why AmAc and formate are good mobile phase modifiers for HPLC–CAD". As described below, Wu et al. [26] demonstrated the analysis of ammonium with CAD by using a higher molar mass acidic mobile phase additive. This is a good example of how, for ionizable species, the choice of mobile phase additive may provide the capability to extend CAD's scope of detection (see Section 6.3.5).

6.4.1.3 Impurity Ions in Pharmaceutical Products

Dai et al. [21] adapted the method of Zhang for analysis of lithium impurities in pharmaceutical products. While lithium carbonate is used to treat mania and bipolar disorders, it has a narrow therapeutic index where higher levels can be severely toxic. Lithium may also occur in pharmaceutical products as a residual impurity due to the common use of lithium-containing reagents for drug compound synthesis. To ensure patient safety, it is critical to accurately quantify trace amounts of lithium in pharmaceutical products. While IC has been used for lithium analysis, it requires aqueous mobile phase for separation and cannot tolerate the injection of high amounts of hydrophobic APIs often required for trace analysis. Using the same trimodal column, the method was optimized for separation of lithium from other matrix components. The LOD was determined to be 0.08 ng, which is significantly (ca. 25-fold) lower than that obtained for sodium ion with the original method. The developed technique was described as a platform method for accurate and sensitive analysis of lithium in a range of pharmaceutical product matrices requiring minimal sample preparation.

6.4.1.4 Parenteral Electrolyte-Amino Acid Solution

Li et al. [23] described an HPLC-CAD method for quality control of potassium aspartate and magnesium aspartate (PAMA) intravenous solution. They used the method of Zhang et al. [12] as a starting point to approach simultaneous analysis of the main components (K^+, Mg^{2+} and Asp) and impurities (Na^+). The method used the same Acclaim Trinity P1 column and was optimized using Response Surface Methodology where the effects of organic content, column temperature, buffer types, pH and concentrations were evaluated. The optimized method was successfully validated with good specificity, linearity, accuracy, and repeatability. The method was described as useful for quality control in PAMA injection and other similar drugs. The authors noted that ions such as Na^+, Mg^{2+}, and Ca^{2+} were commonly observed contaminants in the blank sample solvents and the sample results were calculated after deducting the blank peak area.

6.4.1.5 Intracellular Ions

Winterberg and Kirk [22] developed an HPLC-CAD method for rapid, high-sensitivity measurement of intracellular sodium and potassium ion concentrations. These measurements and the screening of chemical agents to identify molecules targeting ion transport have traditionally involved low-throughput techniques. They used an Acclaim Trinity P1 column and isocratic conditions with mobile phase components of water, AmAc buffer and ACN. The method allowed use of a single low-volume (<10 μL) cell extract, prepared from as few as 10^5 cells. A power regression model was used for calibration. From method validation, LOQs were determined to be 0.9 ng for Na^+ and 2.0 ng for K^+. Selectivity was demonstrated by analysis of cell lysis and cell wash solutions for which no interferences were found. Precision was determined by analyzing three different quality control samples (0.1, 1 and 10 nmol of each of Na^+ and K^+) in independent assays, measured on different days ($n = 4$), revealing a total-assay variation of 1.5% (two-way ANOVA). The method was applied to uninfected human erythrocytes and yielded Na^+ and K^+ content values consistent with those well-established in literature. The utility of this technique was demonstrated by monitoring the ionic changes induced in the intracellular malaria parasite by a new spiroindolone antimalarial KAE609. The authors reported that the method is readily applied to a broad range of cell-types and may be adapted for use with other ions and offers significant advantages compared to presently available methods including simplicity of sample preparation, high-sensitivity, multi-ion detection, and adaptability to a high-throughput multi-well plate format.

6.4.1.6 Excipients Ions in Monoclonal Antibody (mAb) Formulations

Huang et al. [24] described a method for direct analysis of excipients in mAb formulations. During process and formulation development for mAbs and other biologics, operations such as dialysis are employed to achieve the desired protein and excipient composition. Routine assessment is needed during development to ensure that the desired composition doesn't deviate from expectations. A main challenge is that there is a lack of direct and robust methods for analysis of solutions containing high protein concentrations. The approach was based on use of a wide-pore C18 guard column in-line with a HILIC (SeQuant ZIC-pHILIC or ZIC-HILIC) analytical column. The C18 column was used to trap proteins and prevent fouling of the analytical column. Polar excipients, which were unretained on the guard column, were separated on the HILIC column under isocratic conditions and detected by CAD. This simple coupling of a C18 cartridge and a HILIC column was found to be suitable as a general method for direct analysis of several commonly used excipients such as amino acids, carboxylic acids, simple carbohydrates, and inorganic ions across multiple drug formulations containing different therapeutic antibodies.

6.4.1.7 Excipient Ions in Antisense Oligonucleotide Drugs

Li et al. [25] described a method for quantitation of excipient ions in antisense oligonucleotide (ASO) drug products. Among the FDA approved direct delivery (e.g., intrathecal) vehicles for ASOs, artificial cerebrospinal fluid contains the most inorganic ions (Na^+, PO_4^{-3}, Cl^-, K^+, Mg^{+2}, and Ca^{+2}). During process development it is important to ensure that the ionic composition does not change during purification or drug product compounding and that final drug product targets are achieved. They used an Acclaim Trinity P2, which provided strong anion-exchange, weak cation-exchange (CEX) and HILIC interactions within a single column. A binary gradient with mobile phase components of water and AmF was used. The evaporation temperature of CAD was optimized to enhance the signal to noise ratio, especially for those ions at low concentration. The power function setting was optimized to a value of 1.4, which yielded a linear response for ion levels from 25% to 150% of the nominal aCSF concentrations. The qualified range for all compounds, derived from accuracy and linearity data, was 25–150% aCSF with up to 5 mg/mL ASO matrix. Method ruggedness was evaluated by another analyst using a different system in a separate lab with different reagents and on different days and demonstrated by comparable recoveries and precision results. The method was also compared to an orthogonal

ICP-OES method (based on USP general chapter <720>) for cation analysis and the sodium concentrations were within 4%. The method was reported to be specific, accurate, precise, linear, robust, rugged and capable of simultaneously measuring monovalent and multivalent cations and anions.

6.4.1.8 Ions in Drug-Loaded Liposomal Products

Wu et al. [26] described a straightforward LC-CAD method optimized to quantify total, internal, and external ions in drug-loaded liposomal products. Liposomes are one of the most widely used nanoscale drug delivery platforms and are used for a variety of therapeutic agents. Doxil, a form of doxorubicin, was the first liposomal drug product approved by the US FDA. The encapsulation of doxorubicin is based on a novel loading method using a transmembrane ammonium sulfate gradient as the driving force. In addition to ammonium sulfate, other salts are also used for active loading of weak bases in liposomes, including sodium and ammonium salts of acetate, citrate and phosphate as well as a variety of metal salts (e.g., $MnSO_4$, $CuCl_2$). Ion concentrations within a liposomal formulation are one of the most important criteria for establishing a stable and efficient drug encapsulation process and quantification of residual ions in the drug product is a critical aspect of their characterization. These measurements are very challenging due to their non-chromophore/fluorophore nature and low analyte level in limited sample volumes.

This study focused on measurement of ammonium and sulfate ions in Doxil. Separate methods were used for ammonium and sulfate. The method for sulfate used an Acclaim Trinity P2 column and a binary gradient with mobile phase components of water and AmF buffer. The method for ammonium used a SIELC Primesep 100 column which has acidic ion-pairing groups embedded within long hydrophobic ligand chains to separate cations by ion-exchange. Analytes were separated under isocratic conditions with mobile phase components of ACN, water and TFA. As noted above in the study by Wigman et al. [20] and shown in Table 6.1, ammonium is a most common mobile phase additive for use with CAD due to its volatile behavior. In this method, the use of TFA as mobile phase additive enabled the detection by CAD of ammonium in the form of a less volatile TFA salt.

Analytical measurements were confirmed using simple stoichiometry based on the drug crystallization of doxorubicin within the liposome interior. The method was described as quick, highly accurate, with simple preparation requiring low sample volumes with significantly improved detection limits over traditional methods. The authors noted that the technique is easily amenable to quantification of a variety of cations and anions used for remote loading of liposomes. Furthermore, the sample preparation and separation method may allow quantification of other components in Doxil (e.g., sucrose, histidine, sodium, and chloride ions), further improving the characterization methodologies for the "nonliposome inactive ingredients," a requirement in FDA's guidance for industry of liposome drug products.

6.4.1.9 Identification and Quantitation of Ionic Impurities in an Investigational Drug

Ahirrao et al. [27] described their use of CAD as part of a comprehensive characterization of Zidebactam (ZID), a new chemical entity (NCE) under development for treatment of bacterial infections. As an NCE, it is necessary to find and quantify every component and arrive at mass balance and true purity of the drug substance and to prepare and fully characterize a primary reference standard. The drug substance was analyzed using detectors such as photodiode array (PDA), refractive index (RI), MS and CAD using different stationary phases like ODS, HILIC, ion-exchange (anion and cation) and mixed mode along with appropriate mobile phases and conditions. Out of all these analyses, two new unknown peaks were found only in the MMLC-CAD analysis. After being found by CAD, modification of ESI ionization conditions (polarity) allowed the detection by MS, where spectral analysis combined with a thorough investigation of the synthetic procedure led to their identification as sulfate and monomethyl sulfate. MMS and sulfate content was determined in four lots of ZID drug substance by MMLC-CAD as well as the IC method, with comparable results. The MMLC-CAD method used an Obelisc-N column, which has a bonded phase with outer positive and inner negative charges. A binary gradient in HILIC mode was used with mobile phase components of ACN, water and AmF buffer. It was observed in the MMLC-CAD method that analyte determination can be inaccurate when using only single-point calibration. The default power function (PF) value of 1.0 was reported to provide optimal results although only testing at PF 0.67 was mentioned. The sensitivity for sulfate was similar to IC, but better detection limits for MMS were obtained by IC. With CAD, additional peaks were observed that were not seen by IC.

6.4.2 Analysis of Ions in Dietary Supplements

Rmandić et al. [30] described a method for the determination of magnesium, pyridoxine and thiamine in a dietary supplement. Method development utilized analytical quality by design methodology. A ZIC-HILIC PEEK column was used for isocratic separation with mobile phase components of ACN, acetone, and AmF aqueous buffer. The optimal settings of CAD were defined with the one-factor-at-time approach where an evaporation temperature of 50 °C was found to be optimal. The analysis of solutions for selectivity evaluation indicated co-elution of thiamine and lactose, an excipient with an intensive CAD response.

This was resolved by using UV detection at 220 nm. The developed method met all the defined validation criteria set by the ICH Q2(R1).

An HPLC-CAD method for measurement of content and purity of choline bitartrate, choline citrate, and choline chloride was developed by the USP Compendial Development Laboratory, and further evaluated by Lovejoy et al. [31]. A mixed mode SeQuant™ ZIC™ pHILIC column was used for gradient separation with mobile phase components of AmAc buffer, ACN and water. This enabled the separation of choline, its counterions, and several impurities with direct detection by CAD. Several impurities were identified based on retention time comparison to individual standards or by MS including Na+, K+, Cl−, and *O*-(2-hydroxyethyl)choline (Figure 6.7). Choline content was measured over the range of 0.08–0.12 mg/mL with precision ≤3.0% RSD, recovery from 96.7% to 100.8%, and correlation coefficient (r^2) for linear regression of >0.997. An impurity found in choline citrate samples (RT A ~ 6.0 minutes) was further studied using a single-quadrupole mass detector (Figure 6.8). Based on its mass spectrum, the impurity was determined to be *O*-(2-hydroxyethyl)choline, a known by-product formed during the production of choline hydroxide.

6.4.3 Bisphosphonate

Bisphosphonate is a class of drugs that prevent the loss of bone mass. The analysis of bisphosphonate has been challenging because these molecules are polar and ionic and most of them have no UV chromophore. Pre-column derivatization, indirect UV analysis, ion pair, and CE methods were reported to separate and detect these molecules. Etidronate, (1-hydroxyethylidene) bisphosphonate, and its impurities phosphate and phosphite have no UV chromophore and are highly ionic, so it is difficult to separate these compounds by conventional HPLC columns and to detect by ordinary spectrophotometric methods. Liu et al. [28] developed a stability-indicating method for the analysis of etidronate disodium and its related substances by using a mixed mode column (Primesep SB) and CAD. The method was also validated in terms of specificity, linearity, accuracy, precision, sensitivity, and stability. The correlation of peak area versus concentration of etidronate was not linear. The authors used a logarithmic calibration equation and found that the recoveries of etidronate at different levels were in the range of 95.9–102.6%. Similarly, the recoveries of phosphate and phosphite were found in the range of 94.5–103.6% and 96.5–107.9%, respectively. Good accuracy and precision were demonstrated for etidronate and the impurities. The LOQ and LOD for etidronate were 144 and 50 ng, with %RSD of 3.1% and 14.7%, respectively. The LOQ for phosphate and phosphite was 75 ng with %RSD of 2.0%. The LOD for phosphate and phosphite is 50 ng, with %RSD of 4.4% for phosphate and 10.7% for phosphite. It was noted that although CAD is reportedly a universal detector with response

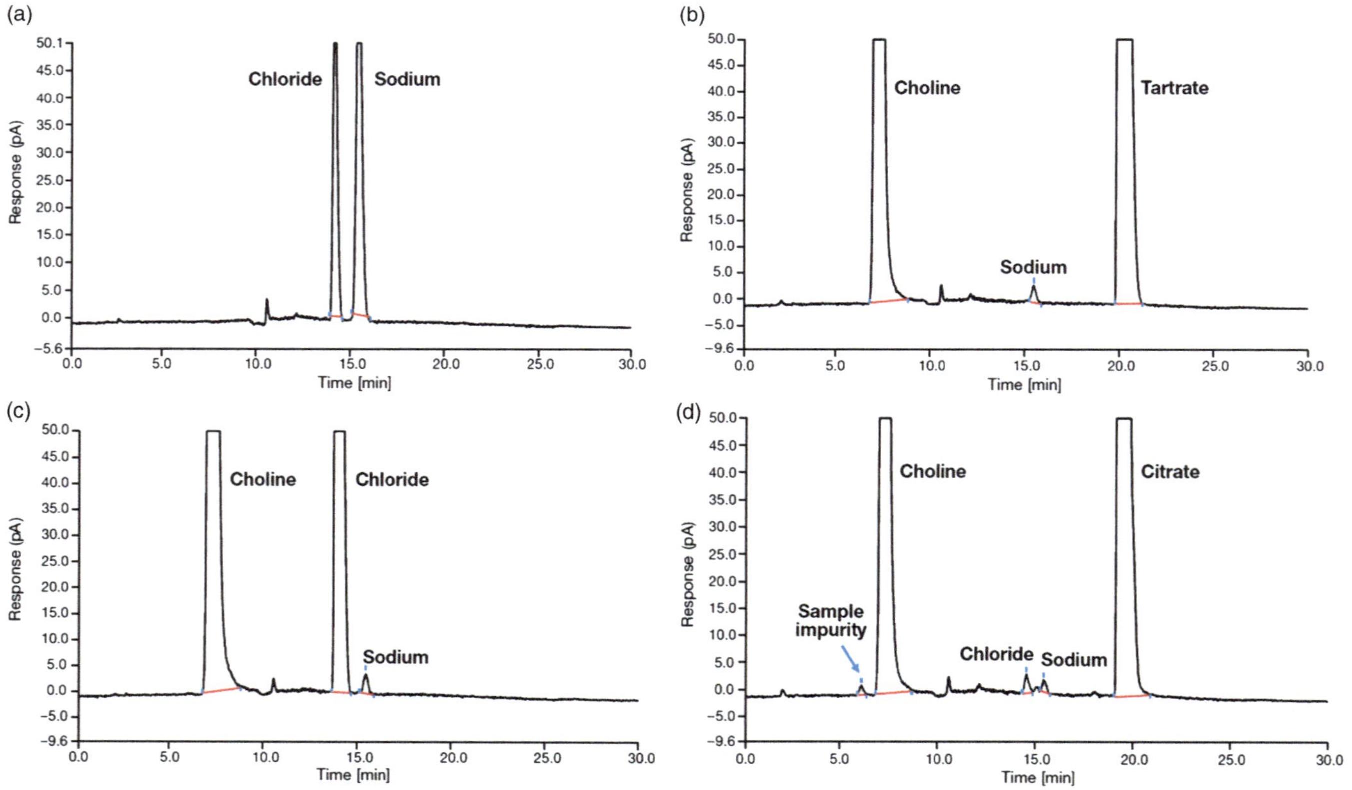

Figure 6.7 (a) Chromatogram of sodium chloride (0.1 mg/mL)—sodium detected as its acetate salt, chloride as its ammonium salt; (b) Analysis of choline bitartrate (2.0 mg/mL) showing sodium impurity; (c) Analysis of choline chloride (2.0 mg/mL) showing sodium impurity; (d) Analysis of choline citrate (2.0 mg/mL) showing sodium, chloride, and a choline-related sample impurity (eluting at 6.0 minutes).

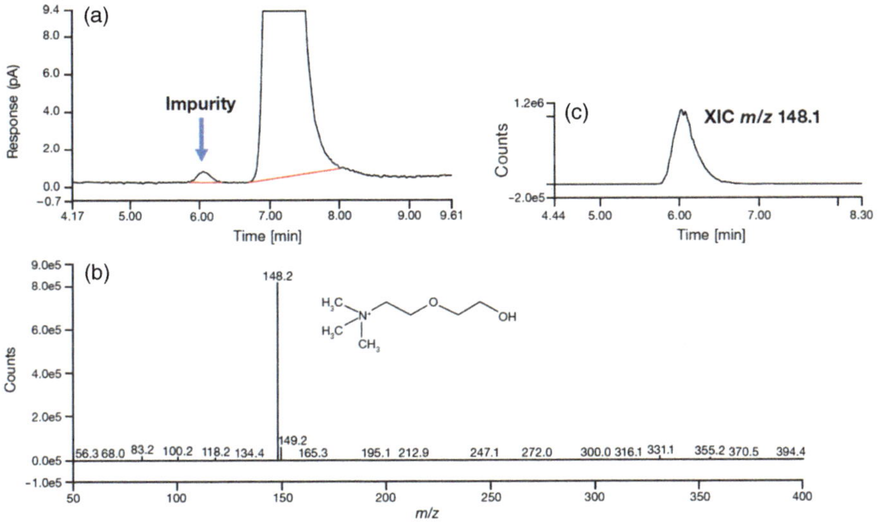

Figure 6.8 (a) CAD chromatogram of choline citrate (2.0 mg/mL) showing an impurity at RT ~ 6 minutes; (b) mass spectrum at the peak apex of the impurity peak at RT ~ 6 minutes; (c) extracted ion chromatogram (XIC) of *m/z* 148.1.

magnitude independent of the analyte chemical properties, Liu et al. observed different recovery responses of phosphate and phosphite by using etidronate standard at impurity levels.

6.4.4 Ionic Liquids

Ionic liquid, also called liquid salt, is a class of organic salts with melting point below the boiling point of water. The interest in ionic liquid is rapidly growing because of the variety of industry applications. Ionic liquids with UV-absorbing groups like imidazolium- and pyridinium-based molecules can be analyzed by UV detector, but for the ionic liquids such as ammonium- and phosphonium-based molecules that lack chromophoric groups, CAD is very attractive. Stojanovic et al. applied the coupling of CAD for the non-chromophoric aliphatic cations with diode array detection for the aromatic anions to simultaneous analysis of a set of new ionic liquids derived from either tricaprylmethylammonium chloride (Aliquat 336) or trihexyltetradecylphosphonium chloride [29]. In this study, a gradient method was developed using Gemini C18 column and 0.1% TFA in water as mobile phase A and 0.1% TFA in ACN as mobile phase B. The authors established a unified calibration function for the quantitative analysis of the quaternary ammonium cations of the ionic liquids, due to lack of single-component standards. The method sensitivity LOD is at the ng level for all tested ions, which is in agreement with the reported figures for other compounds.

References

1 Jackson, P.E. and Pohl, C.A. (1997). Advances in stationary phase development in suppressed ion chromatography. *TrAC, Trends in Analytical Chemistry* 16 (7): 393–400.

2 Weiss, J. (2016). *Handbook of Ion Chromatography, 3 Volume Set*. John Wiley & Sons.

3 Zhang, K. and Liu, X. (2016). Reprint of "mixed-mode chromatography in pharmaceutical and biopharmaceutical applications". *Journal of Pharmaceutical and Biomedical Analysis* 130: 19–34.

4 Pohl, C. (2010). Recent developments in ion-exchange columns for inorganic ions and low molecular weight ionizable molecules. *LC-GC North America* 28 (4): 24–31.

5 Russell, J.J., Heaton, J.C., Underwood, T. et al. (2015). Performance of charged aerosol detection with hydrophilic interaction chromatography. *Journal of Chromatography. A* 1405: 72–84.

6 Lantz, M.D., Risley, D.S., and Peterson, J.A. (1997). Simultaneous resolution and detection of a drug substance, impurities, and counter ion using a mixed-mode HPLC column with evaporative light scattering detection. *Journal of Liquid Chromatography and Related Technologies* 20 (9): 1409–1422.

7 Huang, Z., Richards, M., Zha, Y. et al. (2009). Determination of inorganic pharmaceutical counterions using hydrophilic interaction chromatography coupled with a Corona® CAD detector. *Journal of Pharmaceutical and Biomedical Analysis* 50 (5): 809–814.

8 Pack, B.W. and Risley, D.S. (2006). Simultaneous determination of positive and negative counterions using a hydrophilic interaction chromatography method. *LCGC North America* 24 (8): 776–785.

9 Crafts, C., Bailey, B., Plante, M., and Acworth, I. (2009). Evaluation of methods for the simultaneous analysis of cations and anions using HPLC with charged aerosol detection and a zwitterionic stationary phase. *Journal of Chromatographic Science* 47 (7): 534–539.

10 Liu, X., Pohl, C., Woodruff, A., and Chen, J. (2011). Chromatographic evaluation of reversed-phase/anion-exchange/cation-exchange trimodal stationary phases prepared by electrostatically driven self-assembly process. *Journal of Chromatography. A* 1218 (22): 3407–3412.

11 Liu, X. and Pohl, C.A. (2010). HILIC behavior of a reversed-phase/cation-exchange/anion-exchange trimode column. *Journal of Separation Science* 33 (6–7): 779–786.

12 Zhang, K., Dai, L., and Chetwyn, N.P. (2010). Simultaneous determinatieon of positive and negative pharmaceutical counterions using mixed-mode chromatography coupled with charged aerosol detector. *Journal of Chromatography. A* 1217 (37): 5776–5784.

13 Liu, X. and Pohl, C. (2011). Is HILIC the best way for determination of active pharmaceutical ingredients and counterions?: Thermo Scientific.

14 Walter, T.H., Alden, B.A., and Berthelette, K. (2022). Evaluation of the base stability of hydrophilic interaction chromatography columns packed with silica or ethylene-bridged hybrid particles. *Separations* 9 (6): 146.

15 Jenkins, K. (2015). HILIC separation of carbohydrates using BEH amide particle technology. *Chromatography Today* 14–18.

16 Qian, K., Yang, Z., Zhang, F. et al. (2018). Low-bleed silica-based stationary phase for hydrophilic interaction liquid chromatography. *Analytical Chemistry* 90 (15): 8750–8755.

17 Mitchell, C.R., Bao, Y., Benz, N.J., and Zhang, S. (2009). Comparison of the sensitivity of evaporative universal detectors and LC/MS in the HILIC and the reversed-phase HPLC modes. *Journal of Chromatography B* 877 (32): 4133–4139.

18 Gorecki, T., Lynen, F., Szucs, R., and Sandra, P. (2006). Universal response in liquid chromatography using charged aerosol detection. *Analytical Chemistry* 78 (9): 3186–3192.

19 Ilko, D., Nap, C., Holzgrabe, U., and Almeling, S. (2014). Validation and application of an HPLC-CAD-TOF/MS method for identification and quantification of pharmaceutical counterions. *Pharmeuropa Bio & Scientific Notes* 2014: 81–91.

20 Wigman, L., Scott, B., and Zhang, K. (2013). Material identification by HPLC with charged aerosol detection. *LC-GC North America*. 2013 Sep 1; 31 (9): S37.

21 Dai, L., Wigman, L., and Zhang, K. (2015). Sensitive and direct determination of lithium by mixed-mode chromatography and charged aerosol detection. *Journal of Chromatography. A* 1408: 87–92.

22 Winterberg, M. and Kirk, K. (2016). A high-sensitivity HPLC assay for measuring intracellular Na+ and K+ and its application to *Plasmodium falciparum* infected erythrocytes. *Scientific Reports* 6 (1): 29241.

23 Li, P., Sun, W., Zuo, L. et al. (2018). Fast simultaneous determination of main components and impurity sodium ion in PAMA injection by mixed-mode chromatography. *Journal of Pharmaceutical and Biomedical Analysis* 161: 407–413.

24 Huang, J.Z., Lin, S., Huang, Z., and Bolgar, M.S. (2017). Enablement of the direct analysis of excipients in monoclonal antibody formulations through the incorporation of a wide pore C18 protein trap with hydrophilic interaction liquid chromatography. *Journal of Chromatography B* 1068-1069: 131–135.

25 Li, J., Stolee, J.A., and Meda, A. (2021). Simultaneous quantitation of inorganic ions in oligonucleotides using mixed-mode liquid chromatography coupled with a charged aerosol detector. *Journal of Pharmaceutical and Biomedical Analysis* 204: 114244.

26 Wu, J., Crist, R.M., McNeil, S.E., and Clogston, J.D. (2019). Ion quantification in liposomal drug products using high performance liquid chromatography. *Journal of Pharmaceutical and Biomedical Analysis* 165: 41–46.

27 Ahirrao, V.K., Rane, V.P., Patil, K.R. et al. (2020). Identification of monomethyl sulfate and sulfate impurities in zidebactam using LC–MS and application of mixed-mode liquid chromatography with charged aerosol detection and ion chromatography for quantification. *Chromatographia* 83 (2): 219–228.

28 Liu, X.-K., Fang, J.B., Cauchon, N., and Zhou, P. (2008). Direct stability-indicating method development and validation for analysis of etidronate disodium using a mixed-mode column and charged aerosol detector. *Journal of Pharmaceutical and Biomedical Analysis* 46 (4): 639–644.

29 Stojanovic, A., Lämmerhofer, M., Kogelnig, D. et al. (2008). Analysis of quaternary ammonium and phosphonium ionic liquids by reversed-phase high-performance liquid chromatography with charged aerosol detection and unified calibration. *Journal of Chromatography. A* 1209 (1): 179–187.

30 Rmandić, M., Radjenović, M., Stanković, J. et al. (2021). PDA-CAD method for the determination of magnesium, pyridoxine and thiamine in a dietary supplement supported by analytical quality by design methodology. *Archives of Pharmacy* 71 (Notebook 5): 378–392.

31 Lovejoy, K., Gamache, P. (n.d.). Quality control of choline as a dietary supplement by high performance liquid chromatography coupled to a charged aerosol detector. Thermo Fisher Scientific Application Note 73917.

32 Cohen, R.D., Liu, Y., and Gong, X. (2012). Analysis of volatile bases by high performance liquid chromatography with aerosol-based detection. *Journal of Chromatography. A* 1229: 172–179.

33 Bailey, B., Plante, M., Thomas, D. et al. (2017). Practical use of CAD: achieving optimal performance. In: *Charged Aerosol Detection for Liquid Chromatography and Related Separation Techniques* (ed. P.H. Gamache), 163–189. John Wiley and Sons.

34 Stoll D.R. and Makey D.M. (2019). Mobile phase buffers in LC: effect of buffer preparation method on retention repeatability. *LC-GC North America.* Jul 1; 37 (7): 444–449.

35 Muellner, T., Acworth, I., and Gamache, P. (n.d.). Getting the most out of your charged aerosol detector. Thermo Fisher Scientific Technical Guide.; 73914.

36 Hodge, K., Ten Have, S., Hutton, L., and Lamond, A.I. (2013). Cleaning up the masses: exclusion lists to reduce contamination with HPLC-MS/MS. *Journal of Proteomics* 88: 92–103.

37 Toussaint, B., Immame Hassane Beck, T., Surget, E. et al. (2023). Exploration of the effects of chloride ions on the analysis of polar compounds at low concentrations by hydrophilic interaction liquid chromatography coupled to a charged aerosol detector: application to tromethamine. *Journal of Separation Science* 46 (6): e2200766.

38 Walter, T.H., Alden, B., Belanger, J. et al. (2022). Modifying the metal surfaces in HPLC systems and columns to prevent analyte adsorption and other deleterious effects. *Recent Developments in HPLC and UHPLC, LCGC Supplement* 40: 28–34.

39 Heaton, J.C. and McCalley, D.V. (2016). Some factors that can lead to poor peak shape in hydrophilic interaction chromatography, and possibilities for their remediation. *Journal of Chromatography. A* 1427: 37–44.

7

Analysis of Carbohydrates with Charged Aerosol Detection

Paul H. Gamache

Consultant, Nashua, NH, USA

CHAPTER MENU

7.1 Summary

The use of high performance liquid chromatography (HPLC) with charged aerosol detection (CAD) for carbohydrate analysis continues to grow exponentially (see Chapter 2). Many early applications involve the adoption of CAD in methods that traditionally employed refractive index detection (RID), short-wavelength UV detection and evaporative light scattering detection (ELSD). Improved sensitivity, linearity, dynamic range, and compatibility with gradient elution are commonly cited reasons for choosing CAD. Several more recent papers describe the use of CAD for analyses that have historically employed pre-column derivatization for HPLC with fluorescence detection (HPLC-FLD) and/or direct analysis by high performance anion exchange chromatography with pulsed amperometric detection (HPAEC-PAD). These newer HPLC-CAD methods have been facilitated by ongoing developments in column technologies, an improved understanding of their characteristics and their use with volatile mobile phases that are also compatible with mass spectrometry (MS). Commonly cited reasons for these approaches include the versatility of HPLC, the desire for simplified workflows and for both faster and more comprehensive analysis, which often includes peak identification with MS. In many cases, these HPLC-CAD methods are viewed as complementary to well-established methods. In other examples, CAD is shown to provide unique and highly relevant capabilities that may warrant its use as a preferred technique. This chapter provides a brief overview of HPLC-CAD applications with a focus on recent literature that describes method development and performance evaluation. Within this context, advantages, limitations, and opportunities for addressing some of the challenges encountered in the use of HPLC-CAD for carbohydrate analysis are discussed.

7.2 HPLC Analysis of Carbohydrates

Carbohydrates are key constituents of biological systems with many important functions including energy production and storage, structural support, signal transduction, and immune functions. Many beneficial properties have been associated with carbohydrates including antidiabetic, anti-inflammatory, antioxidant, and neuroprotective effects. Glycoconjugates such as monoclonal antibodies (mAbs) are also well established for their therapeutic use. Glycation of both proteins and lipids and Maillard reactions with amino acids are also examples of some detrimental properties of carbohydrates. Characterization and quantification of carbohydrates is highly relevant to agriculture, biofuels, biopharmaceuticals, biotechnology, herbal medicines, food and beverages, pharmaceuticals, and other disciplines. The analysis of carbohydrates by HPLC, however, can be exceptionally challenging. Carbohydrates are highly polar and are typically not retained with reversed-phase columns. They also lack strong chromophores, which hinders their detection by UV–vis absorbance. The structures of oligosaccharides and more complex carbohydrates include various linkages between monomers along with chirality at different carbons leading to a nearly endless number of isomers that mainly differ in the spatial orientation of atoms. These more subtle differences can make them more difficult to separate compared to the functional group variations that are more common with other analyte classes. Even for simple carbohydrates, isomers of reducing sugars can exist in a mutarotation equilibrium between α and β anomers. Reducing sugars are those with free aldehyde or ketone groups and include all monosaccharides and many other saccharides. Chromatographic separation of reducing sugars can be complicated by their propensity to undergo on-column mutarotation [1]. A slow interconversion rate compared to the on-column mass transfer rate may result in split peaks or partially merged peaks often compromising resolution and quantitation. Anomer separations are therefore usually suppressed. Common methods for analysis of reducing sugars use pre-column derivatization to both block the anomeric interconversion and introduce a chromophore or, more commonly, a fluorophore such as 2-aminobenzoic acid or 2-aminobenzamide [2, 3]. These labels provide low to mid fmol sensitivity with HPLC-FLD [4]. However, the process of derivatization is often costly, tedious, and time-consuming and can introduce significant variability. This variability is mainly related to derivative stability and differences in derivatization efficiency, which for example can be especially poor for fructose [1, 5–7]. While the presence of the label usually facilitates chromatographic separation [8], in some cases it can reduce chromatographic selectivity for different isomers [9]. For these reasons, it is often preferable to separate and detect carbohydrates directly without derivatization. HPAEC-PAD is a well-established and very effective method for direct, label-free, analysis of carbohydrates with high fmol to low pmol sensitivity [10–12]. Carbohydrates are separated at high pH where they exist as anions using high pH-compatible anion exchange columns and base-compatible chromatography systems. Common methods include separation with aqueous sodium or potassium hydroxide eluents and sodium acetate gradients in a strong sodium hydroxide solution (≥100 mM). Analytes are detected by oxidation on a gold working electrode (WE) where a continuous pulsed potential waveform is used to both detect the analytes and restore the WE surface. A main disadvantage of HPAEC is its limited compatibility with conventional HPLC systems and, perhaps more significantly, difficulties in coupling with MS [13]. The high pH of the eluent may also result in unwanted on-column epimerization and chemical modification of analytes [10, 14, 15]. Given the above limitations, there has been great interest in using HPLC with its high versatility and widespread availability for direct analysis.

Common approaches for direct HPLC analysis of carbohydrates include separation by hydrophilic interaction chromatography (HILIC), porous graphitic column (PGC) chromatography, ion exchange chromatography (IEC), and size exclusion chromatography (SEC). A wide range of column technologies has been developed for HILIC-mode separations of carbohydrates. Most early applications with CAD used amino-bonded stationary phases but recent papers more commonly describe the use of amido-bonded phases. Representative methods are reviewed in Section 7.4. When developing separation methods for reducing sugars, a main consideration is to suppress the above-described anomer separation. This can be accomplished by increasing the column temperature (T_c) and/or eluent pH to increase the interconversion rate until the peaks coalesce [9]. Amino-bonded stationary phases can directly suppress anomeric conversion due to the higher localized pH of the amine group. While this may be advantageous to avoid peak splitting and shouldering, several issues are associated with amino-bonded phases. First, reducing sugars can react with the bonded amine group to form Schiff bases leading to low column recovery of analytes and loss of column efficiency [16]. Also, with silica-based amino columns, the creation of a localized high pH by the bonded amine group can self-dissolve the silica leading to prohibitive column bleeding [17–19]. Polymer-based amino columns are more stable under alkaline conditions, but separation efficiency is typically reduced compared to that of silica-based columns [9]. Perhaps more importantly, Schiff base formation remains an issue with polymeric amino columns. Several papers have described the use of nearly neutral amido-bonded silica stationary phases for carbohydrate separations. The bonded amide groups are less prone to irreversible Schiff base column adsorption. However, it is usually necessary to operate at high column T and/or high pH to suppress anomer separation. This may in turn lead to increased column bleeding as discussed in Section 7.3.

RID, ELSD, CAD, and MS are the main detection techniques used for direct analysis of carbohydrates. MS provides a specific and sensitive technique for HPLC, but detection is highly dependent on analytical conditions, and the instrumentation can be costly and operationally complex. RID has considerable limitations in sensitivity, is incompatible with gradient elution and is significantly affected by ambient temperature and system backpressure leading to slow equilibration and baseline instability. ELSD, while compatible with gradients, has significant limitations in precision, sensitivity, dynamic range, and the complex nature of calibration curves. Many detector comparison studies have consistently shown significant advantages of CAD over ELSD. The main benefits of CAD include universal detection of nonvolatile and semi volatile analytes, response for nonvolatile analytes that is largely independent of chemical properties, broad dynamic response range with detection limits spanning from low ng to high μg, good precision for a diverse range of analytes, and simple and reliable operation. These advantages are consistent with theory as discussed in Chapter 1 where the downstream mechanism of diffusional charging of aerosols employed in CAD underlies its more uniform response, lower detection limits, wider dynamic range, and less complex response curves than ELSD. With CAD, like ELSD, response is not perfectly linear, and its response magnitude varies with mobile phase composition; it lacks analyte-specific information such as with diode array or MS; it requires the use of volatile mobile phases and is limited to detection of analytes that behave as nonvolatiles or semi volatiles with the latter having poorer sensitivity. Various approaches to address some of these limitations are discussed throughout this book, especially in Chapter 3. Section 7.3 describes some key considerations relevant to the use of CAD specifically for carbohydrate analysis and is largely based on an overview of recent papers described in Section 7.4.

7.3 Practical Considerations with CAD

As detailed in Chapter 3, the best performance with CAD is obtained when the concentration of less volatile impurities in the eluent is minimized by using solvents specified to have low evaporation residue, columns and conditions that produce minimal bleed, LC systems, and operational practices that minimize contamination, and low concentrations of low molar mass additives. Table 7.1 summarizes the methods described in Section 7.4 and includes information on stationary phase and mobile phase components, which are key variables affecting performance.

7.3.1 Column Bleed

The topic of column bleed is detailed in Chapter 3 and its relevance to specific analyte classes is discussed in Chapter 5 (Lipids), Chapter 6 (Ions), and here for carbohydrates. Column bleed is often more readily observed with CAD than other detection techniques, including ELSD and MS [19]. This can be attributed to its behavior as a mass-flow-dependent universal detector with higher sensitivity to smaller aerosol residue particles than ELSD. Immediate issues with column bleeding are an increase in baseline signal, noise, and drift with CAD, ELSD, and MS and, with MS, the creation of gas-phase ion adducts. It should be noted that while chromatographic effects may not be immediately obvious, bleeding is a symptom of column instability. Chromatographic performance may therefore eventually degrade where changes in retention, decreased column efficiency and peak tailing may become evident. Column bleed is a particular concern with the above analyte classes since many of the separation methods use HILIC and mixed-mode columns. Dissolution of the silica support at higher pH is a main source of bleed with any silica-based column but the issue appears to be more pronounced with columns used for normal phase (NP) and HILIC separations (see also Chapters 5 and 6). While most modern silica-based reversed-phase columns exhibit low bleed when used within their recommended pH and temperature range, there have been reports of significant bleed observed with CAD or MS when using silica-based HILIC columns even when operated within their recommended ranges [18]. Possible factors contributing to higher column bleed with HILIC and normal phase separations include: less protection of the support by an organic bonded-phase layer (e.g., C18); more facile hydrolysis of attached functional groups and/or dissolution of the base silica substrate by the water-enriched layer on the stationary phase surface; greater exposure of the support to higher organic eluents, which shifts the pH higher relative to the aqueous pH; and the above-described self-dissolution of silica-based amino columns [17–19]. The latter columns are now very rarely used with CAD or MS. Qian et al. [19] described a poly(vinyl alcohol)-coated benzylthioethyl-silica (PVA-Bte-Sil) column and studied column bleed using CAD, ELSD, MS, and inductively couple plasma optical emission spectroscopy (ICP-OES). Figure 7.1 compares CAD baseline currents and noise obtained with no column, with the PVA-Bte-Sil column and with two commercially available columns under

Table 7.1 Representative HPLC-CAD applications.

First author [reference #]	Application	Stationary phase	Elution	MP components	T_c °C	CAD T_e °C	LOD (ng)
Grembecka et al. [20]	Simple carbohydrates in beverages	Polymeric amine	Isocratic	Water, ACN	25	N/A	0.5–1.6
Yan et al. [1]	Monosaccharides from oligo and polysaccharides after acid hydrolysis (TFA) in plants	Amide	Isocratic	ACN, AmAc pH 10.8	30	N/A	15–40
Ghosh and Kline [21]	Monosaccharides from hyaluronic acid (GAG) after acid (TFA) hydrolysis	Amide	Binary gradient	ACN, AmAc, TEA	50	N/A	50–83
Hu et al. [22]	Oligosaccharides in traditional medicines	Amide	Binary gradient	Water, ACN	60	N/A	0.24–1.35
Xie et al. [23]	Monosaccharides and oligosaccharides in traditional medicines	Amide	Binary gradient	ACN, TEA	N/A	N/A	15–81 and 304
Ji et al. [5]	Synthetic heparosan oligosaccharides	PGC	Isocratic	Water, ACN, MeOH, TFA	25	N/A	Not reported
Westereng et al. [13]	Oligosaccharides from cellulose by controlled enzyme catalyzed reactions	PGC	Binary gradient	AmAc pH 6.5 ACN	Ambient	N/A	Low ng
Knihtila et al. [6]	Oligosaccharides released from mAbs by PNGase	PGC	Binary gradient	Water, ACN, TEA	60	50	ca. 20-fold lower sensitivity than HPLC-FLD
Hetrick et al. [24]	Glucose impurity in pharmaceutical excipient	Bare silica	Binary gradient	ACN, Water	55	N/A	4[a)]

Xie et al. [25]	Sucrose and fructose impurities in dextran 40	Bare silica	Isocratic	ACN, Water	40	50	0.5
Corrado et al. [26]	Polysaccharide vaccine antigens	Diol	Isocratic	ACN Water AmF pH 3.5	30	55	Not reported
Wang et al. [7]	Sialic acids from therapeutic proteins by acetic acid or enzyme hydrolysis	HILIC, SAX, WCX	Isocratic	Water, ACN, AmF pH 4.0	30	N/A	<7.5
Wang et al. [27]	Partially acid (TFA) hydrolyzed polysaccharides in plants/herbal medicines	HILIC, SAX, WCX	Isocratic	ACN water AmF	40	40	Not reported
Chen et al. [28]	Polysaccharides in herbal medicines by SEC	SEC	Isocratic	AmAc	40	60	Not reported
Corrado et al. [26]	Polysaccharide vaccine antigens	C4	Binary gradient	Water, ACN, TFA	50	55	Not reported

MP, mobile phase; T_c, column or column compartment temperature; T_e, CAD evaporation temperature; N/A, not available.

a) Estimated based on a SNR of 73 for ca. 90 ng.

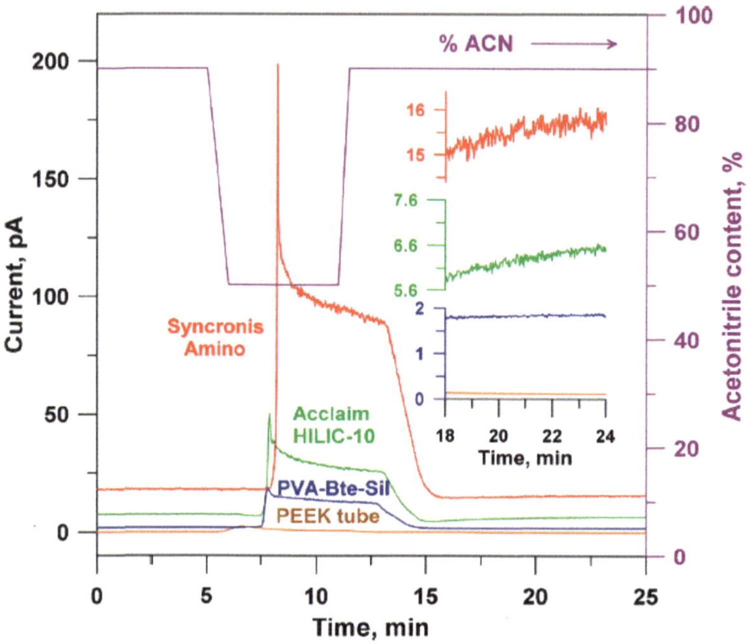

Figure 7.1 Bleed measurement of various HILIC columns as seen by CAD. Time ($MeCN/H_2O$) 0–5 minutes (90 : 10), 6–11 minutes (50 : 50) 11.5–25 minutes (90 : 10). CAD $T_e = 50\,°C$. *Source:* [19] / with permission of the American Chemical Society.

ACN/water gradient conditions. It is notable that all columns exhibited some level of bleed even without any basic mobile phase additive. Also notable, for all tested columns, is the transient baseline spike and overall increase in baseline level and noise associated with a gradient of increasing water. This has significant implications for gradient methods and may be attributed to the above-described more facile dissolution of the silica base in water. The highest column bleed was observed with the silica-based amino column, as expected, while the lowest was with the PVA-Bte-Sil column. An Acclaim™ HILIC-10 column (Thermo Fisher Scientific), which contains proprietary polar groups covalently bonded to silica by hydrolysis-resistant linkages exhibited an intermediate level of column bleed consistent with that of a more recent study [29]. It is interesting that ICP-OES analysis indicated a silicon concentration of 7.7 and 0.3 µg/mL in the respective effluents of a bare silica column and the PVA-Bte-Sil column. Assuming that the relationship between nonvolatile impurity concentration and detector signal is roughly similar to that shown in Figure 1.14 (Chapter 1), the bare silica column would be expected to produce >20-fold higher baseline currents and noise than the PVA-Bte-Sil column.

For carbohydrate separations, the common use of higher T_c and higher pH eluents may further exacerbate the issue with column bleed. Fortunately, HILIC and mixed-mode columns designed for higher pH and T_c stability are now more readily available. This includes pH and T-stabilized amide columns, mixed-mode columns, and PGC columns some of which are further described in Section 7.4. Hopefully, further technological improvements to reduce column bleed will soon become available. When choosing a column for use with CAD, a good practice is to examine baseline currents, noise and drift and to compare the results to those obtained without a column. This should be done over the expected range of T_c and mobile phase variables such as pH, solvent, and additive composition and should include gradient elution. For carbohydrates, higher pH and T_c might be particular areas on which to focus.

7.3.2 Additives and Eluent pH

Elevated baseline currents, drift and noise are common observations when using higher pH (> ~7) eluents with CAD. First generation CAD instruments were even described as incompatible with basic conditions. In some cases, higher background current and noise can be attributed to increased column bleed under alkaline conditions. Another potential cause is the formation, through spray-drying, of nonvolatile carbonate salts. While several labs have observed prohibitively high background currents when using ammonium carbonate buffers, this issue can also occur with other additives. Freshly prepared AmAc buffers, for example, typically produce low background currents and noise. However, this is often found to significantly increase over a relatively short period of time (e.g., within a day or so) and is attributed to ammonium carbonate salt formation that results from absorption of atmospheric CO_2 by the eluent. With newer CAD instruments, the use of an optimally higher evaporation temperature (T_e), for example 50 °C, can be very effective at reducing background signal and noise that is attributable to ammonium carbonate salt formation [30]. Use of a higher T_e may, however, lead to unwanted volatilization of analytes and therefore should be used with caution (see below). Additional practices to minimize carbonate formation include more thorough degassing, sparging, or blanketing of the mobile phase with an inert gas and use of freshly prepared mobile phases. These practices may reduce or obviate the need for use of a higher T_e.

The type and concentration of mobile phase additive and the technique used for buffer preparation are also very important considerations when using CAD. Volatile ionic mobile phase additives can, during spray-drying, combine with other ionic solutes (e.g., analytes, impurities, sample matrix components and other oppositely charged additives) to form less-volatile salts. These effects can be significant especially considering that additive concentrations are often several orders of magnitude higher than typical analyte concentrations within the eluent. The extent to which additives contribute to CAD signal significantly depends on their molar mass where higher molar mass additives, through salt formation, contribute more mass to the aerosol particles. Use of ammonium hydroxide to increase pH is therefore preferred over triethylamine (TEA). Combinations of higher molar mass acids (TFA) and higher molar mass bases (e.g., TEA) within an eluent can lead to particularly high background signal. Also, ionic matrix components such as TFA, which is often used to hydrolyze carbohydrates, can combine with oppositely charged additives to produce large peaks and peak artifacts. It is also recommended to restrict the concentration of additives to the minimum required to achieve the desired separation since higher concentrations typically produce higher background signal. The method of buffer preparation may also be a considerable source of problems. Here, it is highly recommended to prepare any buffers gravimetrically rather than to titrate to the desired pH with a pH electrode. This is intended to ensure greater reproducibility and avoid contamination from pH calibration buffers [31].

7.3.3 Sensitivity, Linearity, and Calibration

Table 7.1 shows reported limits of detection (LODs) in the low ng range, the lowest generally observed with the simplest eluents (e.g., no additives). This is consistent with other analyte classes. The LODs reported with HPLC-FLD and HPAEC-PAD are commonly expressed in units of moles, and it appears that there are few direct comparisons with CAD. Since CAD is a mass flow sensitive device, better sensitivity when expressed on a mole basis should be observed for larger molecules. For example, an LOD of 5 ng for glucose (180.16 g/mol) would correspond to 27.8 pmol while that of a 1000 g/mol analyte would correspond to 5 pmol. An estimated low pmol LOD range for CAD indicates lower sensitivity than HPLC-FLD (low to mid-fmol) and HPAEC-PAD (high fmol to low pmol) as generally reported in literature. In one comparison, Bailey et al. estimated an LOD for simple sugars of 100 pg with HPAEC-PAD and a 5 ng LOD with HILIC-CAD [32]. Knithila et al. [6], based on their adjustment of sample load to match sensitivity, suggested an approximately 20-fold higher sensitivity with HPLC-FLD for labeled *N*-glycans compared to their direct analysis with HPLC-CAD. For this and other applications described in Section 7.4, either adequate sensitivity was achieved with CAD for the intended purpose, or the sample amount was increased to allow comparable sensitivity to established methods. Limitations to increasing the volume and/or concentration of sample analyzed may be encountered when total sample volume is limited or when there is an incompatibility between the sample solvent and the separation technique. This can be a particular problem with HILIC methods when the required sample solvent has a higher elution strength (e.g., water) than the mobile phase (e.g., high ACN) at time of injection.

Several strategies such as using alternative solvents or pre-injecting a small plug of a stronger solvent have been described to address this issue [33].

Many of the applications discussed in Section 7.4 involved the use of first-generation CAD instruments, which did not have options for setting a T_e (indicated by "N/A" in Table 7.1) or a power function value (PFV) setting. These are very useful instrument parameters to optimize performance as detailed in Chapter 3. Briefly, an optimal T_e can improve signal-to-noise ratios (SNR) by altering the extent of vaporization of solutes. This is usually a compromise between maximizing the vaporization of eluent impurities to reduce baseline noise and minimizing vaporization of target analytes. It is generally recommended to use the lowest T_e setting that consistently produces the required sensitivity limits and where the ideal goal is for all analytes to behave as nonvolatiles. For calibration, most of the described carbohydrate methods used linear regression over relatively small mass ranges. However, some used linear regression after log transformation of peak response versus analyte amount. Here, it is highly recommended to optimize the detector output with respect to linearity using the PFV. An optimal PFV extends the linear range and enables straightforward quantitation while often improving accuracy, SNRs, precision and peak shapes. The use of log transformation is generally not recommended. Log–log plots can be visually deceptive as small deviations on the plot can represent large absolute differences. For example, during method development it is important to know whether an analyte behaves as a nonvolatile or semi volatile, since the latter is associated with poorer precision and sensitivity. This is readily seen as a supralinear-shaped response curve especially at lower mass levels. This may go unnoticed in log–log plots. Also, linear regression calculations such as correlation coefficients, are valid only when the experimental uncertainty of replicate y values is not related to the values of x or y. This assumption may not be fully valid after data have been transformed. Chapter 3 describes several options to determine an optimal PFV with minimal experimentation. These tools can be used with any chromatography platform and require only the generation of response curve data at one PFV. This optimal PFV can then be implemented as an instrument setting, which is the recommended approach as discussed below. The PFV setting is a data acquisition parameter like data collection rate or filter time constant that directly affects the signal output and is used to optimize the response with respect to linearity for a given method and range of quantitation.

7.4 Representative Applications

This section provides an overview of HPLC-CAD methods reported in literature. A previous review by Rohrer and Kitamura [34] described the adoption of CAD for analyses that traditionally used RID, short-wavelength UV or ELSD. Commonly cited reasons for choosing CAD over short-wavelength UV and RI include its better stability, sensitivity and compatibility with gradient elution. As with other analyte classes, CAD is often chosen over ELSD due to its wider dynamic range and better linearity, precision and sensitivity. A majority of early methods for simple carbohydrates used either polymeric amino-bonded or silica-based amido-bonded phases. Common applications include quantitation in food, beverages, plants and herbal medicines and monitoring of enzyme reactions for synthesis of desired materials (e.g., bioactives, surfactants, sweeteners) and catalysis of complex carbohydrates (e.g., for compositional analysis, quality control, and biofuel production). Other highlighted methods include the use of a metal loaded cation exchange separation to determine monosaccharide anhydrides in biomass combustion smoke [35]; reversed-phase methods for quality control of aminoglycoside antibiotics [36] and for separation of cycloamyloses up to a degree of polymerization (DP) of about 80 [34]. The use of a PGC stationary phase with CAD was also described for analysis of *N*-linked and *O*-linked glycans released from monoclonal antibodies (mAbs) [32]. mAbs are complex glycoproteins and are the most common therapeutic product of the biopharmaceutical industry. Glycans are a very important aspect of the critical quality attribute (CQA) panel for development and release testing of therapeutic mAbs. Figure 7.2 shows the

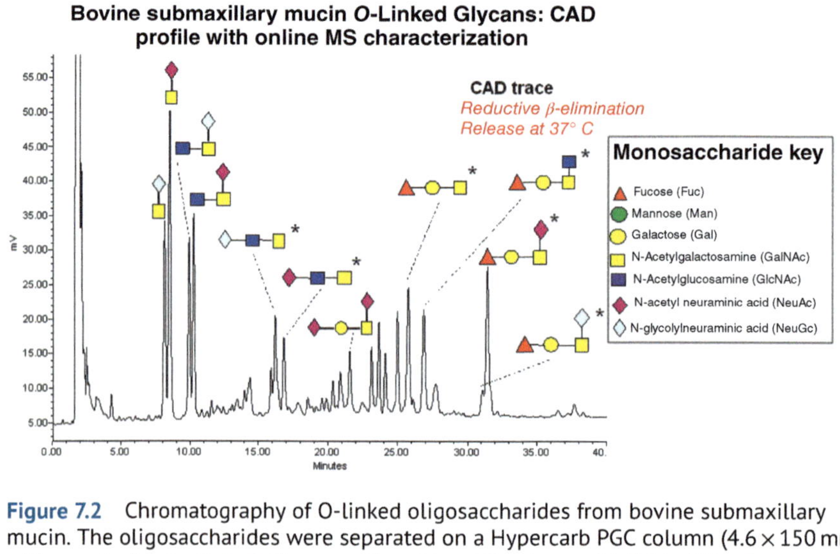

Figure 7.2 Chromatography of O-linked oligosaccharides from bovine submaxillary mucin. The oligosaccharides were separated on a Hypercarb PGC column (4.6 × 150 mm) with a water/acetonitrile/trifluoroacetic acid mobile phase with 90% of the flow going to the CAD and 10% going to a mass spectrometer for oligosaccharide identification. Identified structures are shown on the figure. Peaks marked with * depict proposed structures based on online accurate mass determination and literature. *Source:* Adapted from [37].

O-linked glycans released from bovine submaxillary mucin separated on a PGC column using a water/ACN/TFA mobile phase with detection by CAD. Peaks were identified by MS. Release of *O*-linked glycans is usually conducted under reductive conditions to prevent oligosaccharide degradation. Because the reduced oligosaccharides are not amenable to reductive amination, the typically used fluorophore-labeling reactions are not possible. This makes CAD ideal for their detection when there is enough sample to achieve the required sensitivity and has the important advantage of mobile phase compatibility with MS.

The following discussion includes more recent HPLC-CAD methods (see Table 7.1). These were selected with priority given to papers published within the last 10 years that include details of method development and performance evaluation.

7.4.1 Therapeutic Glycoproteins and Vaccines

Knihtila et al. [6] described their development of a novel, label-free method for analysis of *N*-linked glycan moieties of mAbs. Glycosylation is an important CQA, requiring accurate measurement of released *N*-linked glycans of many therapeutic proteins. Traditionally, *N*-linked glycans are released from glycoproteins after denaturation and enzymatic digestion. The liberated glycans are then derivatized by a process that often involves the use of toxic reagents and can be costly, tedious, time-consuming and variable. This is followed by highly sensitive analysis using HILIC-FLD. In pursuing a label-free HPLC analysis, several stationary phase options for separation of unlabeled glycans were evaluated and the best performance was obtained with PGC. Separation with PGC is based on a combination of hydrophobic interactions, size and shape interactions with the planar stationary phase surface and a polar retention effect, which is facilitated by the high polarizability of graphite. Initial separation conditions used a water/ACN gradient with 0.1% TFA additive. Additive type, gradient slope and T_c were among the variables studied. Interestingly, with an acidic pH (0.1% formic acid) and T_c settings ranging from 50 to 80 °C, there was no evidence of peak splitting or shouldering due to anomer separation. The authors proposed that this may be due to the lack of undesirable interactions between the reducing end of the glycan and the PGC surface. While anomer separation of simple sugars on PGC has been previously demonstrated at a T_c of 5 °C [38], the lack of separation at 50 °C in acidic conditions suggests that PGC may be less prone to this issue than other stationary phases. Since three sialic acid-containing glycans were found to be unretained when using TFA as acidic additive, TEA was investigated and found to retain these analytes. The final conditions used a water/ACN binary gradient with 0.1% TEA and a CAD T_e of 50 °C. The initial sample load was adjusted to achieve the same sensitivity of the labeled glycans and the results demonstrated acceptable comparability to the existing platform methods. The method was therefore able to eliminate the requirement for derivatization without sacrificing peak resolution. This provides an advantage of increased efficiency with significant reduction in analysis time while keeping the reagent cost significantly lower.

Wang et al. [7] described a novel label-free method for quantification of total sialic acids in therapeutic protein products. The terminal monosaccharide of *N*-linked glycans is typically end-capped by sialic acids (SAs), with two primary forms being N-acetylneuraminic acid (NANA) and N-glycolylneuraminic acid (NGNA). Sialylation is often closely monitored as a CQA. A common approach for determination of total sialic acids involves releasing SAs from glycoproteins through hydrolysis, followed by derivatization of the released SAs with 1,2-diamino-4,5-methyleneoxybezene (DMB) followed by HPLC-FLD. Although DMB labeling offers desirable sensitivity and selectivity, the derivatization process is time-consuming and derivatized SAs are not stable, leading to poor method precision. A label-free method was developed using an Acclaim Trinity™ P2 column, which provides HILIC, strong anion exchange, and weak cation exchange interactions. A pre-column C18 cartridge was used to remove proteins, preventing protein interference and analytical contamination enabling direct sample injection. Peak broadening and splitting were observed with a mobile phase of pH 3 and attributed to charge heterogeneity as commonly observed when the eluent pH is close to the analytes' pK_a (NANA pK_a 2.60 and NGNA pK_a 2.92). Optimal peak shape and resolution were obtained with a pH 4 AmF buffer. The final method used isocratic elution with a mobile phase of 65%/25%/10% (v/v/v) of ACN/water/100 mM AmF and T_c of 30 °C The method demonstrated satisfactory specificity, linearity ($R > 0.999$), precision (RSD between 0.4% and 2.1%), sensitivity (QL < 25 ng), and accuracy (recovery between 93% and 102%). The method demonstrated SNRs >5-fold higher than short-wavelength UV, superior precision to DMB-labeling workflows and was successfully applied to total sialic acid determination in a model protein drug product.

Corrado et al. [26] described the development of a universal strategy, termed UniQS (Universal Quantification of polySaccharides), for direct quantification of polysaccharide (PS) and glycoconjugate (i.e., protein-linked saccharide) antigens. Analysis of these antigens is critical for quality evaluation of vaccines used against bacterial infections. Traditional methods include GC–MS, HPLC-FLD after derivatization and direct analysis with HPAEC-PAD. The analysis of long polysaccharide antigens, however, requires significant sample pre-treatment to purify the saccharide and cleave the glycosidic bonds to release the monosaccharide units. These procedures and the chromatographic methods are usually PS or serotype-specific since they depend on the repeating unit structure, properties and glycosidic linkage. The goal of this work was to develop a universal UHPLC-CAD approach for direct quantification of intact and untouched PS. Several bacterial serotypes

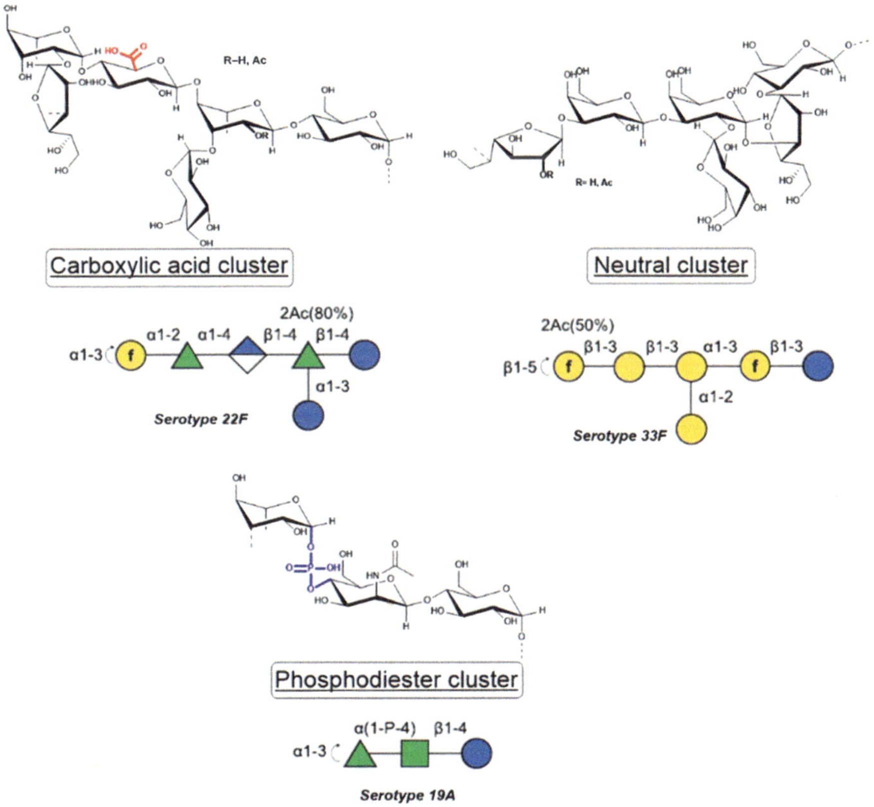

Figure 7.3 Serotypes cluster subdivision examples: carboxylic acid highlighted in red (left), phosphodiesters highlighted in blue (right), and neutral (middle). *Source:* [26] / Springer Nature / CC BY 4.0.

differing in monosaccharide composition, anomeric linkages, linkage positions, and the presence of charges were used for development and evaluation. Based on these characteristics, the PS was classified into three clusters: carboxylic acid; phosphodiester and neutral (see Figure 7.3). To evaluate the universality of the method and its applicability to a wide range of matrices, non-purified PS and process intermediates from the fermentation broth to the final purification steps were tested.

Two methods were developed. The first was a binary gradient reversed-phase (RP) method using a wide-pore C4 column and mobile phase components of water, ACN and trifluoroacetic acid and CAD T_e of 55 °C. Non-purified PS samples were simply centrifuged, filtered and diluted to the target concentration before injection. The PS component of each sample was completely separated from other components (e.g., DNA, fermentation matrix, salts, surfactants, etc.). To preserve long-term detector functionality, the column eluent was diverted to waste at the beginning and end of each run where salts and other high-level matrix components eluted. Each PS produced linear response over the required concentration range. This RP method was found suitable for neutral and carboxylic acid PS clusters but the phosphodiesterase PS clusters were not sufficiently retained. A second method was adapted from the HILIC method described in Ghosh and Kline [21] and used a silica-based diol stationary phase with isocratic separation and mobile phase components of AmF 100 mM pH 3.5, water, and acetonitrile (ACN). The method was optimized for elution and detection of glycans containing a phosphodiesterase moiety and was found to be selective with linear response over the targeted range. The method used the same sample treatment as the RP method and was also found applicable to carboxylic acid but not neutral PS clusters. The authors reported that, to their knowledge, this represents the first report of an underivatized PS chromatographic method for production monitoring of a multivalent PS-based vaccine. The approach was described as simple and automatic with the potential to have an impact within the field of vaccines and biologics research and development because it can be applied to other substrates with similar characteristics. In addition, it could lead to material, economic and time savings in an industrial production process setting, supporting the development of ambitious multivalent vaccines that require a flexible analytical control strategy.

7.4.2 Pharmaceuticals

Ji et al. [5] studied the retention behavior of heparosan oligosaccharides on PGC. Heparosan or N-acetyl heparosan (NAH) is a member of the glycosaminoglycan (GAG) family and is a precursor in the biosynthesis of therapeutic heparin and heparan sulfate (HS). Traditional methods for analysis of NAH involve pre-column derivatization, which can result in analyte loss or compromises in the sensitivity of MS, or HPAEC-PAD, which cannot be conveniently coupled to MS because of the high salt content used within the typical eluents. PGC has attracted much attention as a unique stationary phase with high stability to T and pH and increased retention for hydrophilic analytes and hydrophobic interactions that are quite distinct from conventional RP columns. The overall retention behavior of PGC columns has been described as a combination of hydrophobic interactions, size, and shape interactions with the planar graphite surface and interactions induced by the charge of polar analytes [39]. Starting with ACN as the primary and most widely used organic modifier with PGC, variables including T_c, water, TFA, and MeOH concentrations were studied. TFA was used as an acidic modifier to aid in the separation of acidic oligosaccharides and the addition of small amounts of MeOH had a significant beneficial effect on selectivity. Final conditions used isocratic elution with mobile phase components of water, ACN, MeOH, and TFA and had excellent compatibility with CAD and MS. The combined PGC CAD/MS approach demonstrated high efficiency and excellent sensitivity to elucidate the structures of NAH oligosaccharides in one sample. The authors noted that this easy method has been employed in their lab for monitoring NAH intermediates in the production of well-defined bioengineered heparin and that the method should be promising for investigating the biological function of heparosan and the related enzymes involved in HS biosynthesis.

Hetrick et al. [24] described a comprehensive evaluation of polar stationary phases for separation of simple sugars and sugar alcohols. They screened 16 stationary phases for their ability to separate various mixtures of 26 sugar and sugar alcohol analytes in HILIC mode with ACN/water or acetone/water mobile phases. Stationary phases included unbonded silica and ethylene-bridged hybrid silica along with several bonded phases—amide, amine, diol, pentahydroxy, triazole, and zwitterionic. While the study did not evaluate column bleed, the findings related to column selectivity and chromatographic performance were used to create a design space for use in method development. From this, a HILIC-CAD method was developed for quantitative analysis of glucose, which is a potential impurity in the common pharmaceutical excipient trehalose. The method used a superficially porous unbonded silica-based HILIC column, which was one of only seven columns tested that produced, as desired, a single peak for glucose anomers. A water/ACN binary gradient was used with a column T of 55 °C. The method was described as providing excellent accuracy, linearity, precision and specificity with more than ample sensitivity to quantify glucose at the targeted 0.0009% w/w level.

Ghosh and Kline [21] adapted the method of Yan et al. [1] to the compositional analysis of acidic, amino and neutral monosaccharides derived from carbohydrate-based therapeutics (e.g., hyaluronic acid). A BEH™ amide column (Waters Corp.) was used and the effects of T_c, buffer type and concentration, and TEA additive on chromatographic separation were evaluated. A T_c of 50 °C and a TEA concentration of 0.2% was able to adequately separate acidic and amino sugars along with select neutral sugars without formation of split peaks. Higher concentrations of TEA (>0.2%), high pH (>10), and high AmAc (>25 mM) caused substantial detector noise and an unstable baseline. The optimized method used a binary gradient with mobile phase components of 88% ACN, 0.2% TEA, and 25 mM AmAc, which were able to separate and detect amino sugars (GlcN, GalN, and GlcNAc) and acidic sugars (GlcA, GalA, Neu5Ac, and LIdoA). Calibration was based on linear regression after double logarithmic transformation of peak area versus amount. The LODs were reported to be higher than those reported in Yan et al. [1], and it was suggested that this was due to the higher concentrations of AmAc and TEA in the mobile phase. This is consistent with the expected contribution of TEA-Ac salts to residue particles that are formed during the spray-drying process of CAD (Chapter 3). In addition to monosaccharide compositional analysis by HPLC-CAD, a previously described SEC-CAD method [28] was used to evaluate homogeneity and molecular weight of polysaccharides. The authors described HPLC-CAD as a multipurpose platform for rapid QC of carbohydrate-based therapeutics without the need for laborious, time-consuming sample preparation (e.g., derivatization) and multiple complementary analytical techniques. The risks of epimerization and degradation of carbohydrate and requirement for a dedicated base-compatible system for HPAEC-PAD is also avoided. The authors note, however, that the platform can only be used as a routine QC tool for sufficiently pre-characterized carbohydrate samples, since some samples may contain impurities that might interfere with compositional analysis. Also, some neutral sugar residues were not separated, and complementary separations would be needed if a more comprehensive analysis is required.

Xie et al. [25] described HILIC-CAD analysis of fructose and sucrose residues in Dextran 40 and compared its performance to HILIC-RID and HILIC-MS. Dextran 40, a glucose polysaccharide prepared from sucrose by fermentation, is an active pharmaceutical ingredient (API) primarily used intravenously for increasing plasma volume and as an anti-shock treatment. The presence of residual sucrose and carbohydrate by-products, such as fructose, has been associated with adverse reactions. Sensitive, precise, and accurate methods are therefore required for quality control. The HILIC-CAD method used a superficially porous bare silica column (T_c 40 °C) and isocratic elution with an ACN-water (90 : 10) mobile phase and a CAD T_e of 50 °C. Three other columns were evaluated (2 amino and 1 amido) and found to produce noisy baselines. Table 7.2 summarizes the detection capabilities of the three methods where all demonstrated good linearity but differed in sensitivity. For CAD and RID, the on-column mass LODs (calculated from the listed concentration LODs and injection volumes) were similar for both analytes (CAD ~0.5 ng, RID ~200 ng) while for MS the LODs differed between analytes (2.4 ng for sucrose, 8.4 ng for fructose). The authors reported that the HILIC-MS method was limited by sample matrix effects where injection of higher volumes or higher sample concentrations led to significant baseline fluctuations. The methods using CAD and RID allowed for the analysis of higher sample amounts. As a result, method sensitivity in terms of ppm of the starting material was >20-fold better with HILIC-CAD than with either of the other methods. Validation results of the HILIC-CAD method indicated good repeatability, recovery, durability, and linearity with rapid quantification of fructose and sucrose within 10 minutes.

Table 7.2 Comparison of HILIC-RID, HILIC-MS and HILIC-CAD methods.

Carbohydrate	**HILIC-RID**	**HILIC-MS**	**HILIC-CAD**
Fructose	$Y = 4.774 \times 10^5 X + 2728$	$Y = 0.8666X + 5.026$	$Y = 5.911 \times 10^6 X + 4.308 \times 10^6$
Sucrose	$Y = 4.914 \times 10^5 X + 1471$	$Y = 0.8900X + 2.220$	$Y = 6.110 \times 10^6 X + 4.480 \times 10^6$
Fructose	0.9999	0.9996	0.9991
Sucrose	0.9999	0.9991	0.9990
Fructose	100–1500	1.25–100	0.33–50
Sucrose	100–1500	0.12–2.54	0.33–50
Fructose	50	20	5
Sucrose	50	20	5
Dextran 40	100	6	100
Fructose	9.91 ± 0.40 μg/mL (99.12 ± 4.02 ppm)	1.27 ± 0.02 μg/mL (211.49 ± 2.89 ppm)	0.33 ± 0.01 μg/mL (3.27 ± 0.08 ppm)
Sucrose	10.06 ± 0.26 μg/mL (100.57 ± 2.63 ppm)	0.36 ± 0.01 μg/mL (59.52 ± 1.50 ppm)	0.34 ± 0.01 μg/mL (3.39 ± 0.08 ppm)
Fructose	4.00 μg/mL (40.00 ppm)	0.42 μg/mL (70.42 ppm)	0.10 μg/mL (1.01 ppm)
Sucrose	4.25 μg/mL (42.46 ppm)	0.12 μg/mL (20.13 ppm)	0.11 μg/mL (1.08 ppm)

Source: [25] / with permission of the Royal Society of Chemistry.

7.4.3 Food, Beverages, Plants and Traditional Herbs, and Medicines

Grembecka et al. [20] described a method for quantitative analysis of fructose, glucose, sucrose, and inositol in beverages for purposes of routine quality control. They used a polymeric amino column with T_c of 25 °C and isocratic elution with 75% ACN/25% water. Erythritol was used as an internal standard. Carbonated beverages were degassed by sonication. Otherwise, samples were simply filtered and diluted with water prior to analysis. The method was applied to 85 nonalcoholic beverages including energy drinks (e.g., Figure 7.4), soft drinks, fruit drinks, nectars, and juices. The method was characterized as providing linearity, precision and accuracy that is comparable to an industry standard method that uses RID while also providing high sensitivity with LODs ranging from 0.5 to 1.6 ng on column. The authors reported that the use of CAD allowed greater flexibility to use gradient elution, which is important to enable measurement of other relevant ingredients that are often present in samples (e.g., artificial sweeteners, acidity regulators, caffeine, preservatives).

Westereng et al. [13] compared HPAEC-PAD and HPLC with PGC-based separations with CAD or MS to study the product profile of lytic polysaccharide monooxygenases (LPMOs). LPMOs are key enzymes involved in the degradation

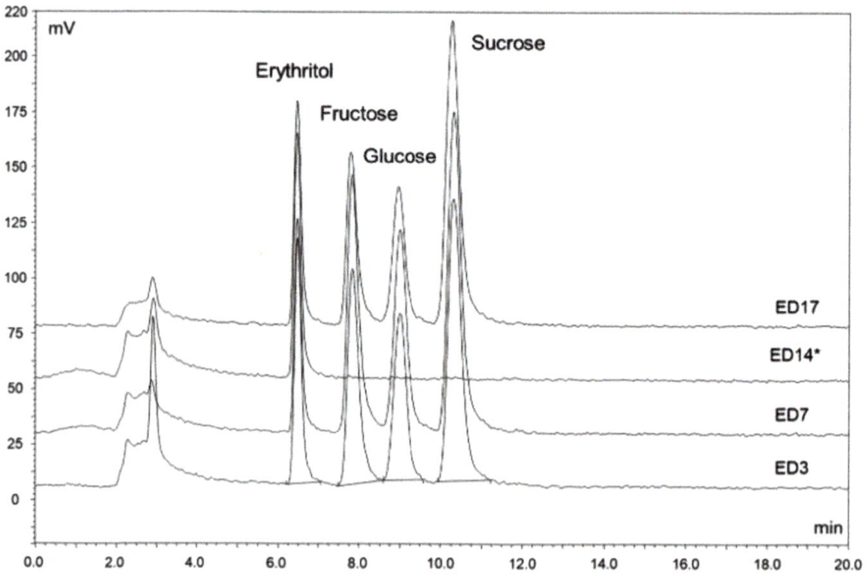

Figure 7.4 HPLC-CAD chromatograms of selected energy drink samples. An aliquot of 10 µL of sample solution was injected. * - sugar-free product. *Source:* [20] / MDPI / CC BY 4.0.

of polysaccharides such as cellulose and chitin. An understanding of substrate specificity and product profile of LPMOs is of particular interest to the biofuel industry. LPMO activity leads to a complex mixture of products, often in small amounts, with various oxidation patterns (C1 or C4) and with varying stabilities. This makes accurate analysis of product profiles a major challenge. HPAEC-PAD has been the method of choice for this analysis providing high resolution and excellent sensitivity but has limitations that include poor compatibility with MS. The HPLC method used a PGC with binary gradient elution with 10 mM AmAc, pH 8.0, and ACN. The HPLC separations showed lower resolution than HPAEC, but this was compensated by easy peak identification with MS. Their studies revealed that peaks observed with HPAEC, and commonly believed to be LPMO products, are in fact produced by on-column decomposition, attributed to the strong alkaline eluents. As a consequence, chromatograms with HPAEC are undesirably complex, some of the measured oligosaccharides are altered from their native forms and some components cannot be detected. In contrast to HPAEC, the HPLC methods are directly compatible with MS and CAD, which enables online peak validation and quantification with LOD levels in the low ng range. Notably, PGC-based separation allowed isolation of C4-oxidized standards for the first time. The authors commented that CAD was incompatible with high pH mobile phases and therefore lowered the pH to 6.5 when using CAD. As discussed in Section 7.3, this study used a first generation CAD instrument which did not have the option to increase T_e in order to extend its pH compatibility range.

Yan et al. [1] described a novel method for neutral monosaccharide compositional analysis of plant-derived oligo and polysaccharides using HILIC-CAD. The objectives were to investigate the use of this approach to aid in structural elucidation of bioactive PSs and for routine quality control of herbal medicines. Method development included evaluation of column type, additives, pH, and T_c. An amino column was first investigated and, although anomer separation could be prevented, they observed greatly varying responses between analytes attributable to the formation of Schiff bases. A HILIC BEH Amide column was chosen since it is not susceptible to Schiff base formation and is specified as stable over a broad pH range [2–11] and resistant to high temperature up to 90 °C. While higher column T produced single sharp peaks for individual monosaccharides, there were some coelutions. Better separation was obtained at higher pH (10.8) and, at this pH, anomer separation was not observed even at lower T_c. Resolution could also be improved by increasing either buffer or ACN concentration, albeit at the expense of longer run times. At higher buffer concentrations, solubility was also cited as a potential issue as was an increase in CAD baseline noise. The optimized method used isocratic elution with a mobile phase of 22 mM ammonium acetate (pH 10.8) in 88% ACN as the mobile phase with a T_c of

30 °C. The method provided good linearity ($r^2 \geq 0.9981$), high precision (RSD ≤4.43%) and acceptable accuracy (94.02–103.37% recovery). The obtained LODs (15–40 ng), while higher than the pg level obtained by HPAEC-PAD [40] and GC–MS, were adequate for the purpose of plant-derived oligo- and polysaccharide analysis. The monosaccharide compositions of the polysaccharide from *Eclipta prostrata* L. and stachyose were successfully profiled. The method could also distinguish potential sample matrix impurities including chloride, sulfate and sodium. This newly developed method was described as convenient, efficient, and reliable for monosaccharide analysis in a single injection while avoiding steps of derivatization and associated sample clean-up. The authors noted that, while HPAEC is another powerful tool for direct sugar analysis, coelutions especially for closely eluting pairs (Rha/Ara and Xyl/Man), remain a critical issue. While many other methods cannot easily measure the common monosaccharide, fructose, this was easily measured with the described method. The authors also noted that the flexibility of HPLC, its widespread availability and the method's compatibility with MS provide significant advantages over other methods that require specialized reagents and instrumentation.

Hu et al. described a method for oligosaccharides in *Pseudostellaria heterophylla*, a well-known Chinese medicine. A BEH Amide column was used with T_c 60 °C and binary water/ACN gradient elution. A quadratic curve fit was used for quantitation of eight oligosaccharides (DP2–DP7) over a concentration range of ca. three orders of magnitude with LODs ranging from 0.24 to 1.35 ng on column. Wang et al. [27] described a heart cutting 2D HPLC-CAD method for analysis of polysaccharides found in plants. The main objectives were to evaluate its ability to provide detailed structural information and to allow QC of polysaccharides in Traditional Chinese Medicine (TCM). The workflow is depicted in Figure 7.5. SEC was used in the first dimension for initial separation and to provide molar mass-related information and the second dimension used a mixed mode separation to generate polysaccharide fingerprints. Polysaccharide samples were subjected to partial acid hydrolysis to simplify their complex structures before analysis.

Samples were injected on the SEC column, which used water as mobile phase with isocratic elution. Three heart cuts were taken at specific time points. The second dimension column was an Acclaim Trinity P2 trimodal column, which provided strong anion exchange, weak cation exchange, and HILIC interactions. Components were separated using a binary gradient with mobile phase components of ACN, water, and AmF buffer. ACN was also added after the second dimension column to provide a more consistent and higher ACN concentration to the detector. This was found to provide a higher signal and more stable baseline with CAD. A known concentration of dextran (molar mass 44,100 g/mol) was used as a reference substance for method development and to evaluate performance. Precision and repeatability of peak areas for six peaks from the second dimension

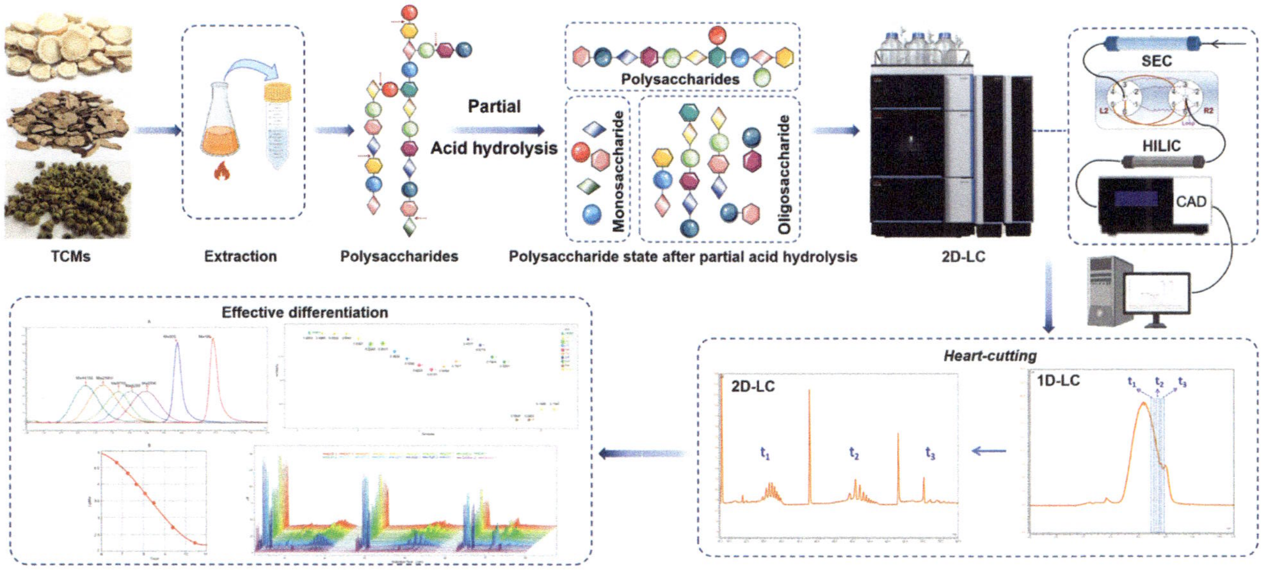

Figure 7.5 Graphical summary of natural polysaccharides analyzed by 2D-LC. *Source:* [27] / MDPI / CC BY 4.0.

were < 2%. The authors noted that the method can distinguish different varieties of TCM polysaccharides and can also be used as a basis for identifying TCMs. Additional work, however, is needed to improve sample purification to reduce the interference of small molecules and to incorporate MS to aid in peak identification.

References

1 Yan, J., Shi, S., Wang, H. et al. (2016). Neutral monosaccharide composition analysis of plant-derived oligo-and polysaccharides by high performance liquid chromatography. *Carbohydrate Polymers* 136: 1273–1280.

2 Anumula, K.R. (2006). Advances in fluorescence derivatization methods for high-performance liquid chromatographic analysis of glycoprotein carbohydrates. *Analytical Biochemistry* 350 (1): 1–23.

3 Jeong, Y.R., Kim, S.Y., Park, Y.S., and Lee, G.M. (2018). Simple and robust N-glycan analysis based on improved 2-aminobenzoic acid labeling for recombinant therapeutic glycoproteins. *Journal of Pharmaceutical Sciences* 107 (7): 1831–1841.

4 Thermo Scientific (2016). Analysis of 2-aminobenzamide (AB) labeled glycans using HPLC with fluorescence detection. Thermo Scientific Technical Note 109.

5 Ji, X., Hu, G., Zhang, Q. et al. (2016). Simultaneous analysis of heparosan oligosaccharides by isocratic liquid chromatography with charged aerosol detection/mass spectrometry. *Carbohydrate Polymers* 152: 337–342.

6 Knihtila, R., Chemmalil, L., Sawant, P. et al. (2022). Development of a novel, label-free N-glycan method using charged aerosol detection. *Journal of Chromatography B* 1212: 123502.

7 Wang, Y., Xu, Y., Peters, B.-H. et al. (2025). Development and validation of a label-free HPLC-CAD method to determine total sialic acid in therapeutic proteins. *Journal of Pharmaceutical and Biomedical Analysis* 263: 116937.

8 Vreeker, G.C.M. and Wuhrer, M. (2017). Reversed-phase separation methods for glycan analysis. *Analytical and Bioanalytical Chemistry* 409 (2): 359–378.

9 Fu, Q., Liang, T., Li, Z. et al. (2013). Separation of carbohydrates using hydrophilic interaction liquid chromatography. *Carbohydrate Research* 379: 13–17.

10 Cataldi, T.R., Campa, C., Angelotti, M., and Bufo, S.A. (1999). Isocratic separations of closely-related mono-and disaccharides by high-performance anion-exchange chromatography with pulsed amperometric detection using dilute alkaline spiked with barium acetate. *Journal of Chromatography. A* 855 (2): 539–550.

11 Rohrer, J.S. (2000). Analyzing sialic acids using high-performance anion-exchange chromatography with pulsed amperometric detection. *Analytical Biochemistry* 283 (1): 3–9.

12 Dionex (1997). Glycoprotein oligosaccharide analysis using high-performance anion-exchange chromatography. Thermo Scientific Technical Note 42.

13 Westereng, B., Arntzen, M.Ø., Aachmann, F.L. et al. (2016). Simultaneous analysis of C1 and C4 oxidized oligosaccharides, the products of lytic polysaccharide monooxygenases acting on cellulose. *Journal of Chromatography. A* 1445: 46–54.

14 Isaksen, T., Westereng, B., Aachmann, F.L. et al. (2014). A C4-oxidizing lytic polysaccharide monooxygenase cleaving both cellulose and cello-oligosaccharides*. *The Journal of Biological Chemistry* 289 (5): 2632–2642.

15 Mariño, K., Bones, J., Kattla, J.J., and Rudd, P.M. (2010). A systematic approach to protein glycosylation analysis: a path through the maze. *Nature Chemical Biology* 6 (10): 713–723.

16 Inoue, Y. and Yamamoto, A. (2014). Stationary phases for the separation of reducing sugars by normal-phase partition chromatography. *Chromatography* 35 (2): 63–72.

17 Jenkins, K. (2015). HILIC separation of carbohydrates using BEH amide particle technology. *Chromatography Today* 14–18.

18 Walter, T.H., Alden, B.A., and Berthelette, K. (2022). Evaluation of the base stability of hydrophilic interaction chromatography columns packed with silica or ethylene-bridged hybrid particles. *Separations* 9 (6): 146.

19 Qian, K., Yang, Z., Zhang, F. et al. (2018). Low-bleed silica-based stationary phase for hydrophilic interaction liquid chromatography. *Analytical Chemistry* 90 (15): 8750–8755.

20 Grembecka, M., Lebiedzińska, A., and Szefer, P. (2024). Application of HPLC coupled with a charged aerosol detector to the evaluation of fructose, glucose, sucrose, and inositol levels in fruit juices, energy drinks, sports drinks, and soft drinks. *Beverages* 10 (4).

21 Ghosh, R. and Kline, P. (2019). HPLC with charged aerosol detector (CAD) as a quality control platform for analysis of carbohydrate polymers. *BMC Research Notes* 12 (1): 268.

22 Hu, D., Han, B., Chen, C. et al. (2020). Determination of seven oligosaccharides and sucrose in Pseudostellaria heterophylla by pressurized liquid extraction and ultra-high performance liquid chromatography with charged aerosol detector and tandem mass spectrometry. *Journal of Chromatography. A* 1609: 460441.

23 Xie, Q., Tian, H., Huan, X. et al. (2022). Quality evaluation of codonopsis radix and processed products based on the analysis of monosaccharides and oligosaccharides by liquid chromatography coupled with charged aerosol detector. *Phytochemical Analysis* 33 (2): 262–271.

24 Hetrick, E.M., Kramer, T.T., and Risley, D.S. (2017). Evaluation of a hydrophilic interaction liquid chromatography design space for sugars and sugar alcohols. *Journal of Chromatography. A* 1489: 65–74.

25 Xie, S., Jin, Z., Huang, Y., and Huang, Q. (2025). Quantitative analysis of carbohydrate residues in dextran 40 from various sources: a comparative study using high-performance liquid chromatography coupled with a charged aerosol detector. *Analytical Methods* 17 (8): 1765–1773.

26 Corrado, A., De Martino, M., Bordoni, V. et al. (2023). A universal UHPLC-CAD platform for the quantification of polysaccharide antigens. *Scientific Reports* 13 (1): 10646.

27 Wang, H., Jin, H., Chai, R. et al. (2024). An analysis of polysaccharides from eight plants by a novel heart-cutting two-dimensional liquid chromatography method. *Food* 13 (8): 1173.

28 Chen, Q.-L., Chen, Y.-J., Zhou, S.-S. et al. (2018). Laser microdissection hyphenated with high performance gel permeation chromatography-charged aerosol detector and ultra performance liquid chromatography-triple quadrupole mass spectrometry for histochemical analysis of polysaccharides in herbal medicine: ginseng, a case study. *International Journal of Biological Macromolecules* 107: 332–342.

29 Li, Z., Li, S., Zhang, F. et al. (2021). A hydrolytically stable amide polar stationary phase for hydrophilic interaction chromatography. *Talanta* 231: 122340.

30 Bailey, B., Plante, M., Thomas, D. et al. (2017). Practical use of CAD: achieving optimal performance. In: *Charged Aerosol Detection for Liquid Chromatography and Related Separation Techniques* (ed. P.H. Gamache), 163–189. John Wiley and Sons.

31 Stoll, D.R. and Makey, D.M. (2019). Mobile phase buffers in LC: effect of buffer preparation method on retention repeatability. *LC-GC North America.* 37 (7): 444–449.

32 Bailey, B., Ullucci, P., Bauder, R. et al. (2014). Carbohydrate analysis using HPLC with PAD, FLD, charged aerosol detection, and MS detectors. ThermoFisher Scientific: Boston, MA, USA.

33 Taylor, M.R., Kawakami, J., and DV, M.C. (2023). Managing sample introduction problems in hydrophilic interaction liquid chromatography. *Journal of Chromatography. A* 1700: 464006.

34 Rohrer, J.S. and Kitamura, S. (2017). Determination of carbohydrates using liquid chromatography with charged aerosol detection. In: *Charged Aerosol Detection for Liquid Chromatography and Related Separation Techniques* (ed. P.H. Gamache), 311–325. John Wiley and Sons.

35 Dixon, R.W. and Baltzell, G. (2006). Determination of levoglucosan in atmospheric aerosols using high performance liquid chromatography with aerosol charge detection. *Journal of Chromatography. A* 1109 (2): 214–221.

36 Joseph, A. and Rustum, A. (2017). Applying charged aerosol detection to aminoglycosides: development and validation of an RP-HPLC method for gentamicin and netilmicin. In: *Charged Aerosol Detection for Liquid Chromatography and Related Separation Techniques* (ed. P.H. Gamache), 393–423. John Wiley and Sons.

37 Chai, W., Hounsell, E.F., Cashmore, G.C. et al. (1992). Neutral oligosaccharides of bovine submaxillary mucin: a combined mass spectrometry and 1H-NMR study. *European Journal of Biochemistry* 203 (1–2): 257–268.

38 Gabbanini, S., Lucchi, E., Guidugli, F. et al. (2010). Anomeric discrimination and rapid analysis of underivatized lactose, maltose, and sucrose in vegetable matrices by U-HPLC-ESI-MS/MS using porous graphitic carbon. *Journal of Mass Spectrometry* 45 (9): 1012–1018.

39 Russo, M., Camillo, M.R.T., La Tella, R. et al. (2024). Principles and applications of porous graphitic carbon stationary phase in liquid chromatography: an update. *Journal of Chromatography. A* 1719: 464728.

40 Zhang, Z., Khan, N.M., Nunez, K.M. et al. (2012). Complete monosaccharide analysis by high-performance anion-exchange chromatography with pulsed amperometric detection. *Analytical Chemistry* 84 (9): 4104–4110.

8

Analysis of Pharmaceutical Excipients with Charged Aerosol Detection

Ulrike Holzgrabe

Institute of Pharmacy, University of Würzburg, Wuerzburg, Germany

CHAPTER MENU

8.1 Summary

Excipients often do not have a chromophore and are complex mixtures which make their characterization and quality control difficult. Currently the quality is controlled by the determination of bulk parameters, such as acid value, hydroxyl value, etc. Only a few HPLC methods are described with regard to the characterization in more detail. Besides mass spectrometric (MS) detection, charged aerosol detection (CAD) plays a major role in HPLC analysis of excipients. The pros and cons of HPLC-CAD analysis are discussed here.

8.2 Introduction

A finished product (FP) of a drug, namely a coated or uncoated tablet, a suspension, a capsule, or an ointment, contains the active pharmaceutical ingredient (API) and typically several excipients of different chemical nature. The latter are necessary in order to form the finished product, to regulate the API release and to guarantee the stability of the API. Since a large portion of the FPs are excipients, the quality of these compounds must be ensured. The quality and its evaluation are regulated by international pharmacopoeias, such as the European Pharmacopoeia (PhEur) [1], the United States Pharmacopeia (USP) [2], and the Japanese Pharmacopoeia (JP) [3]. However, many excipients, mainly polymers, are characterized by the determination of bulk parameters derived from assays for fatty acids (FAs), characterizing the overall molecular weight, the overall composition, the degree of oxidation or saturation, etc., but none provide the detailed composition of the excipient or its purity.

Many excipients are natural products isolated from plants, such as sugars, FAs, fats, and waxes. Hence, they are always mixtures, containing not only one component or one class of components, but also ions including nitrate, nitrite, halogens, to name only a few. Furthermore, many excipients are synthesized (often from natural products) and are mixtures too. In these cases, the PhEur [1] defines the composition by a production statement, e.g., glycerol distearate: "*Mixture of diacylglycerols, mainly distearoylglycerol, together with variable quantities of mono- and triacylglycerols. It is obtained by partial glycerolysis of vegetable oils containing triacylglycerols of palmitic (hexadecanoic) and stearic (octadecanoic) acids or by esterification of glycerol with stearic acid. The fatty acids may be of vegetable or animal origin.*" This statement combines the natural origin (plants or animals) of the fatty acids, which therefore may contain other components besides stearic and palmitic acid, with the synthesis pathway which will undoubtedly result in a mixture of mono-, di-, and triacylglycerols of varying composition. The PhEur simply tests the acid, iodine and saponification values, in addition to free glycerol and the FA composition, the latter determined by size exclusion chromatography (SEC) and gas chromatography (GC), respectively.

Because of their many uses, such as in food, medicine, and cosmetics, excipients are frequently obtained from a few different providers, most of whom offer varying composition and quality. While these may all meet the requirements defined by the PhEur or another pharmacopoeia they may have different functionality-related properties [4]. However, it is clear that a more detailed and more quantitative analysis is necessary to characterize the composition and the quality of such excipients. Due to the multicomponent character of the excipients, their analysis is challenging. Furthermore, many of the aforementioned excipients do not have a chromophore, which makes UV detection difficult or even impossible. Hence, reversed phase (RP) or SEC combined with an aerosol-based detector, either an Evaporative Light Scattering Detector (ELSD) or CAD is the analytical system of choice. MS detection is also an option, but it is more expensive and tends to be difficult to operate.

In the following sections, examples of increasing complexity are presented and discussed for representative excipients, starting with simple polyethylene glycols

(PEGs), up to more complex polysorbates (PSs) and including FAs and their corresponding esters and ethers, acylglycerols, sorbitols, and sorbitol esters.

8.3 Polyethylene Glycols (Macrogols) and Derivatives

PEGs, also called macrogols, are polymers made from ethylene oxide "with the general formula H-[OCH_2-CH_2]*n*-OH where n represents the average number of oxyethylene groups" [1], with a molecular weight range from 300 to 35,000 Da. Besides the application as an excipient, PEG reagents are often used for the pegylation of proteins. PEGylated molecules generally take longer to be cleared from the body, resulting in a longer half-life. This can help to achieve improved efficacy at lower doses or longer dosing intervals. The PhEur determines the average length of the polymers by means of the hydroxyl value but does not provide any information as to the molar mass distribution—neither the broadness of the size distribution nor the concentration of the fractions of the same size. It is not even possible to differentiate between mono- and polydisperse PEGs. Since the composition of a polymer determines its physicochemical properties, a more detailed characterization is required in addition to the evaluation of non-PEG impurities.

Classical analysis of PEGs is based on either RP or SEC with a variety of detectors including differential refractive index (dRI), multi-angle light scattering (MALS), ELSD, or CAD. As presented in Figure 8.1, an Xterra™ RP18, 250–4.6 mm, 5 μm column (Waters Corp.) and a methanol/water gradient could easily separate the components of small PEGs and the ELSD was able to detect all components [5]. Similarly, Theiss et al., were able to separate PEGs from 300 to 3000 Da using a YMC-Pack™ Pro C18, 150 mm × 4.0 mm, 3 μm column (YMC Co. Ltd.) with an acetonitrile (ACN)/water gradient and detection by CAD (see Figure 8.1) [6]. Interestingly, the resolution worsened for larger PEGs. The assignment of all peaks in the chromatogram was achieved using an ion trap-MS. Similar separations of the PEG units was reported using an Acclaim™ Surfactant Plus column (Thermo Fisher Scientific) with an ammonium acetate/ACN gradient and CAD [7]. The combination of SEC and CAD was also able to characterize the quality of PEGs [8]. Recently, the Zhang group performed a comparison of dRI, MALS, and CAD. All of them are appropriate for the intended purpose [9]. Although CAD response was somewhat dependent on the molecular weight of the PEG, especially with high MWs, and was not as linear as the other detectors, CAD was more sensitive for measurement of the PEGs as well as impurities such as ions.

Poloxamers are nonionic surfactants consisting of co-polymers of oxyethylene and oxypropylene that can have different lengths and compositions of the ethylene and propylene oxide units. The PhEur [1] determines the ratio of oxypropylene and oxyethylene by means of quantitative ^1H NMR spectroscopy, but this is limited,

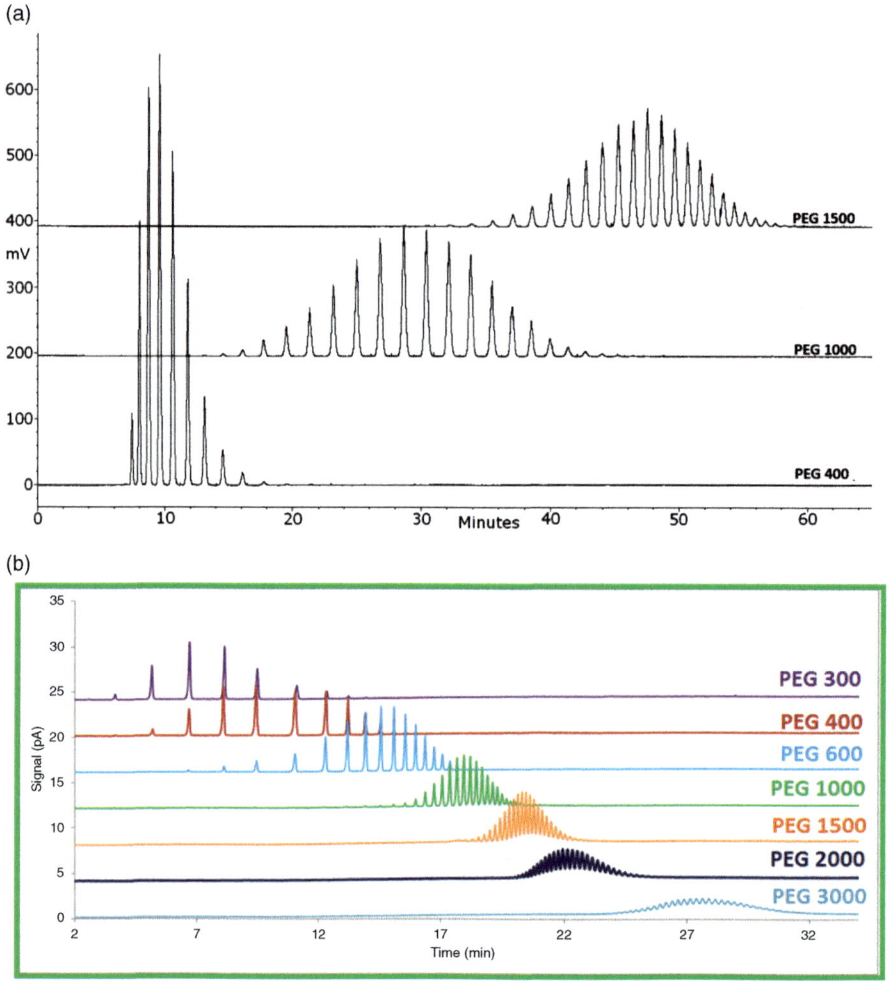

Figure 8.1 (a) Stacked chromatograms of three low molecular weight PEGs (400, 1000, and 1500), obtained with a Xterra RP 18 column (250 mm × 4.6 mm, 5 µm) and MeOH/H_2O gradient in 50 minutes (30–50%), flow rate 0.5 mL/min and an ELSD. *Source:* Brinz and Holzgrabe [5] / with permission of John Wiley & Sons. (b) Stacked chromatogram of PEGs of varying length obtained with a YMC-Pack Pro C18 (150 mm × 4.0 mm, 3 µm) and an ACN/water gradient and CAD. *Source:* Theiss and Holzgrabe [6] / with permission of Elsevier.

$$HO{-}\left[CH_2CH_2O\right]_y{-}\left[CH_2\right]_x{-}CH_3$$

X = 4 – 60
Y 12: lauryl-; 14: myristyl-; 16: cetyl/palmityl-; 18: stearyl-; 18:1 ω-9: oleyl-alcohol

Figure 8.2 General structural formula of macrogol fatty acid ethers.

as this technique is not available in every quality assurance laboratory. Bareford et al. used RP chromatography with a C3 column (Zorbax™ 300SB C3 4.6 mm × 7.5 ca., 5 μm, Agilent Technologies, Inc.) and a water/ACN/trifluoroacetic acid (TFA) gradient to separate the Poloxamer 188 from corresponding high and low molecular weight polymers, which were then measured using CAD [10]. As Poloxamer 188 and PSs are often used for formulations of monoclonal antibodies, it is of interest to characterize both in a single analysis. Recently, Zegota et al. described a high-throughput multidimensional liquid chromatography method in which the protein was removed by means of a combination of cation and anion exchange columns resulting in a mixture of isolated Poloxamer 188, PS 20, and PS 80, which could be simultaneously profiled and quantified on a RP 18 column and CAD [11]. Such methods are also important for the evaluation of polymer degradation processes in formulations. For further information on PSs in bioformulations, see Chapter 12 in this book.

In the PhEur, a couple of PEG ethers consisting of various alcohols are monographed [1]. They are always mixed PEGs with the corresponding linear fatty alcohols, mainly the ones given in the name. They also contain free PEGs (of varying lengths) and free alcohols (as impurities). The production procedure is rudimentary and is described by the number of moles of ethylene oxide reacted per mole of the corresponding alcohol. As shown by the structural formula (Figure 8.2), these nonionic surfactants do not contain a chromophore, but they are readily detectable by CAD. Using a Kinetex™ C18 column (100 mm × 3.0 mm, 2.6 μm, Phenomenex, Inc.) with a water/ACN (15/85) mobile phase, PEG lauryl ether and the free PEGs can be separated. By decreasing the ACN content to 70%, the individual units of the PEGs could be seen (see Figure 8.3) [12]. Interestingly, the PhEur describes the same surfactant as a local anesthetic API, named Lauromacrogol 400, with similar quality requirements.

Additional applications of HPLC-CAD for analysis of polydisperse PEGs and PEG-based excipients are discussed in Theiss et al. [6].

8.4 Fatty Acids and Derivatives

FAs are an important component of many excipients, especially of surfactants. As already mentioned, FAs are of natural origin, either from vegetable oils or

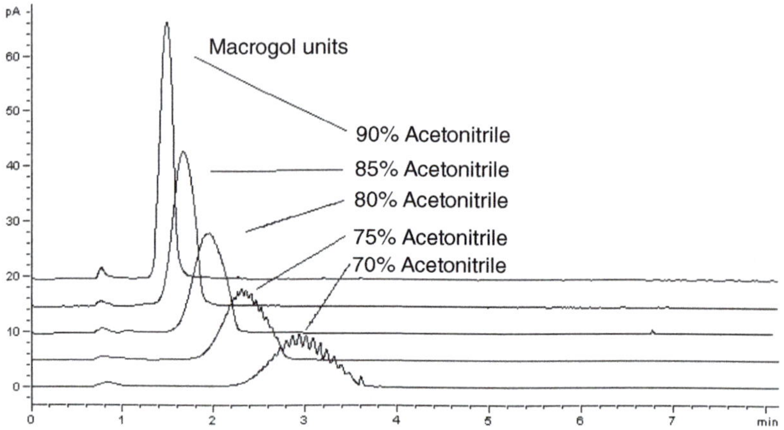

Figure 8.3 Stacked plot of polidocanol, a local anesthetic, showing the macrogol unit separation on a Kinetex C18, 100 mm × 3.0 mm, 2.6 μm column with decreasing ACN content in the mobile phase. *Source:* Ilko et al. [12] / with permission of Elsevier.

fats, or from animals. Hence, they are mixtures. The PhEur contains only four monographs on FAs: (i) caprylic acid, containing 10 FA (ester) impurities; (ii) oleic acid containing eight FAs with an amount between 5% and 18%; (iii) palmitic acid whose monograph gives only stearic acid as an impurity; (iv) stearic acid which may contain different amounts of palmitic acid [consequently, the PhEur differentiate between three qualities depending upon the amount of stearic acid present (50%, 70%, and 95%)]. The compositions are determined by GC as described in a general monograph (2.4.22. PhEur).

The determination of FA composition by means of HPLC-CAD must consider that the short-chain FAs are volatile or semi-volatile depending upon their chain length. While lauric and myristic acids are almost impossible to quantify and detect properly, longer FAs are sufficiently nonvolatile to give a quantifiable signal. The dependence of the signal heights on the evaporation temperature (T_e) was demonstrated by Schilling et al. [13] (see Figure 8.4). As expected, the signal height attenuates with increasing T_e. Since this effect is very pronounced, the T_e must be carefully chosen. Higher T_e settings reduce the baseline noise, because volatile impurities in the mobile phase and other reagents are evaporated but may impact analyte signal. The signal-to-noise ratio can be increased by increasing T_e as long as the analytes continue to behave as nonvolatiles. However, when attempting to measure shorter chain FAs and other more volatile analytes, it is recommended to use the lowest temperature setting that consistently produces the required sensitivity limits. This should provide the best reproducibility and most uniform response among analytes as further detailed in Chapter 3.

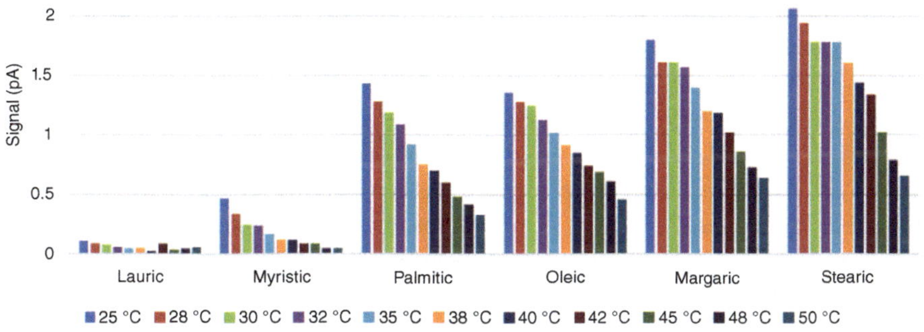

Figure 8.4 Dependence of the peak height of a selection of FAs on the evaporation temperature [Kinetex C18 column (100 mm × 2.1 mm, 2.6 μm), gradient elution with aqueous formic acid and ACN/HCOOH; injection of 10 ng on column]. *Source:* Schilling et al. [13] / with permission of Elsevier.

CAD response depends strongly on the composition of the mobile phase, especially the amount of organic solvent reaching the detector. The higher the concentration of ACN, the better the sensitivity. This holds especially true for the late-eluting acids, because the ACN amount increases due to the gradient. Alternatively, an inverse gradient might be applied in order to enhance the ACN amount before detection (see Chapter 4).

Zhao et al. demonstrated the power of RP HPLC-CAD for oleic acid profiling [14]. Using a gradient from 0.05% glacial acetic acid with ACN, an Acclaim 120 C18 column (250 mm × 4.4 mm, 5 μm) and a T_e of 35 °C, the authors were able to discover far more FAs than with the compendial GC method, especially long-chain FAs, in addition to the GC-undetectable linolenic acid. They applied an external standard thus avoiding the different response factors seen with the GC normalization method, which also suffers from losses upon derivatization and evaporation in the source. Taken together, the HPLC-CAD method extends the range of FAs measured and outperforms the compendial GC method.

For FA analysis, the precise adjustment of separation and CAD conditions is of vital importance. The analysis of quantitative structure–property relationships can be performed by machine learning approaches. The Protić group identified the solvent acetone to be superior regarding detector response compared to the environmentally more harmful ACN [15]. Applying an analytical quality by design approach, the separation of FAs can be easily optimized [16] with a flow rate of 0.575 mL/min, a column temperature of 33 °C, and a gradient with ACN from 70% to 80% within 14.2 minutes are recommended.

FA esters with polyethylene glycol are nonionic surfactants which are used in cosmetics. They are often characterized by GC and more recently by means of RP-HPLC-ELSD. Lee et al., described an HPLC-CAD method, using a Wakosil® 10 C18 column, 250 mm × 4.6 mm, 10 μm (FujiFilm Wako Chemical Corp.) with a water–ACN gradient and ELSD [17]. The separation method was similar to the HPLC-CAD method developed by Theiss et al. [6].

Fats are triacylglycerols and thus esters. They are always a mixture of mono-, di-, and tri-glycerides, "*which may be obtained either by esterification of hydrogenated FAs of vegetable origin with glycerol or by interesterification of hydrogenated vegetable oils*" as defined for hard fat by the PhEur [1]. Fats and oils, depending on whether they are solid or liquid, are often used in capsules, ointments, cosmetics, and food, but also in washing powder, cleaning agents, and chocolate, which is especially true for palm oil.

As early as 2011, de al Mata-Espinosa discriminated between non-olive blended oils and olive oils by means of an HPLC-CAD method and the use of chemometrics. Using a LiChrospher® C18 column (250 mm × 4 mm, 5 μm; Merck KGaA), elution with an ACN and hexane-isopropanol gradient, and CAD, they were able to differentiate the different oils via principal components and linear discriminant analysis [18]. Broader studies by the same group, applying almost the same HPLC method, were aimed at the authentication of the geographical origin of olive and palm oil, respectively [19, 20].

With the introduction of small interfering RNA and messenger-RNA therapeutics, lipid-based nanoparticles are becoming increasingly important. These consist of complex triglycerides such as dimyristoyl-sn-glycero-3-phosphoethanolamine-*N*-[methoxy(polyethylene glycol)-2000] ammonium salt (DMPE-PEG 2000), which is also a mixture. The stability of such excipients can be easily analyzed by means of the RP18-HPLC-CAD method using a Hypersil Gold™ PFP column, 150 mm × 4.6 mm, 3.0 μm (Thermo Fisher Scientific), and a gradient formed by an acidified water/methanol mixture and ACN [21]. Further examples of such drug delivery systems can be found in Chapter 11 on lipid-based formulations.

An even more complex system is polysorbates (PSs), containing PEGs of different length, free FAs, fatty alcohols, sorbitols, and derivatives thereof. In the PhEur [1], they are characterized by the above-mentioned bulk parameters, a GC method for the composition of FAs, and head-space GC for the limitation of ethylene oxide and dioxan. Besides the fact that the mixtures are not clearly separated and characterized, the PhEur keeps the analysts rather busy. However, as already mentioned, the full characterization of PSs can be done in one run [6]. More HPLC-CAD methods for PSs are described in Chapter 12.

8.5 Carbohydrate Polymers and Derivatives

Carbohydrates are a major component of plants and their extracts. The various free sugars, mono-, di-, oligo-, and polysaccharides are often analyzed by means of HPLC-CAD, as some representative studies demonstrate [22–24]. Carbohydrates

Figure 8.5 Structural skeleton of cellulose, which can have a varying length and be substituted by hydrogen, methyl, ethyl, or hydroxypropyl groups.

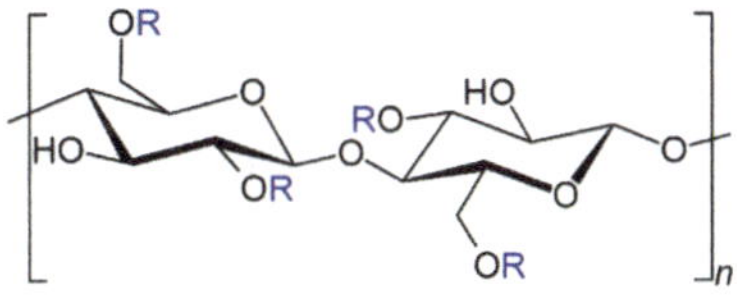

are a kind of hybrid between active pharmaceutical ingredients, food ingredients, and excipients. Hence, a quality control is mandatory. On the one hand, an analysis of mixture compositions must be performed, and on the other hand, the content of various free sugar units must be determined. Hydrophilic interaction chromatography (HILIC) is especially useful for the separation of sugars and is perfect for detection by CAD, because no derivatization is necessary. Recently, some applications applying HILIC-CAD were published, e.g., [25–27]. Further information on carbohydrate analysis is given in Chapter 7 of this book.

Hydroxypropylmethylcellulose (HPMC), also called hypromellose (see Figure 8.5), is always a mixture and can occur in different types, varying in the percentage of methoxy and hydroxypropoxy groups. The PhEur [1] determines the percentage of both groups as iodides by splitting off the groups with hydriodic acid and measuring the isopropyl iodide and methyl iodide content by means of GC. This method considers neither the locus of alkyl substitution nor the composition of hypromellose. It gives only the average substitution. As early as 2011, the Schoenmaker's group tried to characterize hypromellose by 2D-chromatography using RP in one dimension and SEC in the second dimension with CAD [28]. They visualized the data using Matlab® (The MathWorks, Inc.) software. However, later the group studied polymeric excipients by applying MS detection only.

It is worth mentioning that the Acworth group at Thermo Fisher Scientific was able to quantify HPMC in milk products by using a non-porous Presto FF-C18 (2.0 μm, 150 mm × 4.6 mm) column (Imtakt, Corp), using gradient elution with water, ACN, and isopropanol and CAD [29].

Sucrose esters are monographed in international pharmacopoeias. E.g., sucrose stearate (see Figure 8.6) is a "*mixture of sucrose esters, mainly sucrose stearate,*

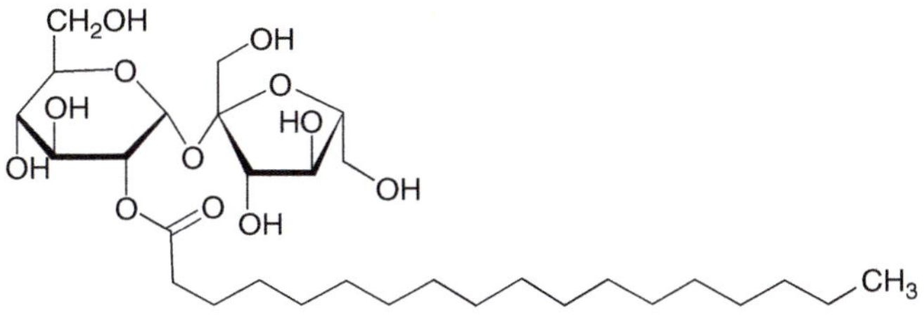

Figure 8.6 Structural formula of a sucrose ester, showing the esterification of sucrose with stearic acid in one position.

obtained by transesterification of stearic (octadecanoic) acid methyl esters of vegetable origin with sucrose." The manufacture of FA methyl esters includes a distillation step. A corresponding monopalmitate is also monographed. The PhEur [1] determines the FA composition by means of GC, free sucrose by HPLC, and the percentage of monoesters, diesters and the sum of tri- and polyesters by means of SEC using dRI, which is characterized by low robustness. However, the Pedersen group was able to separate and annotate the regioisomers of sucrose caprate and laurate by RP-HPLC using a Symmetry® C18, 250 mm × 4.6 mm, 5 μm column (Waters Corp), an isocratic elution with water/ACN acidified with 0.05% TFA, and a CAD detector [30, 31]. This might be a good starting point to revise the laborious pharmacopoeia monographs of sucrose stearate and palmitate.

8.6 Conclusions

Since CAD is a kind of universal detection technique, it is ideal for the qualitative and quantitative analysis of excipients. Both the characterization of excipient composition (being often mixtures of natural products) and impurity profiling [including synthesis by- and starting-products as well as degradation (hydrolysis) products] are easy to perform. In comparison to detection by UV and ELSD, CAD is more sensitive. Only in very few cases CAD cannot be used because of analyte volatility, e.g., short length FAs such as myristic acid. Despite the advantages of CAD, it is not often used in the quality control of excipients, which might be due to the old-fashioned and also outdated test methods of the major pharmacopoeias, which must be urgently replaced by modern HPLC-CAD methods. Moreover, such HPLC-CAD methods would provide an almost holistic picture of the quality of the excipients.

References

1 European Pharmacopoeia, Council of Europe (2023). 11th Ed., European Directorate for the Quality of Medincines and Healthcare., Strasbourg.

2 United States Pharmacopeia (USP) (2024). USP-NF, Maryland, USA. https://www.usp.org/.

3 Japanese Pharmacopoeia (2021). 18th Ed., Pharmaceutical and Medical Agency Tokyo, Japan. https://www.pmda.go.jp/english/rs-sb-std/standards-development/jp/0029.html.

4 Braun, A.C., Ilko, D., Merget, B. et al. (2015). Predicting critical micelle concentration and micelle molecular weight of polysorbate 80 using compendial methods. *European Journal of Pharmaceutics and Biopharmaceutics* 94: 559–568.

5 Brinz, D. and Holzgrabe, U. (2008). Analysis of the molecular size distribution of polyethylene glycols by CE-UV versus HPLC with evaporative light scattering detection. *Electrophoresis* 29 (17): 3605–3611.
6 Theiss, C. and Holzgrabe, U. (2018). Characterization of polydisperse macrogols and macrogol-based excipients via HPLC and charged aerosol detection. *Journal of Pharmaceutical and Biomedical Analysis* 160: 212–221.
7 https://assets.thermofisher.com/TFS-Assets/CMD/Application-Notes/ai-71101-hplc-cad-surfactants-emulsifiers-ai71101-en.pdf (accessed 25 February 2025).
8 Kou, D., Manius, G., Zhan, S., and Chokshi, H.P. (2009). Size exclusion chromatography with corona charged aerosol detector for the analysis of polyethylene glycol polymer. *Journal of Chromatography. A* 1216 (28): 5424–5428.
9 Wang, J., Tang, S., and Zhang, K. (2024). Quantitation of polyethylene glycol by size exclusion chromatography with charged aerosol, differential refractive index, and multi-angle light scattering detectors. *Journal of Pharmaceutical and Biomedical Analysis* 238: 115854.
10 Bareford, L., Peng, H., Ali, A. et al. (2019). Development of a rapid and reliable analytical method for screening poloxamer 188 for use in cell culture process. *Biotechnology Progress* 35 (3): e2792.
11 Zegota, M.M., Schuster, G., De Pra, M. et al. (2024). High throughput multidimensional liquid chromatography approach for online protein removal and characterization of polysorbates and poloxamer in monoclonal antibody formulations. *Journal of Chromatography. A* 1720: 464777.
12 Ilko, D., Puhl, S., Meinel, L. et al. (2015). Simple and rapid high performance liquid chromatography method for the determination of polidocanol as bulk product and in pharmaceutical polymer matrices using charged aerosol detection. *Journal of Pharmaceutical and Biomedical Analysis* 104: 17–20.
13 Schilling, K., Pawellek, R., Lovejoy, K. et al. (2018). Influence of Charged Aerosol Detector instrument settings on the ultra-high-performance liquid chromatography analysis of fatty acids in polysorbate 80. *Journal of Chromatography. A* 1576: 58–66.
14 Zhao, X., He, Y., Chen, J. et al. (2021). Identification and direct determination of fatty acids profile in oleic acid by HPLC-CAD and MS-IT-TOF. *Journal of Pharmaceutical and Biomedical Analysis* 204: 114238.
15 Pawellek, R., Krmar, J., Leistner, A. et al. (2021). Charged aerosol detector response modeling for fatty acids based on experimental settings and molecular features: a machine learning approach. *Journal of Cheminformatics* 13 (1): 53.
16 Walther, R., Krmar, J., Leistner, A. et al. (2023). Analytical quality by design: achieving robustness of an LC-CAD method for the analysis of non-volatile fatty acids. *Pharmaceuticals (Basel)* 16 (4): 478.

17 Lee, Y.H., Jeong, E.S., Cho, H.E., and Moon, D.-C. (2008). Separation and determination of polyethylene glycol fatty acid esters in cosmetics by a reversed-phase HPLC/ELSD. *Talanta* 74 (5): 1615–1620.

18 De la Mata-Espinosa, P., Bosque-Sendra, J., Bro, R., and Cuadros-Rodriguez, L. (2011). Discriminating olive and non-olive oils using HPLC-CAD and chemometrics. *Analytical and Bioanalytical Chemistry* 399 (6): 2083–2092.

19 Obisesan, K.A., Jiménez-Carvelo, A.M., Cuadros-Rodriguez, L. et al. (2017). HPLC-UV and HPLC-CAD chromatographic data fusion for the authentication of the geographical origin of palm oil. *Talanta* 170: 413–418.

20 Vera, D.N., Jiménez-Carvelo, A.M., Cuadros-Rodríguez, L. et al. (2019). Authentication of the geographical origin of extra-virgin olive oil of the Arbequina cultivar by chromatographic fingerprinting and chemometrics. *Talanta* 203: 194–202.

21 Kurmi, M., Suryavanshi, V., Panduranga, N.S. et al. (2021). Development of HPLC-CAD stability indicating assay method for polyethylene glycol-conjugated phospholipid (DMPE-PEG 2000) and identification of its degradation products. *Journal of Pharmaceutical and Biomedical Analysis* 198: 113967.

22 Chen, L., Hu, D., Liang, X. et al. (2018). Preparation and identification of oligosaccharides in lotus seeds and determination of their distribution in different parts of lotus. *Electrophoresis* 39 (15): 2020–2028.

23 Jabłońska-Ryś, E., Sławińska, A., Skrzypczak, K., and Goral, K. (2022). Dynamics of changes in pH and the contents of free sugars, organic acids and LAB in button mushrooms during controlled lactic fermentation. *Food* 11 (11): 1553.

24 Liu, Y., Song, J., Liu, S. et al. (2024). A universal method for profiling and characterization of oligosaccharides in traditional Chinese medicines. *Journal of Pharmaceutical and Biomedical Analysis* 244: 116129.

25 Ghosh, R. and Kline, P. (2019). HPLC with charged aerosol detector (CAD) as a quality control platform for analysis of carbohydrate polymers. *BMC Research Notes* 12 (1): 268.

26 Hetrick, E.M., Kramer, T.T., and Risley, D.S. (2017). Evaluation of a hydrophilic interaction liquid chromatography design space for sugars and sugar alcohols. *Journal of Chromatography. A* 1489: 65–74.

27 Pitsch, J. and Weghuber, J. (2019). Hydrophilic interaction chromatography coupled with charged aerosol detection for simultaneous quantitation of carbohydrates, polyols and ions in food and beverages. *Molecules* 24 (23): 4333.

28 Greiderer, A., Steeneken, L., Aalbers, T. et al. (2011). Characterization of hydroxypropylmethylcellulose (HPMC) using comprehensive two-dimensional liquid chromatography. *Journal of Chromatography. A* 1218 (34): 5787–5793.

29 Plante, M., Bailey, B., and Acworth, I. (2016). Analysis of emulsifiers in foods by high pressure liquid chromatography and corona charged aerosol detection. https://assets.thermofisher.com/TFS-Assets/CMD/posters/PN-70995-AOAC2014-Emulsifiers-Food-PN70995-EN.pdf (accessed 25 February 2025).

30 Lie, A. and Pedersen, L.H. (2013). Elution strategies for reversed-phase high-performance liquid chromatography analysis of sucrose alkanoate regioisomers with charged aerosol detection. *Journal of Chromatography. A* 1311: 127–133.

31 Lie, A., Wimmer, R., and Pedersen, L.H. (2013). Design of experiments and multivariate analysis for evaluation of reversed-phase high-performance liquid chromatography with charged aerosol detection of sucrose caprate regioisomers. *Journal of Chromatography. A* 1281: 67–72.

9

Application of Charged Aerosol Detection in Traditional Herbal Medicines Analysis

Baiping Ma[1], *Ping Hu*[2], *Haizhen Liang*[1], *Yan Jin*[3], *and Guobin Shen*[3]

[1] *Beijing Institute of Radiation Medicine, Beijing, China*
[2] *School of Chemistry and Molecular Engineering, East China University of Science and Technology, Shanghai, China*
[3] *Thermo Fisher Scientific (China) Co., Ltd., Chromatography Mass Spectrometry Division, Shanghai, China*

CHAPTER MENU

9.1 Introduction

THMs have a history spanning thousands of years and have gained considerable global attention in recent years, coinciding with the rising popularity of natural therapies. In China, compound prescriptions consisting of various herbal components, known as Traditional Chinese Medicines (TCMs), constitute a significant part of traditional medicine, featuring unique theoretical systems and extensive clinical applications. Typically, TCM formulas contain numerous herbs, each comprising multiple active compounds. These compounds may synergistically act through different pathways and multiple targets within the human body, enabling the treatment of complex diseases, the management of chronic illnesses, and the prevention of disease through holistic regulation. Although generally recognized for their safety and minimal side effects, herbal medicines are not entirely devoid of toxicity. Therefore, establishing rigorous approval procedures, stringent quality standards, comprehensive safety and efficacy evaluations, and robust quality control systems for THMs/TCMs has become increasingly critical. These measures necessitate advanced analytical technologies.

As previously mentioned, THMs/TCMs contain various bioactive compounds, often present at low concentrations, embedded in complex matrices that significantly complicate their quantification. Thus, analytical methods capable of simultaneously analyzing multiple components with high sensitivity and selectivity, such as chromatographic techniques, are essential for quality evaluation and control. Among chromatographic methods, HPLC is predominantly employed in THMs/TCMs analysis, primarily because most active components are non-volatile. The detector is an essential component of LC systems. UV detectors are widely favored due to their high sensitivity, broad linear range, operational simplicity, and stability [1–3]. However, compounds lacking UV-absorbing chromophores cannot be detected by conventional HPLC-UV setups, necessitating universal detectors. Several universal detectors compatible with HPLC include MS, nuclear magnetic resonance spectrometry (NMR), refractive index detection (RID), chemiluminescent nitrogen detection (CLND), evaporative light scattering detection (ELSD), condensation nucleation light scattering detection (CNLSD), and CAD. RID, based on changes in refractive index, is widely used due to its non-specificity, but suffers from low sensitivity, incompatibility with gradient elution, and susceptibility to minor temperature and pressure fluctuations [4]. MS is highly sensitive, can provide structural information, and has become a broadly applicable detection system [5, 6]. However, due to varying ionization efficiencies among different compounds, the response factors of MS are inconsistent. Additionally, MS is costly, unsuitable as routine instrumentation, and requires highly trained operators. CLND offers exceptionally high sensitivity and is often regarded as a universal detector [7]. However, CLND is only responsive to nitrogen-containing analytes, and the mobile phase must be nitrogen-free; hence, it cannot use acetonitrile as an organic modifier or ammonium-based additives in HPLC eluents. ELSD, CNLSD, and CAD are all aerosol-based universal HPLC detectors suitable for measurement of all non-volatile analytes and many semi-volatile analytes, significantly advancing the detection of UV-transparent compounds [8]. ELSD, introduced earliest among these detectors, has been effectively applied in THMs/TCMs quantitative analysis [9–12]. But it exhibits limited reproducibility, poor sensitivity, narrow linear, and dynamic response ranges under certain conditions, marked inter-analyte response differences and response changes with solvent composition. CNLSD, commercially known as nano-quantity analyte detector (NQAD), operates similarly to ELSD but with superior sensitivity and linear range [13]. Nevertheless, analyte response consistency is relatively poor.

CAD is an aerosol-based LC detector developed for the determination of non-volatile and many semi volatile analytes. As a universal detector, CAD offers greater sensitivity than ELSD, possesses a broad dynamic range (approximately four orders of magnitude), high precision, and response consistency independent of analytes' optical properties or ionization capability in the gas phase [14]. These characteristics, alongside its reliability and ease of operation, make CAD an exceptional detector widely applied in HPLC analysis [15]. Furthermore, CAD is compatible with various chromatographic separation modes, including reversed-phase chromatography [16], mixed-mode chromatography [17], hydrophilic interaction chromatography [18, 19], supercritical fluid chromatography [20], and size exclusion chromatography [21]. It supports both standard and narrow-bore columns and gradient elution, demonstrating versatility in analyzing diverse compounds. While CAD shows consistent response factors under isocratic conditions for non-volatile analytes, response variability under gradient elution conditions may restrict its application scope. However, as discussed below, this response issue can be overcome using inverse gradient compensation (see also Chapter 4). Additionally, the response diminishes significantly for semi-volatile analytes depending on their volatility.

9.2 Basic Ideas for Method Development of HPLC-CAD Analysis

9.2.1 Factors Affecting CAD Sensitivity

9.2.1.1 Effect of Mobile Phase Composition and Gradient on CAD Sensitivity

According to CAD detection principles (see Chapter 1), the response is proportional to particle diameter, which depends on initial droplet size and analyte concentration as shown in Equation 9.1.

$$d_p = d_d\left(\frac{c}{\rho_p}\right)^{1/3} \tag{9.1}$$

where ρ_P is particle bulk density (partly determined by analyte density), c is analyte concentration, d_p is particle diameter, and d_d is droplet diameter. Droplet diameter depends on several factors, including the density and viscosity of the mobile phase, both of which are influenced by its composition.

Higher organic content in the mobile phase increases nebulizer transfer efficiency, thus enhancing particle delivery to the detection chamber and resulting in higher signals. During gradient elution, changes in organic modifiers alter mobile phase composition during different gradient stages, affecting aerosol formation and droplet evaporation. Variations in droplet and particle transmission efficiency within the detector may result in a 5- to10-fold difference in response for individual analytes [22], impacting CAD response consistency.

To evaluate how varying concentrations of mobile phase modifiers affect CAD sensitivity, Eom et al. [23] tested different concentrations of ammonium acetate buffer, ammonium formate buffer, acetic acid, and formic acid. The results showed improved detector sensitivity when the buffer concentration in the mobile phase decreased below a certain threshold.

Thus, using aerosol-based detectors (e.g., CAD or ELSD) in gradient analysis affects response consistency and sensitivity. Employing an inverse gradient compensation technique can address this issue through stabilizing evaporator conditions and enhancing response consistency by counterbalancing the organic solvent and modifier concentrations before the detector. Gorecki et al. [24] employed a second pump to deliver an exact inverse gradient compared to the analytical HPLC system, and both flows were mixed in a tee piece before introduction to the CAD. After applying mobile phase compensation, highly consistent calibration curves were obtained for six sulfonamide drugs. The proposed approach allows quantification of unknown compounds, e.g., in pharmaceutical mixtures. Our lab compared the calibration curves obtained using only the analytical gradient with those generated with the inverse gradient applied. Figure 9.1 illustrates the inverse gradient effectively normalized the peak area relative to the analytical gradient.

9.2.1.2 Effect of Mobile Phase Flow Rate on CAD Sensitivity

CAD is a typical mass-sensitive detector; peak heights increase with higher flow rates of mobile phase within the HPLC ranges (see Chapter 1). However, non-volatile impurities in the mobile phase result in higher baseline currents and noise at higher flow rates. Thus, the influence of flow rate on the signal-to-noise ratio (SNR) is bidirectional and should be experimentally optimized. Eom et al. [23] evaluated flow rates of 0.8, 1.0, and 1.2 mL/min using 0.1 mmol ammonium acetate (pH 4.0)-acetonitrile, finding the highest sensitivity at 1.0 mL/min.

9.2.1.3 Effect of Evaporator Temperature on CAD Sensitivity

Evaporator temperature affects analyte distribution between gas and condensed phases, directly influencing CAD sensitivity. For instance, higher temperatures significantly reduce CAD signals for semi-volatile long-chain fatty acids. Baseline noise is also temperature-dependent; hence, evaporator temperature requires experimental optimization [25]. Temperatures between 35 and 50 °C are suitable for most active components in TCMs, but generally, lower evaporator temperatures enhance reproducibility and response consistency.

9.2.1.4 Effect of Column Bleed on CAD Sensitivity

Column bleed is a common phenomenon in HPLC. Since the substances eluted from an HPLC column are typically non-volatile and CAD generates a universal response independent of the molecular structure, the column bleed contributes to

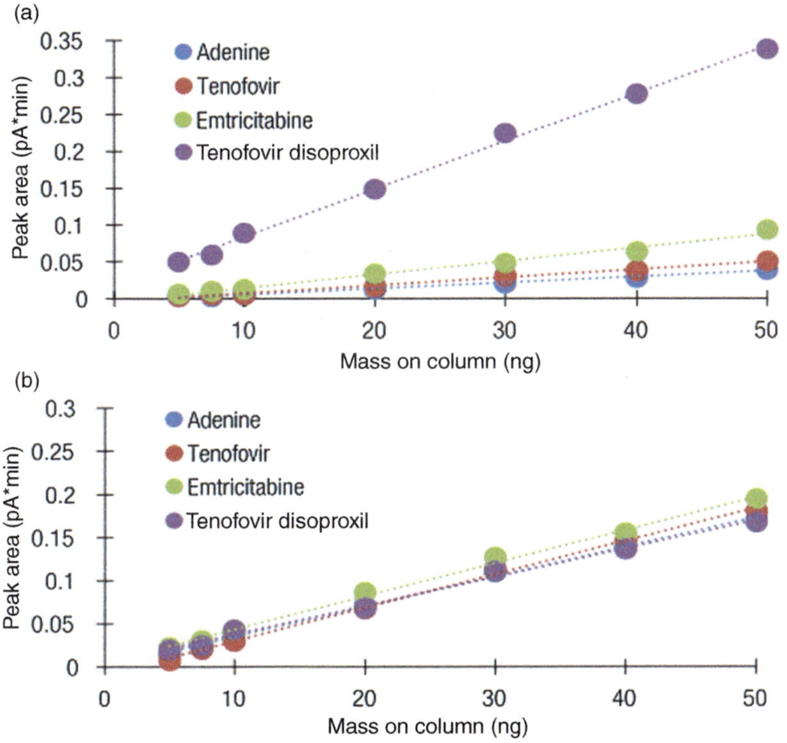

Figure 9.1 Calibration curves for tenofovir disoproxil and impurities either (a) with analytical gradient only and (b) with inverse gradient.

baseline noise and adversely affects the SNR of CAD. Certain stationary phases, such as amino-, amide-, and cyano-bonded silica gel, have been observed to exhibit significant bleed throughout the operating range of the mobile phase when monitored by CAD. Teutenberg et al. [26] conducted experiments to evaluate the applicability of UV and CAD for the online monitoring of the bleed of five HPLC columns (Luna C18, Hypercarb, ZirChrom Carb, PLRP-S, and TiO_2-Carb) at 200 °C. The results showed that the carbon-clad titanium dioxide column exhibited the lowest bleed, whereas the silica-based C18 stationary phase demonstrated the highest bleed due to susceptibility of the silica support to hydrolysis at high temperatures.

9.2.2 Application of Fingerprint Chromatogram and Characteristic Chromatogram of TCM

The fingerprint profile of TCMs systematically characterizes their complex chemical composition using chromatographic, spectroscopic, or other analytical techniques, reflecting their overall chemical characteristics. In contrast, the characteristic chromatogram focuses specifically on key or representative components within TCMs, highlighting distinct chemical markers. Both fingerprint and characteristic chromatograms serve as contemporary quality control techniques for TCMs, addressing challenges arising from their chemical complexity and the difficulty in quality control. These chromatographic profiles facilitate the transition from traditional empirical identification to objective quantitative analysis, ensuring batch consistency, authenticity verification, and standardized production.

HPLC is the most commonly utilized technique for generating fingerprint and characteristic chromatograms of TCMs. Although UV detection is commonly used in both methods, UV-transparent active components in TCMs remain undetectable in HPLC-UV chromatograms, undermining their systematic and comprehensive characteristics. Recently, HPLC-CAD fingerprint and characteristic chromatograms have been increasingly applied to quality evaluation of herbal materials, prepared herbal pieces, formulated granules, TCM compound prescriptions, and natural products. For example, our lab developed a UPLC-UV-CAD dual-detection fingerprint method for Lulutong, the dry and mature fruit of *Liquidambar formosana* Hance, which effectively fused chemical information on triterpenoids, flavonoids, and organic acids (Figure 9.2). The similarity of fusion fingerprint chromatograms from 18 batches of Lulutong ranged between 0.935 and 0.992, indicating high consistency across different origins.

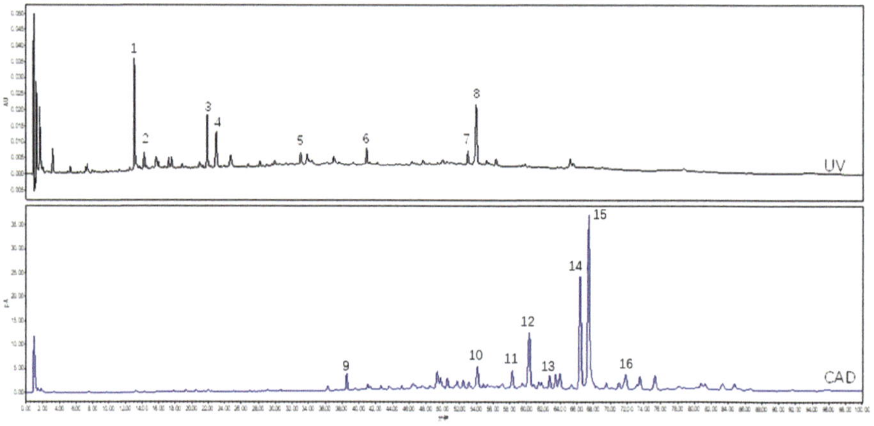

Figure 9.2 UPLC-UV-CAD dual-detection fingerprint chromatogram of Lulutong (Top: UPLC-UV fingerprint chromatogram. 1. Ellagic acid; 7. alisol C-23 acetate; 8. liquidambolide; bottom: UPLC-CAD fingerprint chromatogram, 13. oleanolic acid and ursolic acid; 14. liquidambaric acid; 15. oleanonic acid, and ursonic acid).

Zhang et al. [27] developed HPLC-CAD fingerprint chromatograms for the water-soluble non-saponin fractions of *Panax notoginseng, Panax ginseng, and Panax quinquefolius*. The analysis was performed on a Hypercarb porous graphitized carbon column (3.0 mm × 150 mm, 3 μm) using acetonitrile and 0.1% perfluoropentanoic acid as the mobile phase at a flow rate of 0.8 mL/min. The evaporator temperature and nitrogen pressure for CAD were set at 50 °C and 60.1 psi (1 psi ≈ 6.895 kPa), respectively. The chemical profiles of the water-soluble fractions from these herbs exhibited overall similarities, although the relative contents of dencichine and other specific components varied significantly. The results showed that the systematic characterization of the water-soluble non-saponin fractions is essential for the comprehensive quality assessment of these medicinal herbs.

Our lab studied the chemical profile of a traditional Chinese herbal pair named GX, composed of *Trichosanthis fructus* (Gualou) and Allii macrostemonis bulbus (Xiebai), using HPLC-CAD. A comparison of fingerprint chromatograms acquired by HPLC-CAD and HPLC-ELSD is shown in Figure 9.3. The HPLC-CAD fingerprint chromatogram revealed additional chromatographic peaks that were not detectable with ELSD due to weak signal intensity, demonstrating significantly higher sensitivity of CAD detector, particularly for fatty acids. These results indicate that CAD is more suitable for fingerprint analysis of the GX herbal pair.

TCM formulated granules are modern pharmaceutical preparations produced from traditional herbal slices through processes such as extraction, separation, concentration, drying, granulation, and packaging. These standardized granules are characterized by uniform specifications, dosages, and quality standards. They enable clinical practitioners to conveniently apply individualized treatments without decoction, ensuring safety, hygiene, and ease of use. Establishing quality standards for formulated granules is crucial for ensuring product quality, advancing industry standardization, and promoting the internationalization of TCM. The Chinese Pharmacopoeia Commission has issued national standards for over 300 types of TCM formulated granules. CAD detectors are widely applied in developing HPLC or UHPLC characteristic chromatograms and in quantitative assays of these granules, as detailed in Table 9.1.

In summary, CAD has been recognized as a universal detector with a lower detection limit compared to ELSD. Regardless of whether the components of TCM possess UV absorption, all semi volatile and nonvolatile substances, commonly analyzed by HPLC, can be effectively detected by CAD and visualized in the fingerprint or characteristic chromatograms. This approach allows a more comprehensive representation of the overall chemical profile of TCMs. Furthermore, the combination of CAD and UV detectors in series to generate dual-detection fingerprints was demonstrated to be a promising strategy for the analysis of TCMs.

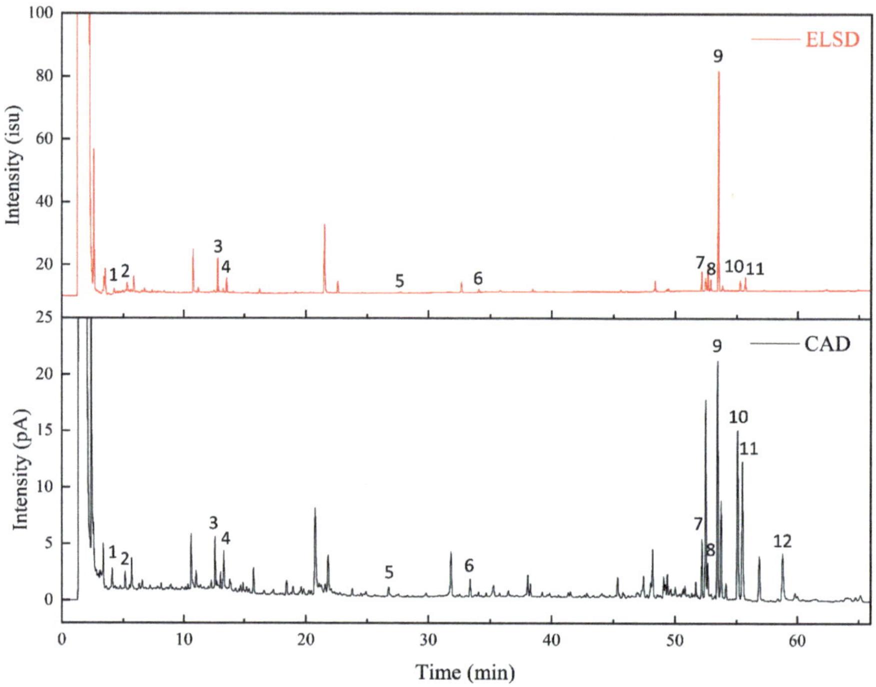

Figure 9.3 Comparison of HPLC-ELSD and HPLC-CAD fingerprints of the GX herbal pair. 1. Phenylalanine; 2. tryptophan; 3. macrostemonoside I; 4. macrostemonoside P; 5. 5β-3-*O*-β-D-glucopyranosyl-furost-26β-ol-3-*O*-acetyl-*O*-β-D-glucopyranosyl-(1 → 2)-*O*-[β-D-glucopyranosyl-(1 → 3)]-*O*-β-D-glucopyranosyl-(1 → 4)-β-D-galactopyranoside; 6. macrostemonoside E; 7. linolenic acid; 8. palmitoleic acid; 9. linoleic acid; 10. palmitic acid; 11. oleic acid; 12. stearic acid.

9.2.3 Quantitative and Semiquantitative Application in TCM

9.2.3.1 Quantitative Analysis of Multi-components with a Single Marker

TCMs contain diverse and complex ingredients, and evaluating their quality based on only one or two markers is often insufficient to reflect their true composition. Therefore, quality evaluation of TCMs is progressively moving toward simultaneous monitoring of multiple components. However, this faces challenges related to the availability and cost of reference standards. Quantitative analysis of multi-components with a single marker (QAMS) was proposed to address this issue, employing internal functional and proportional relationships among active components in TCMs [28–30]. It utilizes one component as an internal reference to simultaneously determine multiple components and has been increasingly adopted for quality assessment of various TCMs, including incorporation into the

Table 9.1 Application of CAD in quality standards of formulated granules.

Formulated granules	Application and detected compounds	Standard level
Ophiopogon japonicus (Chuan Maidong)	Characteristic chromatogram: ophiopogonin ophiopogon saponin, ophiopogonin D, polyphyllin VI	National
Ophiopogon japonicus (Zhe Maidong)	Characteristic chromatogram: ophiopogonin, peak 6 (S2): ophiopogonin D	National
Celosiae Semen	Characteristic chromatogram and quantification: celoside H, celoside I	National
Salt-processed *Morinda officinalis*	Characteristic chromatogram: oligosaccharides, sucrose, kestose, nystose; quantification: nystose	Provincial
Polygonatum odoratum	Characteristic chromatogram: fructose, glucose, sucrose; quantification: fructose	Provincial
Ilex cornuta	Quantification: sanguisorbin I	National
Homalomena occulta	Quantification: sucrose	Provincial

Chinese Pharmacopoeia. For gradient methods, the use of IGC is recommended to normalize response between components (Chapter 4).

Zhu et al. [31] established a QAMS method using HPLC-CAD for simultaneous determination of sucrose and oligosaccharides in *Pseudostellaria heterophylla*. The relative errors (REs) between the improved QAMS (iQAMS, different markers for the gradient elution program) and the external standard method (ESM) were below 3.00%. However, significant differences appeared between iQAMS and conventional QAMS (one marker for the whole gradient program), suggesting the necessity of method improvement, especially for gradient elution affecting analyte responses. The accuracy, precision, reproducibility, and stability of the iQAMS method showed satisfactory validation results, indicating its suitability as a substitute for ESM in quantitative analyses.

Yao et al. [32] developed an HPLC-CAD-based QAMS method for the simultaneous quantification of astragaloside, astragaloside II, astragaloside IV, calycosin-7-*O*-β-D-glucoside, formononetin, and 7,2′-dihydroxy-3′,4′-dimethoxyisoflavan in *Astragalus membranaceus*. Astragaloside II was used as the internal reference, yielding relative correlation factors (RCF) values of 0.561, 0.835, 0.299, 0.796, and 0.799 for the other components, respectively. The results verified the feasibility and reproducibility of the QAMS method, showing no significant difference from ESM. Although all five RCFs deviated from 1, this was possibly due to the lack of inverse gradient compensation techniques. Nonetheless, RCF reproducibility remained unaffected by gradient elution.

Zhao et al. [33] developed an HPLC-CAD method with QAMS to determine four lignans in *Magnoliae Flos* extract. The results showed that RCFs of the analytes were close to 1, indicating that the response signal was independent of the chemical structure.

Wu et al. [34] developed an HPLC-CAD and QAMS approach to assay nine components (Danshensu, protocatechuic aldehyde, calycosin-7-*O*-β-D-glucoside, rosmarinic acid, notoginsenoside R_1, salvianolic acid B, ginsenoside Rg_1, salvianolic acid A, ginsenoside Rb_1) of Qishen Yiqi dripping pills (QSYQ). The mobile phase compensation method was applied to maintain a constant mobile phase composition in the CAD detector. They used an inverse gradient where mobile phases were separately passed through two pumps and two columns and subsequently recombined using a mixing tee before entering the CAD detector, maintaining 50% solvent B throughout the run. Compared to the linear method and slope method, the multi-point method had more accurate calculation results and was selected to calculate RCFs. Except for the relatively volatile protocatechuic aldehyde, the RCFs relative to Rg_1 ranged from 0.898 to 1.381 for other substances, suggesting that mobile phase compensation effectively achieved response consistency. A robustness test showed that the RSDs of the RCFs were all less than 6.87%. Relationships between analyte properties and RCF and response factor (RF) in CAD were studied, indicating preliminarily that RCF and RF had no obvious relationship with molecular weight. Compounds with boiling points above 480.30 °C were found to be more suitable for analysis with QAMS based on HPLC-CAD. Finally, validation results confirmed that the QAMS method based on HPLC-CAD was successfully established, demonstrating satisfactory repeatability, accuracy, and cost-efficiency for multi-component quantitative analysis of TCM.

9.2.3.2 Semiquantitative Method

Zheng et al. [35] established and validated a semiquantitative method using HPLC-CAD for the determination of caffeic acid, vitexin-2″-*O*-rhamnoside, hyperoside, euscaphic acid, and ursolic acid in hawthorn leaves. To evaluate the feasibility of the semiquantitative method, the RF and the RCFs (with hyperoside as the reference) of five analytes were compared before and after inverse gradient compensation. RSD of the RFs for these five components was significantly reduced from 46.87% to 10.69% after applying inverse gradient compensation, indicating improved consistency of CAD responses. However, due to the exceptionally high response of euscaphic acid, its relative correction factor was only 0.747, suggesting the semiquantitative method was unsuitable for this compound when relying solely on inverse gradient compensation. The influence of power function values (PFV) on RFs was further investigated. Within the PFV range of 0.7–1.5, responses gradually approached consistency as PFV decreased. It was speculated that euscaphic acid's three hydroxyl groups might attract more positive charges during aerosol charging, resulting in elevated detector response. Decreasing the PFV improved both linearity and response consistency. Ultimately, a PFV of 0.7 was selected, and the semiquantitative method was successfully applied to analyze hawthorn leaves harvested at different times using HPLC-CAD combined with inverse gradient compensation.

Similarly, Shi et al. [36] established a semiquantitative method using UPLC combined with CAD and inverse gradient compensation. Six representative compounds of different chemical types, including magnoflorine (alkaloid), spinosin (flavone), 6″-feruloylspinosin (flavone), jujuboside A (triterpenoid saponin), jujuboside B (triterpenoid saponin), and betulinic acid (triterpenoid acid), were selected to assess CAD response consistency. The RSD of average RFs for these six compounds was significantly reduced from 61.50% to 7.87% after using inverse gradient compensation. Moreover, each compound demonstrated consistent responses in the fingerprint chromatogram, and peak areas directly reflected their content accurately. Clearly, inverse gradient compensation significantly improved CAD response consistency, providing substantial advantages for semiquantitative and QAMS analyses.

9.2.3.3 External Standard Method

The external standard method (ESM) is widely applied for determining active components in TCMs using CAD, as shown in Table 9.2.

Xu et al. [50] conducted a comprehensive comparative investigation of 12 ginseng products, simultaneously quantifying 15 known marker ginsenosides using a fully validated UHPLC-CAD approach. Despite the chemical complexity of different ginseng products, baseline separation of all 15 ginsenoside analytes was achieved using a CORTECS UPLC Shield RP18 column and optimized CAD parameters. CAD demonstrated higher sensitivity compared to conventional UV detection methods, particularly for chromophore-free saponins like 24(R)-pseudoginsenoside F11. The analysis of 119 batches of ginseng samples using this UHPLC-CAD method provided valuable information for the authentication and quality differentiation of ginseng products. Nevertheless, the responses for the 15 ginsenosides exhibited considerable variability. Using ginsenoside Rg_1 as a reference, the relative peak areas of the other 14 compounds ranged from 0.57 to 1.15, higher than previously reported [51].

Baker and Regg [52] developed a method combining UPLC followed by UV, CAD, and HRMS to analyze standardized extracts of *Ginkgo biloba*. They observed that the CAD response intensity varied by up to sixfold when comparing injections of 100% aqueous versus 100% organic solvents. Furthermore, applying an inverse gradient increased CAD and MS responses for early-eluting polar analytes, improving CAD signals two to threefold across most of the chromatogram without adversely affecting peak shape, as evaluated by asymmetry and tailing factors.

Table 9.2 The application of HPLC-CAD combined with ESM on TCMs.

No	Markers	Type of compounds	Chromatogr-aphic mode	Sample	References
1	Stachydrine hydrochloride	Alkaloid	HILIC	Compound *Leonurus japonicus* capsules	[37]
2	Glycocholic acid, hyodeoxycholic acid, cholic acid, glycodeoxycholic acid, chenodeoxycholic acid, deoxycholic acid	Bile acid	RPLC	Niuhuang Jiedu tablets	[38]
3	Liquiritin, isochlorogenic acid B, platycoside E, platycodin D3, platycodin D, tenuifolin, glycyrrhizic acid, onjisaponin B	Flavonoid glycoside, polyphenol, saponin	RPLC	Jiegeng Donghua tablets	[39]
4	Oleanolic acid, betulinic acid, corosolic acid, maslinic acid, ursolic acid	Triterpene acid	RPLC	*Crataegi frucuts* and its processed products	[40]
5	Rosmarinic acid-3-*O*-β-D-glucoside, rosmarinic acid	Polyphenol	RPLC	*Perilla Frutenscens* seed and fried seed formulations	[41]
6	Gypenoside XLVI and LVI	Saponin	RPLC	*Gynostemma pentaphyllum*	[42]
7	Fructose, glucose, sucrose	Monosaccharide	HILIC	*Imperatae rhizoma*	[43]
8	Ferulic acid, glycyrrhizic acid, coniferly ferulate, senkyunolide A, Z-ligustilide, imperatorin, isoimperatorin	Phenylpropanoids, triterpenoid saponins, phthalein, coumarins	RPLC	Chuanxiong Chatiao granules	[44]
9	Rutin, quercitrin, hesperidin, baicalin, wogonin, kaempferol, baicalein	Flavonoid	RPLC	Xiaoerjiegan tablets	[45]
10	Bilobalide, ginkgolide A, ginkgolide B, ginkgolide C	Terpenoid lactones	RPLC	Ginkgo leaf preparations (ginkgo leaf tablets, ginkgo leaf dispersible tablets, ginkgo leaf capsules, ginkgo leaf dripping pills)	[46]
11	Rhamnose, fructose, mannitol, D-anhydrous glucose, sucrose, raffinose	Monosaccharide	HILIC	Qingzhi Huayu formula	[47]
12	Ginsenoside Rb1, astragaloside IV	Saponins	RPLC	Qizhitongluo capsules	[48]
13	Asiaticoside B, madecassoside	Triterpenoids	RPLC	Sanjin tablets	[49]

Table 9.3 Sensitivity of different detectors for terpene lactones in ginkgo leaf extract.

	Detector	Bilobalide	Ginkgolide C	Ginkgolide A	Ginkgolide B
	CAD	4	6	10	10
LOD/ng	ELSD	67	68	142	243
	RI	296	44	213	146
	CAD	14	14	28	27
LOQ/ng	ELSD	202	206	427	811
	RI	1011	147	708	489

Our lab developed an HPLC-CAD method for determining terpene lactones in ginkgo leaf extracts and compared its sensitivities with that of ELSD and RID (Table 9.3). CAD exhibited superior sensitivity to ELSD and RID for bilobalide, ginkgolide C, ginkgolide A, and ginkgolide B. Additionally, CAD provided more consistent responses compared to ELSD.

9.2.4 Combined Application of 2D-LC and CAD

Two-dimensional liquid chromatography (2D-LC) integrates two chromatographic modes with orthogonal separation mechanisms (see Chapter 13 for greater details). By employing a unique modulation interface, 2D-LC combines two chromatography columns based on different separation principles, such as normal phase (NP), reversed phase (RP), ion exchange (IE), size exclusion (SE), hydrophilic interaction liquid chromatography (HILIC), and chiral separations, achieving high orthogonality. Figure 9.4 illustrates the differences in selectivity among Accucore series columns, which arise from the differences in the interaction forces between analytes and the stationary phase. Highly orthogonal separation modes provide different selectivity for target analytes, minimizing chromatographic peak overlaps and thus obtaining excellent peak capacities and separation performance. For complex samples, appropriate chromatographic column combinations can create tailored separation systems, which, when coupled with suitable detection methods, significantly enhances qualitative and quantitative capabilities. Due to superior selectivity, resolution, peak capacity, and analytical efficiency (the ability to resolve more peaks within a given time) compared to one-dimensional liquid chromatography (1D-LC), 2D-LC has emerged as an effective and reliable analytical technique for complex samples.

2D-LC can be classified according to the extent and strategy of transferring fractions from the first dimension (1D) to the second dimension (2D). Based on this, it can be generally categorized as heart-cutting (HC) 2D-LC, comprehensive

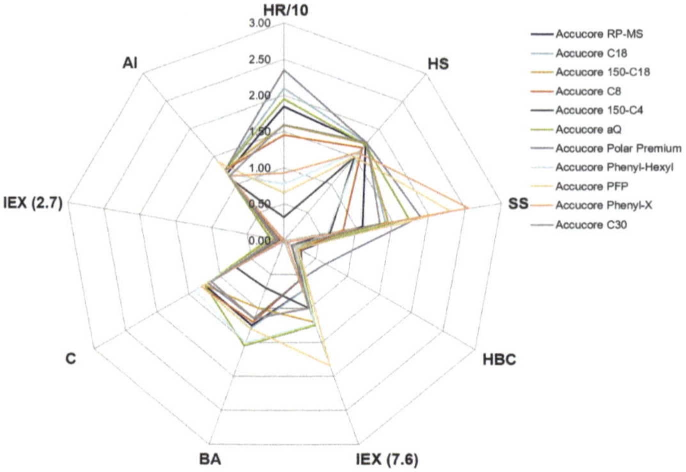

Figure 9.4 The selectivity of Accucore series columns. HR: hydrophobic retention; HS: hydrophobic selectivity; SS: steric selectivity; HBC: hydrogen bonding capacity; IEX (7.6): ion exchange capacity (pH 7.6); BA: base activity; C: chelation; IEX (2.7): ion exchange capacity (pH 2.7); AI: acid interaction.

two-dimensional liquid chromatography (LC × LC), and selective comprehensive two-dimensional liquid chromatography (sLC × LC). The application of 2D-LC in traditional herbal medicines analysis is increasingly widespread, typically used in conjunction with detectors such as UV and MS.

Xu et al. [53] elaborated a multiple heart-cutting two-dimensional liquid chromatography method coupled with charged aerosol detector (MHC-2DLC/CAD) approach to quantitatively assess ginseng from multiple Chinese patent medicines (CPMs). A Poroshell 120 EC-C18 column and an XBridge Shield RP18 column were utilized in the first dimension (1D) and the second-dimension (2D) chromatography, respectively. Compared with UV detectors, CAD can monitor chromophore-free compounds, such as 24(R)-p-F11 (a known chemical marker for *P. quinquefolius*), offering higher sensitivity in detecting ginsenosides using the MHC-2DLC approach. After optimization of the modulation interface, stationary phase, mobile phase, switch time window, and elution program, eight consecutive heart-cuts achieved efficient separation of 16 ginsenosides within 85 minutes. The developed MHC-2DLC/CAD method demonstrated excellent linearity ($R^2 > 0.999$), repeatability (RSD < 6.73%), stability (RSD < 5.63%), inter- and intra-day precision (RSD < 5.57%), recovery (93.76–111.14%), and the LOD and LOQ ranged from 0.45 to 2.37 ng and 0.96 to 4.71 ng, respectively. This study indicated that MHC-2DLC combined with CAD provides powerful online separation and detection capabilities, and a sound reference for developing

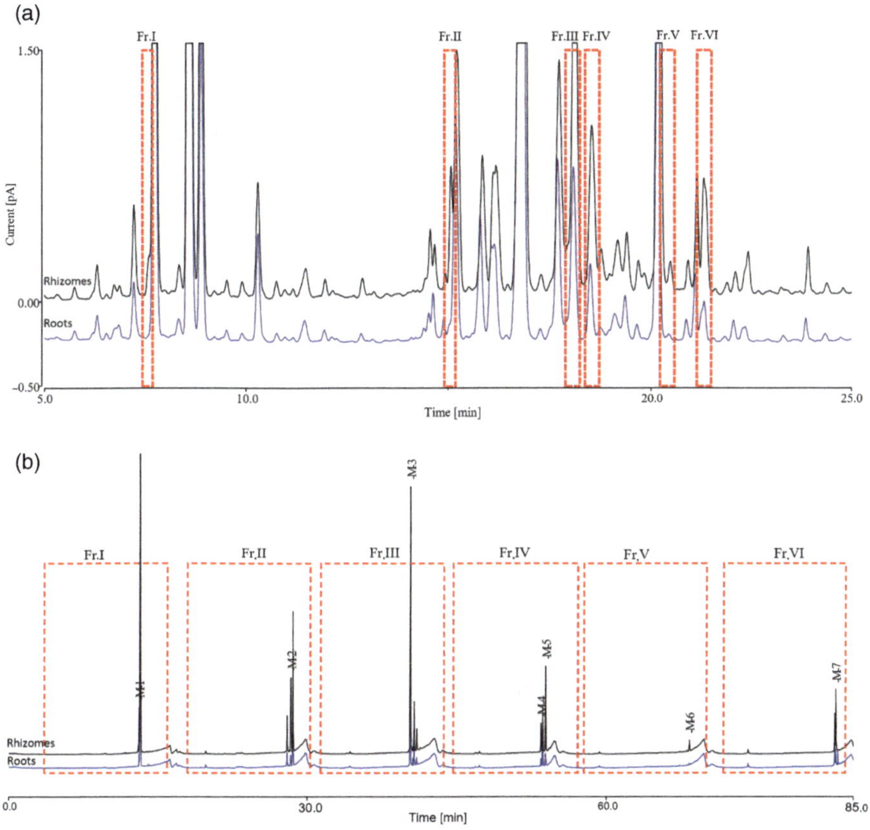

Figure 9.5 First-dimensional chromatogram showing six target fractions (a) and second-dimensional chromatograms (b).

multicomponent quantitation methods applicable to other TCMs beyond the CPMs tested.

Our lab developed a multi-heart-cutting 2DLC/CAD/MS method to identify and confirm intrinsic markers for intuitive discrimination between roots and rhizomes of Notoginseng. The multi heart-cutting 2DLC method selected Acclaim PA C16 as the first-dimensional column and an Accucore C30 as the second-dimensional column, maintaining first-dimensional separation performance to target potential marker compounds based on retention time (Figure 9.5). A method was established based on MHC-2DLC/CAD for quantifying five major active ginsenosides and three differential markers. The method validation results confirmed the method's stability, accuracy, and reliability, demonstrating its applicability in quality evaluation of actual medicinal materials.

9.2.5 Fraction Collection Triggered by CAD

For structural identification, pharmacological research, and quality evaluation of chemical constituents in natural medicines, obtaining purified standard compounds is essential. Due to the complexity of natural product constituents, including wide molecular weight ranges, diverse compound types, structural similarities (common homologues and isomers), significant polarity differences and wide-ranging contents, advanced separation techniques are required. Preparative chromatography, capable of separating compounds with minimal physical and chemical differences, is a highly effective method for complex systems and is widely used, particularly in the final purification stages of TCM constituents. In preparative chromatography, target fractions can be manually collected upon detecting peaks eluting from the column. While flexible, this method is labor-intensive and inefficient. To overcome these limitations, automatic fraction collectors have been developed. Modern multifunctional automated fraction collectors trigger fraction collection based on real-time signals from instrument software, predefined time windows, or detection signals such as peak height, threshold, or slope.

UV detectors are commonly used in preparative LC. However, for compounds lacking UV absorption, CAD detector provides a variable alternative. Song et al. [54] isolated a UV-inactive heptasaccharide from an active fraction of *Trillium tschonoskii,* a traditional medicinal plant to treat dizziness and headaches in the Tujia ethnic group, using high-temperature semi-preparative chromatography on a PGC Hypercarb column. The target heptasaccharide was successfully obtained using a Thermo Vanquish Core HPLC system with a CAD detector and a Vanquish Fraction Collector. The accurate structure of the heptasaccharide was elucidated by ^{1}H and ^{13}C NMR spectra combined with a "de novo sequencing" strategy, and the heptasaccharide was identified as maltoheptaose.

Our lab employed a Thermo Fisher's Vanquish analytical preparative LC system and multifunctional automated fraction collector for purifying saponins from Allii macrostemonis bulbus, widely used in treating chest disorders. The active ingredients—steroidal and furostanol saponins lack UV absorption and exhibit considerable structural complexity and isomerism. To address this, a fraction concatenation strategy was employed. A series of fractions were collected in arithmetic sequences and recombined into new simplified samples. The HPLC-CAD chromatogram of total saponin extract is shown in Figure 9.6, with highly overlapping peaks between 12.5 and 27.5 minutes. This 15-minute window was divided into 15 groups (1 minutes each), further subdivided into six fractions (10 seconds each). Fractions with identical timing across all groups (e.g., the first 10 seconds of each minute) were pooled by color code, yielding six concatenation fractions (fractions 1–6). Re-analysis by HPLC-CAD indicated significant simplification of the chromatograms compared to the original, validating

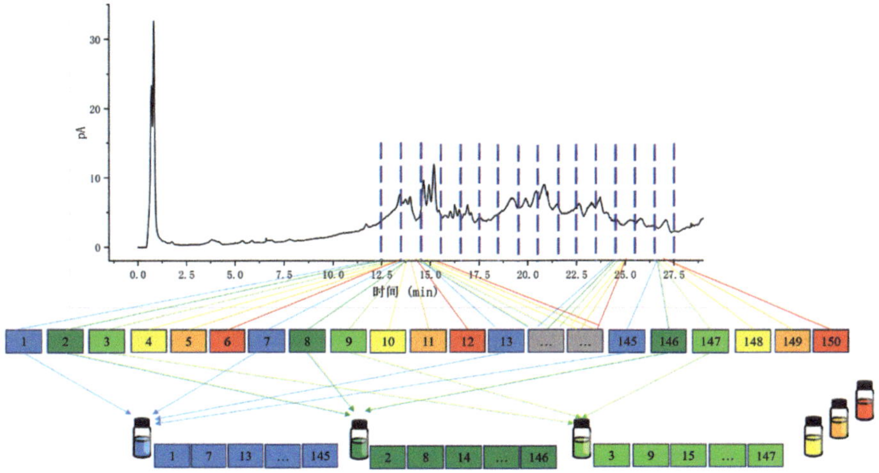

Figure 9.6 Fraction concatenation of Allii macrostemonis bulbus's total saponins.

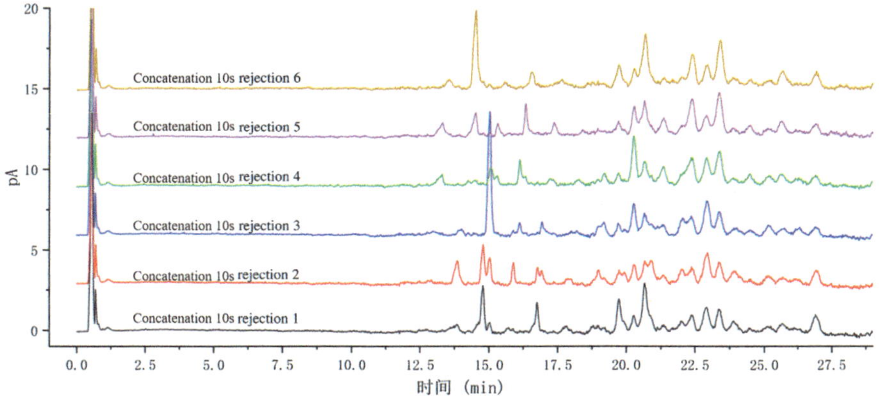

Figure 9.7 HPLC-CAD reanalysis chromatogram of concatenated fractions of Allii macrostemonis bulbus's total saponins.

the effectiveness of the concatenation strategy for simplifying complex natural product samples (Figure 9.7).

9.3 Application of CAD in Quality Analysis of THMs

The ability of THMs to prevent and treat diseases is fundamentally based on their chemical constituents. However, one single herb often contains a variety of components with different types, structures, and concentrations, along with significant matrix interference. This complexity is further compounded in the primary clinical formulation of traditional herb-compound preparations. As a highly sensitive universal detector, CAD has demonstrated considerable advantages in analyzing components within complex systems in recent years. Here we use the analysis of different types of chemical constituents as examples to illustrate the technical characteristics of CAD in the quality analysis of THMs.

9.3.1 Terpenoids/Saponins

Terpenoids, widely distributed in nature, are a diverse class of constituents characterized by complex skeletons and a broad spectrum of biological activities. They can be classified as monoterpenes, sesquiterpenes, diterpenes, and triterpenes based on the number of isoprene units in the structure. Monoterpenes and sesquiterpenes are the primary components of the volatile oils found in THMs, while diterpenes are mainly found in plants from families such as *Araliaceae, Asteraceae, Euphorbiaceae, Lamiaceae*, and *Rubiaceae*. Triterpenes, composed of 30 carbon atoms, exist in nature either freely or as glycosides formed by their combination with sugars. Some commonly used THMs, such as Ginseng Radix et Rhizoma, Notoginseng Radix et Rhizoma, Glycyrrhizae Radix Et Rhizoma, and Ziziphi Spinosae Semen, contain triterpenes and their saponins. Due to the diverse variations of the basic skeleton and the absence of stable aromatic rings in terpenoids, their absorption in the UV region relies primarily on functional groups such as double bonds and ester bonds. Consequently, only weak absorption is observed in the range of 200–250 nm, making it challenging to achieve ideal baseline separation in practical applications. Additionally, because sugar structures lack conjugated groups, saponins exhibit similar behavior under UV. Although the sensitivity of evaporative light-scattering detector (ELSD) for detecting terpenoids and their saponin compounds is higher than the low-wavelength detection by UV, its sensitivity for detecting low-abundance components is limited, restricting its further application in herbal studies. In contrast, CAD, which does not depend on the molecular structure of compounds and exhibits higher sensitivity than ELSD, has been widely applied in the investigations of terpenoids and their saponins in THMs.

Our lab established a HPLC-CAD method for the chromatographic fingerprint analysis of 20 batches of Toosendan Fructus (TF) from different resources and compared the chromatograms obtained using three detectors (UV, ELSD, and CAD). The results showed that the CAD fingerprint exhibited higher sensitivity than the ELSD fingerprint, revealing more chromatographic peak information for TF. Additionally, the CAD fingerprint was more suitable for detecting compounds with no or weak UV absorption, particularly the characteristic

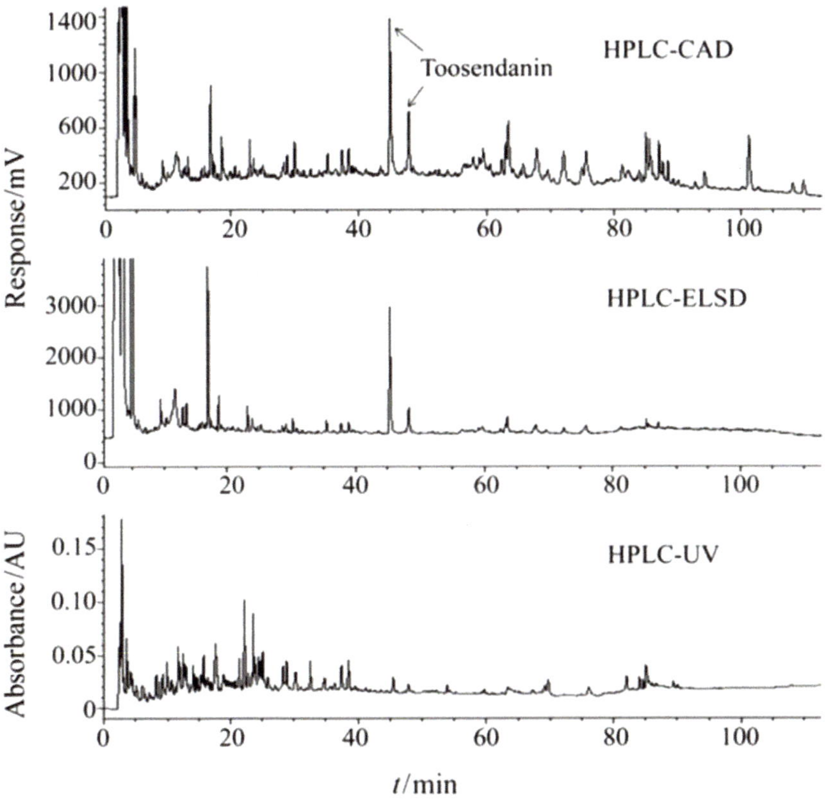

Figure 9.8 Comparison of the chromatographic fingerprints by HPLC-CAD, HPLC-ELSD, and HPLC-UV with the same sample.

compound-toosendanin (one triterpenoid compound), which was well-separated using HPLC-CAD (Figure 9.8). Subsequently, a quantitative method utilizing HPLC-CAD to determine the content of toosendanin in TF samples was also developed in our lab and the limits of detection (LODs) and quantification (LOQs) of CAD and ELSD were compared. As shown in Figure 9.9, CAD method exhibited a lower LOD and LOQ than ELSD, being 1/30 and 1/25 of the latter, respectively. Furthermore, the method validation showed that the RSD values of intra-day precision, inter-day precision, reproducibility, stability, and recovery test were all less than 3%, indicating that this HPLC-CAD method is sensitive, stable, and reliable.

Lu et al. [42] developed a method utilizing UHPLC-CAD for the determination of two triterpenoid saponins, gypenoside XLVI and LVI, in *Gynostemma pentaphyllum* sourced from Fujian Province. The method demonstrated good

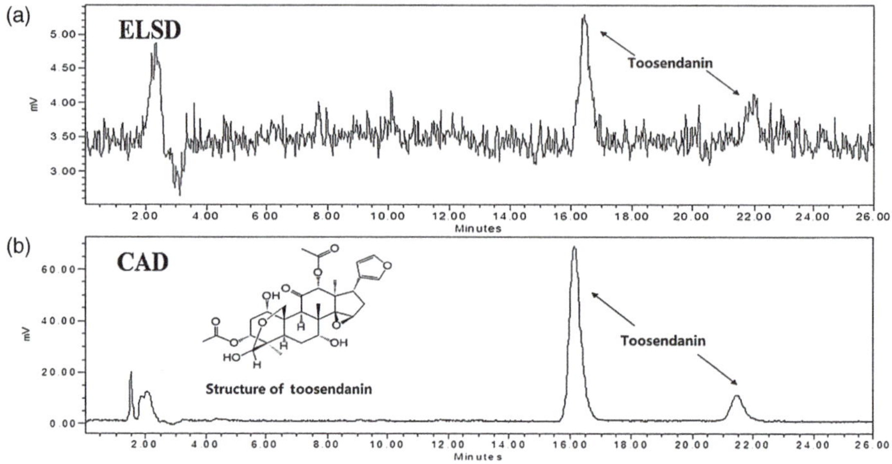

Figure 9.9 ELSD chromatogram (a) and CAD chromatogram (b) of the same toosendanin standard solution (30 μg/mL).

precision, reproducibility, stability, and accuracy, with RSDs of less than 3%. Gypenoside XLVI and LVI exhibited good linear relationships within the concentration ranges of 9.94–318.94 μg/mL and 12.78–409.00 μg/mL, with correlation coefficients of 0.9993 and 0.9995, respectively. Nguyen et al. [55] established an HPLC-UV-CAD method to simultaneously quantify nine triterpenoid saponins in Vietnamese ginseng (VG). Among them, five characteristic ocotillol-type saponins, including majonoside-R1 (M-R1), majonoside-R2 (M-R2), vina-ginsenoside-R2 (V-R2), vina-ginsenoside-R11 (V-R11), and pseudoginsenoside-RT4 (p-RT4), lack conjugated units such as double bonds in their structures, leading to a low response even at 198 nm and making them prone to interference from complex sample matrices. The universal nature of CAD effectively mitigates this issue, with LOD and LOQ for M-R1 at 7.6 and 22.9 μg/mL, respectively, which are four times lower than those obtained using UV, see Table 9.4.

In addition to the aforementioned triterpenoid saponins, saponin compounds also include a class of steroid saponins with spirostanol-aglycones structures, which exhibit little to no UV absorbance or only minimal absorbance by UV detection. Yang et al. [56] simultaneously determined the contents of three major steroid saponins (dehydrocorybulbine VII, 3β-*O*-α-L-pyranoglucosyl-(1 → 4)-[α-L-pyranoglucosyl-(1 → 2)]-*O*-β-D-glucose, and dehydrocorybulbine VI) in the rhizoma and fibrous roots of *Trillium tschonoskii* and *T. kamtschaticum* using HPLC-CAD. A comparison of LODs and LOQs for UV (203 nm), ELSD, and CAD was conducted, revealing LODs of 0.32, 0.21, and 0.04 μg, and LOQs of 0.56, 0.38, and 0.10 μg, respectively, demonstrating the superior sensitivity of CAD. Wang et al. [57] developed a method for the determination of the steroid

Table 9.4 Calibration curve, concentration range, LOD, and LOQ of nine triterpenoid saponins.

Compounds	Calibration curve	R^2	Range (mg/mL)	LOD (ng)	LOQ (ng)
UV					
N-R1	$y = 31.0544x + 0.095$	0.9997	0.008–0.53	48	146
M-R1	N.D.	N.D.	N.D.	N.D.	N.D.
G-Rg1	$y = 89.4402x + 0.2917$	0.9999	0.019–1.24	48	146
M-R2	$y = 1.4089x + -0.0086$	0.9965	0.238–1.24	624	1890
V-R11	N.D.	N.D.	N.D.	N.D.	N.D.
p-RT4	N.D.	N.D.	N.D.	N.D.	N.D.
V-R2	N.D.	N.D.	N.D.	N.D.	N.D.
G-Rb1	$y = 66.7414x + 0.1563$	1.0000	0.015–0.98	26	82
G-Rd	$y = 71.789x + 0.2697$	1.0000	0.21–1.36	42	130
CAD					
N-R1	$\log(y) = 0.8795 \times \log(x) + 0.8698$	0.9983	0.008–0.53	106	318
M-R1	$\log(y) = 0.8053 \times \log(x) + 1.1228$	0.9985	0.018–1.13	184	556
G-Rg1	$\log(y) = 0.8136 \times \log(x) + 1.1613$	0.9981	0.019–1.24	186	562
M-R2	$\log(y) = 0.8565 \times \log(x) + 1.0409$	0.9975	0.019–1.24	152	458
V-R11	$\log(y) = 0.8254 \times \log(x) + 1.3789$	0.9961	0.010–0.67	90	276
p-RT4	$\log(y) = 0.8649 \times \log(x) + 1.0157$	0.9970	0.018–1.13	142	428
V-R2	$\log(y) = 0.7981 \times \log(x) + 1.0384$	0.9970	0.016–1.00	128	390
G-Rb1	$\log(y) = 0.8161 \times \log(x) + 1.2291$	0.9983	0.015–0.98	156	472
G-Rd	$\log(y) = 0.8165 \times \log(x) + 1.2531$	0.9987	0.21–1.36	158	480

Source: Data from [55].

saponin 25-*epi*-officinalisnin II in Asparagi Radix using UHPLC-CAD. The results indicated an LOD of 12.5 ng and an LOQ of 25.0 ng, with RSD values for precision, reproducibility, and 24-hour stability all less than 3.0%. The average recovery rate was 98.63%. Content determination results of 11 botanical samples from different regions ranged from 0.36 to 6.86 mg/g, successfully quantifying low-content, non-UV-absorbing steroid saponins in Asparagi Radix.

9.3.2 Carbohydrates

Carbohydrates are widely distributed in THMs, constituting 80% to 90% of the dry weight of plants, and can be classified into monosaccharides, oligosaccharides, and polysaccharides based on their hydrolysis ability and molecular weight.

Oligosaccharides are sugars formed by the polymerization of 2–9 monosaccharides linked by glycosidic bonds, while polysaccharides are composed of more than 10 monosaccharides, resulting in high molecular weight species. Compared with saponins, there are a higher number of hydroxyl groups and fewer chromophore units in the structures of carbohydrate derivatives, making their detection difficult using conventional HPLC conditions. A common approach for detection of carbohydrates by UV is to derivatize the carbohydrates with a chromophore. This both complicates the procedure and reduces reproducibility. Underivatized carbohydrates can be detected using ELSD or RID, but RID cannot perform gradient elution and is easily affected by factors such as the composition of the mobile phase, temperature, and flow rate, leading to unstable baselines. ELSD offers low sensitivity and is primarily suitable for analyzing carbohydrates with high concentrations. CAD provides versatility and high sensitivity, allowing for the detection of carbohydrates without derivatization. Consequently, its application in detecting carbohydrate components in THMs has gradually increased alongside the rapid development of sugar science.

Our lab developed a sensitive and universal method for the analysis of oligosaccharides in THMs using UHPLC-CAD by systematically optimizing chromatographic columns, mobile phases, column temperatures, and other analytical conditions. This method does not require derivatization, and the samples obtained through simple ultrasound extraction could be directly injected for analysis. The proposed separation and detection procedures were successfully used to evaluate six THMs, leading to the discovery of oligosaccharides in Menthae Haplocalycis Herba and Chrysanthemi Indici Flos for the first time, as well as the measurement of the abundance of oligosaccharides in Morindae Officinalis Radix, as shown in Figure 9.10. Xu et al. [58] simultaneously determined the content of monosaccharides (fructose, D-mannose, and D-glucose) and seven oligosaccharides (sucrose, maltose, maltotriose, maltotetraose, maltopentaose, maltohexaose, and maltoheptaose) in Dendrobii Officinalis Caulis using HPLC-CAD without pre-column derivatization. The addition of a modifier, triethylamine, to the mobile phase, eliminated the double-peak behavior of fructose and glucose observed in aqueous solution (due to mutarotation, which forms stereoisomers), and produced single peaks in the chromatogram. The established method provided good separation for the 10 carbohydrates, with low LODs and LOQs. The RSDs for precision, stability, and repeatability were all less than 3.0%. Chen et al. [59] employed HPLC-CAD to simultaneously quantify monosaccharide and oligosaccharide components in Glycyrrhizae Radix et Rhizoma (including sucralose, rhamnose, arabinose, fructose, mannose, glucose, sucrose, lactulose, maltose, lactose, and lupeose), achieving good separation, and LODs and LOQs below 102 and 204 ng, respectively. Xie et al. [60] established an efficient HPLC-CAD method for the simultaneous determination of two monosaccharides (D-fructose and D-glucose),

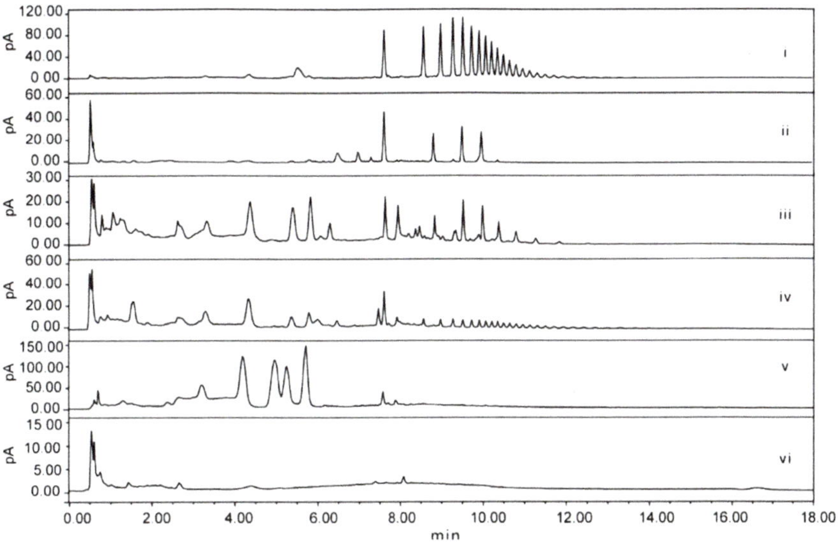

Figure 9.10 Profiling the oligosaccharides in different THMs by UHPLC-CAD. (i) Morindae Officinalis Radix. (ii) Ziziphi Spinosae Semen. (iii) Menthae Haplocalycis Herba. (iv) Chrysanthemi Indici Flos. (v) Crataegi Fructus. (vi) Commelinaeherba.

sucrose, and fructo-oligosaccharides (FOS) (GF2-GF6) in Codonopsis Radix, which demonstrated a good correlation over three orders of magnitude within the range of tested concentrations (Table 9.5), with coefficients greater than 0.999.

9.3.3 Organic Acids

Bile acids, derivatives of cholanic acid, belong to the class of steroid compounds and are predominantly found in animal bile, mainly present in animal-derived medicines, such as bear bile powder and ox bile. This class of substances comprises many isomers with similar polarities, complicating their analysis and detection. Similar to saponins and carbohydrates, bile acids lack chromophores and only exhibit terminal UV absorption, resulting in poor sensitivity and major interference when measured using a UV detector. Moreover, derivatization approaches are often cumbersome. In recent years, several studies have utilized CAD technology to investigate bile acids. Notably, a CAD detection method for deoxycholic acid has been included in the 40th edition of the United States Pharmacopeia.

Our lab established a method for the simultaneous determination of five bile acids (cholic acid, ursodeoxycholic acid, hyodeoxycholic acid, chenodeoxycholic acid, and deoxycholic acid) in Pulvis Fellis Suis, Pulvis Billis Bovis, Pulvis Fellis Caprinus, and Pulvis Fellis Galli using HPLC-CAD. The results showed that the

Table 9.5 Quadratic regression data and limits of detection and quantitation for eight selected standards.

Analytes	Regression equation	R^2	Quantitation range (μg/mL)	LOQ (ng)	LOD (ng)
D-Fructose	$y = 5.6915x^2 + 36.6075x + 5.7280$	0.9996	15.05–2006.00	30.0	15.0
D-Glucose	$y = 16.4167x^2 + 53.3824x + 3.7229$	0.9996	11.34–1134.00	56.8	17.0
Sucrose	$y = 15.0054x^2 + 48.5290x + 3.1570$	0.9998	10.63–1063.00	212.6	31.8
GF2	$y = 58.1231x^2 + 67.9977x + 0.3984$	0.9998	6.06–303.20	121.2	60.6
GF3	$y = 74.5278x^2 + 68.1328x + 0.0672$	0.9999	10.41–208.20	208.2	41.6
GF4	$y = 52.2779x^2 + 50.4754x - 0.0661$	0.9998	15.47–206.20	309.4	61.8
GF5	$y = 40.9951x^2 + 47.1042x - 0.0552$	0.9996	15.15–202.00	303.0	80.8
GF6	$y = 11.6395x^2 + 35.5965x + 0.0126$	0.9999	20.28–202.80	405.6	304.2

Source: Data from [60].

method achieved a LOQ of 30 ng for each bile acid and a LOD of 6 ng, compared to an LOD of 30 ng for ELSD. For a 3 mg/L standard solution, the signal-to-noise ratios (S/N) of the five bile acids in CAD were 9.6, 14.6, 18.1, 17.9, and 26.0, respectively, while the S/N ratios in ELSD were 4.6, 4.7, 4.9, 7.4, and 7.0, indicating that the sensitivity of CAD is up to three times greater than that of ELSD (Figure 9.11). Additionally, the quantitation range for the bile acids tested with CAD spanned from 0.6 to 300 mg/L, covering three orders of magnitude, whereas the linear range for ELSD was limited to just one order of magnitude (3–60 mg/L), as shown in Table 9.6. Furthermore, an important characteristic of the CAD is that the detection signal is independent of the structure of the analyte, meaning that compounds with different structures exhibit consistent responses. The results showed that the RSD of the response factors for the five bile acids using CAD was 4.9%, compared to an RSD of 15.3% for the response factors obtained with ELSD, indicating that CAD provides greater consistency in response factors than ELSD (see Table 9.7).

Moreover, an HPLC-CAD quantitative method for six types of bile acids in Niuhuang Jiedu tablets was established in our lab. The results indicated that the method is sensitive, accurate, and reliable, demonstrating a strong linear relationship between the peak areas and concentrations of the six bile acids, with R^2 values greater than 0.999 across three orders of magnitude. Furthermore, the responses of CAD are typically regarded as approximately linear or quasi-linear within a range of 1.5–2.0 orders of magnitude, but exhibit non-linearity over a broader range, represented by the equation $A = am^b$ (where A is the peak area, m is the sample amount, and a is the sensitivity). As noted in the previous study by *our lab* (see Table 9.6), the quantitation range for the five bile acids involved three orders of magnitude, and each standard curve was quadratic ($y = x^2 + x + c$). The

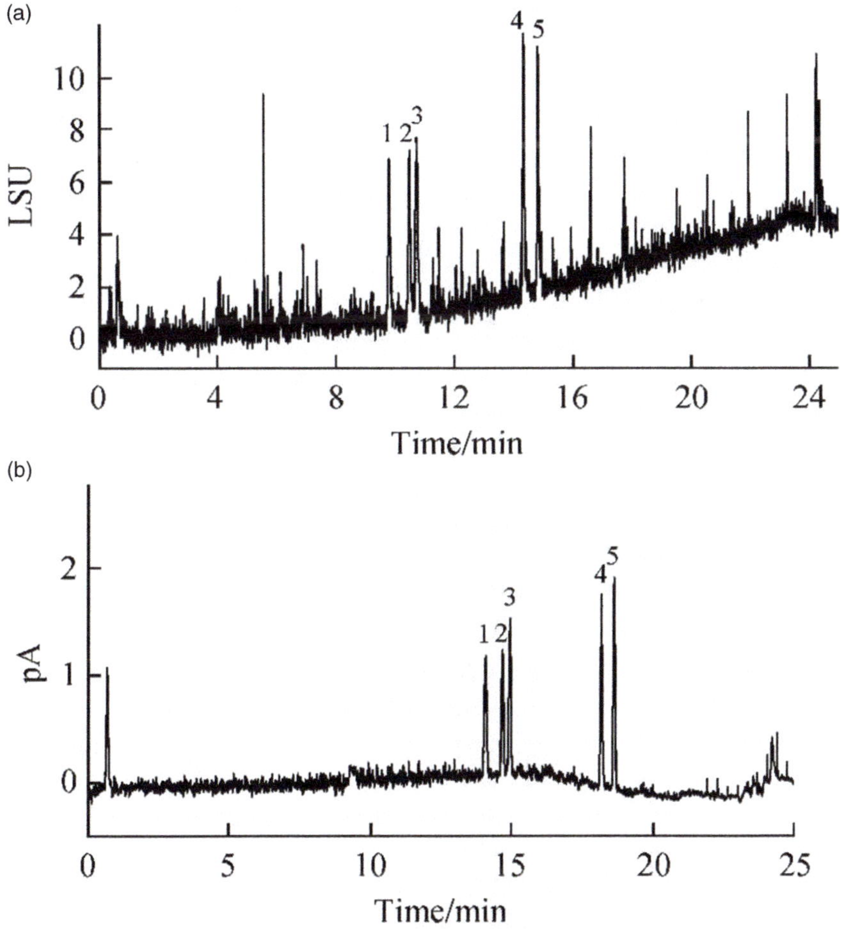

Figure 9.11 ELSD chromatogram (a) and CAD chromatogram (b) of five kinds of standard bile acids. 1. Cholic acid, CA; 2. ursodeoxycholic acid, UDCA; 3. hyodeoxycholic acid, HDCA; 4. chenodeoxycholic acid, CDCA; 5. deoxycholic acid, DCA.

power function value (PFV) is a built-in setting parameter of the CAD instrument used to adjust the exponent *b*. When optimally set, the PFV can enhance the linearity of the CAD response curve across a wide range of mass concentrations, thereby simplifying data processing. This study found that when the PFV value was set to 1.35, the standard curve for the six target bile acids exhibited the best linearity, resulting in a linear standard curve across three orders of magnitude ($y = x + c$), as shown in Table 9.8.

Table 9.6 Linear ranges, regression equations, correlation coefficients of five kinds of standard bile acids.

Bile acid	Quantitation range (mg/mL)	Regression equation	Correlation coefficient
CA	0.5888–294.4	$y = -29.2830x^2 + 27.3466x + 0.1098$	0.9994
UDCA	0.5908–295.4	$y = -24.0867x^2 + 25.9913x + 0.1408$	0.9988
HDCA	0.6296–314.8	$y = -22.3867x^2 + 24.7532x + 0.1562$	0.9989
CDCA	0.5868–293.4	$y = -35.5282x^2 + 29.1901x + 0.1894$	0.9988
DCA	0.5828–291.4	$y = -37.2348x^2 + 31.0860x + 0.1933$	0.9988

Table 9.7 Response factor consistency of five kinds of standard bile acids.

Response factor	CA	UDCA	HDCA	CDCA	DCA	RSD (%)
CAD	0.3847	0.3929	0.3699	0.3926	0.4226	4.9
ELSD	0.2663	0.3073	0.3305	0.3620	0.4000	15.3

Table 9.8 Calibration equation, LOD, LOQ, repeatability and precision of six bile acids ($n = 6$).

Bile acid	Linear range (μg/mL)	Regression equation	R^2	LOD (ng)	LOQ (ng)	Repeatability (RSD, %)	Precision (RSD, %)
GCA	2.10–972	$y = 0.80x + 0.18$	0.9997	13.2	21.0	2.1	1.8
HDCA	2.09–1046	$y = 0.85x + 0.24$	0.9997	12.6	20.9	1.9	1.5
CA	2.05–978	$y = 0.91x - 0.02$	0.9997	12.8	20.5	1.7	1.9
GDCA	2.56–1022	$y = 0.76x - 0.08$	0.9991	15.3	25.6	1.2	2.0
CDCA	1.25–658	$y = 1.01x + 0.12$	0.9997	8.6	12.0	1.7	1.6
DCA	2.05–1025	$y = 0.89x + 0.06$	0.9998	11.4	19.1	1.5	1.9

Fatty acids are also an important class of active components in THMs, particularly in fruit and seed-based herbs, with their core structure being carboxylic acids that possess long hydrocarbon chains. These compounds typically have relatively low polarity, often necessitating gas chromatography for analysis after saponification and methylation. Additionally, fatty acids exhibit minimal UV absorption, resulting in low responses and high baseline noise when analyzed using UV detectors—these issues can be really addressed with CAD. Wu et al. [61]

established methods for determining the content of fatty acids in *Brucea javanica* (L.) Merr. seeds and conducting fingerprint analysis using HPLC-CAD. The LODs and LOQs for five fatty acids (linolenic acid, linoleic acid, palmitic acid, oleic acid, and stearic acid) ranged from 3.68 to 19.58 ng and 10.80 to 68.00 ng, respectively. The analytical method avoided the complicated pretreatment process typically required for fatty acid detection and demonstrated good sensitivity, making it applicable for determining fatty acids in other THMs. Subsequently, the team applied the developed HPLC-CAD method to detect fatty acids in *Agastache rugosa* seeds and found the abundance of linolenic acid [62].

9.3.4 Alkaloids

Most alkaloids have complex cyclic structures with nitrogen atoms incorporated into the rings, exhibiting significant and unique biological activities. Notable examples include morphine, berberine, and ephedrine, making them an important class of constituents in THMs. The structures of alkaloids are diverse, with some alkaloid compounds, such as terpenoid and steroid alkaloids, along with partial simple heterocyclic derivatives, posing challenges for analysis using conventional detectors due to the absence of conjugated systems. Such issues can be effectively solved by CAD.

Our lab developed a non-derivative method for the qualitative and quantitative analysis of isosteroidal alkaloids from *Fritillaria thunbergii* using HPLC-CAD and compared the performance of CAD with that of ELSD. The CAD exhibited a good linear relationship within a dynamic range of four orders of magnitude (0.06–44 mg/L), with correlation coefficients exceeding 0.999. It demonstrated good precision with overall intra- and inter-day RSD values for peak area of less than 1.71% and 2.63%, respectively, whereas the RSD values for ELSD were more than 3.09% and 9.94%. Moreover, CAD was found to be 30–55 times more sensitive than ELSD, enabling the detection of low levels of reference compound impurities that could not be detected by ELSD (see Figure 9.12). For example, for the reference alkaloid, peimissine, the peak area percent detected by ELSD was 91.66%, while the same sample showed a peak area of only 50.1% under CAD detection. Similar results were observed for the other four alkaloids, indicating that CAD is suitable for estimating the presence of impurities. Zhu [63] determined the contents of five alkaloids in eight *Fritillaria* plants using HPLC-ELSD and HPLC-CAD methods. It was found that CAD exhibited superior sensitivity compared to ELSD, with both LODs and LOQs lower than those of ELSD, enabling the detection of low concentrations of alkaloids. Consequently, cluster analysis results demonstrated that different *Fritillaria* medicinal herbs could be distinctly distinguished using the dataset obtained from CAD, which provides a reliable means for quality control of *Fritillaria* plants.

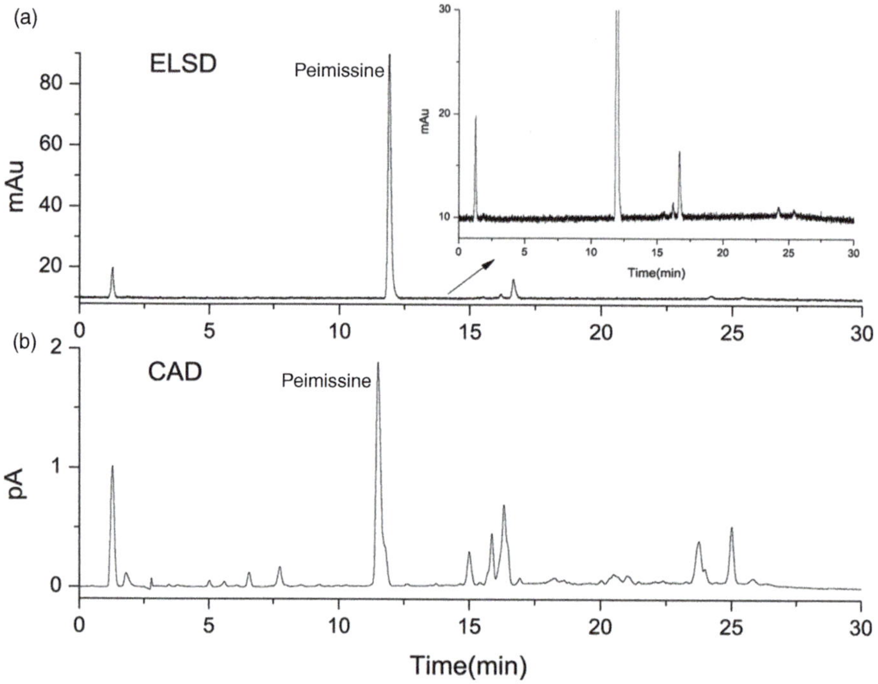

Figure 9.12 The analysis of peimissine standard compounds using ELSD (a) and CAD (b).

9.3.5 Other Applications of CAD

Traditional Chinese prescriptions (TCPs) are the primary clinical application form of THMs. Currently, existing quality control methods that rely on single indicator compounds are inadequate for ensuring the safety and efficacy of these complex prescriptions, severely limiting the market expansion and internationalization of TCPs. Given the diverse structures of chemical constituents in TCPs and the pronounced differences in their concentrations, UV and ELSD methods face limitations that prevent them from fully elucidating the chemical composition of TCPs. In recent years, research using CAD has increased due to its versatility and high sensitivity in the quality evaluation of TCPs.

Yuanhu Zhitong capsules have demonstrated proven clinical efficacy. The Chinese Pharmacopoeia (2015 version) specifies only dl-tetrahydropalmatine, one alkaloid, as a quality indicator for this prescription. Deng et al. [64] employed HPLC-CAD to simultaneously quantify four alkaloids and two coumarins in Yuanhu Zhitong capsules and compared the performance of CAD with DAD. The results indicated that four alkaloid derivatives exhibited maximum absorption at 280 nm, while two coumarins displayed peak absorption at 248 and 301 nm, respectively. When comparing the DAD spectra of samples at these wavelengths with data obtained by CAD, it was found that CAD detected significantly more chromatographic peaks than DAD, demonstrating higher sensitivity and a more stable baseline. Additionally, the linear range for the six indicators by CAD analysis spanned from 0.013 to 2.52 μg, and the method was proven to be accurate, convenient, and highly sensitive, providing a more comprehensive approach for the quality control of multicomponent in Yuanhu Zhitong capsules. Luo et al. [65] established an HPLC-CAD method for the determination of eight ingredients in Fengshiding tablets, which included various types of compounds such as alkaloids (anabasine), triterpenes and their saponins (glycyrrhizic acid and glycyrrhetinic acid), sesquiterpenes (chinenside A), coumarins (imperatorin and isoimperatorin), and organic acids (salicin and paeonol). For these compounds with different characteristic absorption wavelengths, CAD exhibited good responses with stable baselines and high sensitivity. Compared to the single indicator of paeonol recorded in Chinese Pharmacopoeia, this method provides a more comprehensive approach for the quality control of Fengshiding tablets. Xie et al. [66] developed a method using HPLC-CAD to determine six structurally diverse compounds (catalpol, puerarin, salvianolic acid B, ginsenoside Rg1, ginsenoside Rb1 and ginsenoside Rd) in Bushen Huoxue prescription. The regression equations were calculated in three forms of $y = a\ x + b$, $y = a\ x^2 + b\ x + c$, and $\mathrm{Log}(y) = a\ \mathrm{Log}(x) + b$ (x represents the concentration of the standard solution in mg/mL, and y the corresponding peak area of injected compounds). The results indicated that both linear function and quadratic function were more suitable for the tested analytes ($R^2 > 0.9990$), compared with log–log function (Table 9.9). Considering the convenience and relatively narrow concentration range of the six compounds, the linear regression was ultimately adopted for the subsequent studies. The HPLC-CAD method established in this study demonstrated high accuracy, good repeatability, and ease of operation, providing a reference for the improvement of quality standard of the Bushen Huoxue prescription.

Liu et al. [67] established a versatile "Monomethod-Multiuse" analytical strategy using HPLC-CAD that can simultaneously evaluate and control the quality of raw materials and different Shengmai preparations (SMF) after optimizing the column, column temperature, flow rate, and CAD parameters (see Figure 9.13). The "Monomethod-Multiuse" strategy can simultaneously characterize 50 components from SMF and distinguish the raw materials from different origins, and qualities. Fourteen components in 19 batches samples of different formulations (oral liquids, capsules, and injections) of SMF from various manufacturers were quantitatively determined, including 11 ginsenosides and 3 lignans. The universal analytical strategy established based on HPLC-CAD is sensitive, simple, and convenient, effectively overcoming the selectivity limitations of UV and providing valuable guidance for improving the quality control

Table 9.9 The calibration curves (R^2) of six tested analytes.

Analyte	Linear function	Quadratic function	Log–log function
Catalpol	$y = 17.649x + 0.1951$ (0.9995)	$y = -2.0699x^2 + 18.613x + 0.1365$ (0.9995)	$\mathrm{Log}(y) = 0.8693\mathrm{Log}(x) + 1.1892$ (0.9993)
Puerarin	$y = 23.811x + 0.6453$ (0.9991)	$y = -1.5577x^2 + 25.164x + 0.4921$ (0.9994)	$\mathrm{Log}(y) = 0.8342\mathrm{Log}(x) + 1.3494$ (0.9989)
Salvianolic acid B	$y = 31.070x + 1.1178$ (0.9993)	$y = -1.9833x^2 + 33.493x + 0.7324$ (0.9998)	$\mathrm{Log}(y) = 0.8418\mathrm{Log}(x) + 1.4889$ (0.9988)
Ginsenoside Rg1	$y = 51.463x + 0.0358$ (0.9995)	$y = 20.873x^2 + 48.381x + 0.0950$ (0.9998)	$\mathrm{Log}(y) = 0.9151\mathrm{Log}(x) + 1.6164$ (0.9989)
Ginsenoside Rb1	$y = 68.032x + 0.0990$ (0.9997)	$y = 1.6684x^2 + 67.687x + 0.1083$ (0.9997)	$\mathrm{Log}(y) = 0.9788\mathrm{Log}(x) + 1.8226$ (0.9996)
Ginsenoside Rd	$y = 63.483x + 0.1406$ (0.9993)	$y = -11.23x^2 + 66.349x + 0.0453$ (0.9995)	$\mathrm{Log}(y) = 1.0573\mathrm{Log}(x) + 1.8738$ (0.9969)

Source: [66] / with permission of Elsevier.

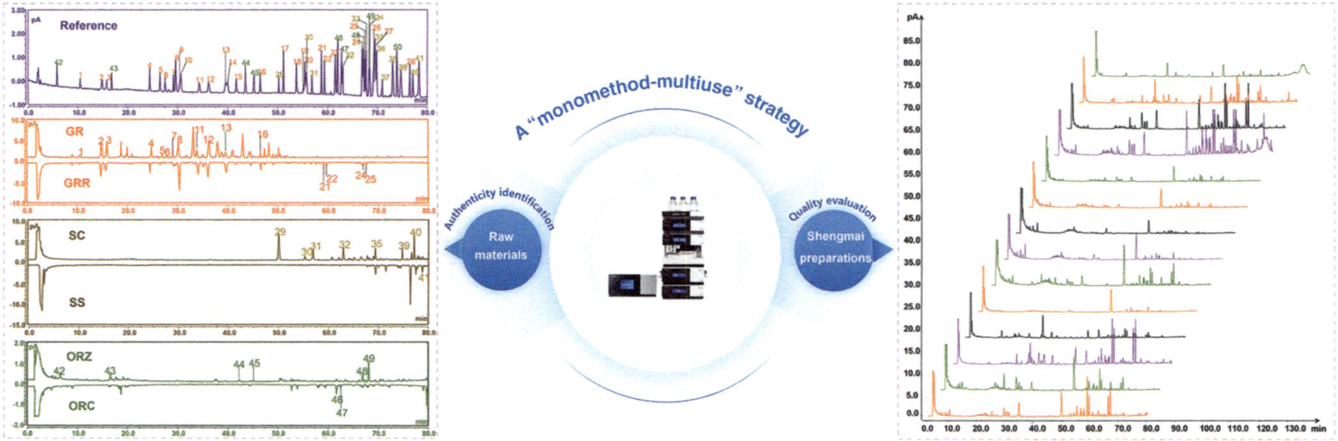

Figure 9.13 A versatile "Monomethod-Multiuse" analytical strategy based on HPLC-CAD. *Source:* [67] / with permission of Elsevier.

of SMF. Cao et al. [68] developed a UHPLC-CAD fingerprinting method and a UPLC–MS/MS multi-component quantitative analysis method for Zishen Yutai pills. They found that there were more peaks obtained by CAD than those by UV, with a stable baseline and high sensitivity. The combination of UPLC-CAD fingerprinting and multicomponent quantification has proven to be an effective and reliable quality control strategy for Zishen Yutai pills.

9.4 Conclusion

The rapid development of modern instrumental technology has not only offered new opportunities for the study of THMs but also opened up technical pathways for the modernization. CAD, with its detection principles that do not rely on the structure of compounds, as well as its high sensitivity and ease of operation, addresses the challenge of analyzing primary constituents in THMs (such as saponins, carbohydrates, bile acids, etc.) that are difficult to detect through conventional means, thus providing valuable solutions and means for research in these areas. In recent years, researchers have increasingly utilized CAD to design and establish strategies for the quality evaluation of THMs and even complex preparations, showcasing significant advantages. Future efforts should focus on enhancing the application of CAD in quality evaluation, thereby facilitating the transition from academic research to practical quality applications.

References

1 Jones, K. and Malcolme-Lawes, D.J. (1988). High-performance ultraviolet absorption detector for liquid chromatography. II. A prototype detector. *Journal of Chromatography. A* 441 (2): 387–393.

2 Shi, Y., Sun, C., Zheng, B. et al. (2010). Simultaneous determination of nine ginsenosides in functional foods by high performance liquid chromatography with diode array detector detection. *Food Chemistry* 123 (4): 1322–1327.

3 Lee, H.J. and Kim, C.Y. (2010). Simultaneous determination of nine lignans using pressurized liquid extraction and HPLC-DAD in the fruits of Schisandra chinensis. *Food Chemistry* 120 (4): 1224–1228.

4 Westerbuhr, S.G. and Rowlen, K.L. (2000). On-column refractive index detector for flash chromatography. *Journal of Chromatography. A* 886 (41): 9–18.

5 Vogeser, M. and Seger, C. (2008). A decade of HPLC–MS/MS in the routine clinical laboratory-goals for further developments. *Clinical Biochemistry* 41 (9): 649–662.

6 Korfmacher, W.A. (2005). Foundation review: principles and applications of LC-MS in new drug discovery. *Drug Discovery Today* 10 (20): 1357–1367.

7 Lucy, C.A. and Harrison, C.R. (2001). Chemiluminescence nitrogen detection in ion chromatography for the determination of nitrogen-containing anions. *Journal of Chromatography A* 920 (1–2): 135–141.

8 Hutchinson, J.P., Li, J., Farrell, W. et al. (2011). Comparison of the response of four aerosol detectors used with ultra high pressure liquid chromatography. *Journal of Chromatography. A* 1218 (12): 1646–1655.

9 Almeling, S. and Holzgrabe, U. (2010). Use of evaporative light scattering detection for the quality control of drug substances: influence of different liquid chromatographic and evaporative light scattering detector parameters on the appearance of spike peaks. *Journal of Chromatography. A* 1217 (14): 2163–2170.

10 Kong, W., Jin, C., Liu, W. et al. (2010). Development and validation of a UPLC-ELSD method for fast simultaneous determination of five bile acid derivatives in Calculus Bovis and its medicinal preparations. *Food Chemistry* 120 (4): 1193–1200.

11 Chai, X.Y., Li, S.L., and Li, P. (2005). Quality evaluation of Flos Lonicerae through a simultaneous determination of seven saponins by HPLC with ELSD. *Journal of Chromatography. A* 1070 (1–2): 43–48.

12 Kim, S.N., Ha, Y.W., Shin, H. et al. (2007). Simultaneous quantification of 14 ginsenosides in Panax ginseng C.A. Meyer (Korean red ginseng) by HPLC-ELSD and its application to quality control. *Journal of Pharmaceutical and Biomedical Analysis* 45 (1): 164–170.

13 Koropchak, J.A., Heenan, C.L., and Allen, L.B. (1996). Direct comparison of evaporative light-scattering and condensation nucleation light-scattering detection for liquid chromatography. *Journal of Chromatography. A* 736 (1–2): 11–19.

14 Vervoort, N., Daemen, D., and Török, G. (2008). Performance evaluation of evaporative light scattering detection and charged aerosol detection in reversed phase liquid chromatography. *Journal of Chromatography. A* 1189 (1–2): 92–100.

15 Vehovec, T. and Obreza, A. (2010). Review of operating principle and applications of the charged aerosol detector. *Journal of Chromatography. A* 1217 (10): 1549–1556.

16 Joseph, A. and Rustum, A. (2010). Development and validation of a RP-HPLC method for the determination of gentamicin sulfate and its related substances in a pharmaceutical cream using a short pentafluorophenyl column and a Charged Aerosol Detector. *Journal of Pharmaceutical and Biomedical Analysis* 51 (3): 521–531.

17 Zhang, K., Dai, L., and Chetwyn, N.P. (2010). Simultaneous determination of positive and negative pharmaceutical counterions using mixed-mode chromatography coupled with charged aerosol detector. *Journal of Chromatography. A* 1217 (37): 5776–5784.

18 Nováková, L., Solichová, D., and Solich, P. (2009). Hydrophilic interaction liquid chromatography - charged aerosol detection as a straightforward solution for simultaneous analysis of ascorbic acid and dehydroascorbic acid. *Journal of Chromatography. A* 1216 (21): 4574–4581.

19 Huang, Z., Richards, M.A., Zha, Y. et al. (2009). Determination of inorganic pharmaceutical counterions using hydrophilic interaction chromatography coupled with a Corona® CAD detector. *Journal of Pharmaceutical and Biomedical Analysis* 50 (5): 809–814.

20 Takahashi, K., Kinugasa, S., Senda, M. et al. (2008). Quantitative comparison of a corona-charged aerosol detector and an evaporative light-scattering detector for the analysis of a synthetic polymer by supercritical fluid chromatography. *Journal of Chromatography. A* 1193 (1–2): 151–155.

21 Kou, D., Manius, G., Zhan, S., and Chokshi, H.P. (2009). Size exclusion chromatography with Corona charged aerosol detector for the analysis of polyethylene glycol polymer. *Journal of Chromatography. A* 1216 (28): 5424–5428.

22 Cobb, Z., Shaw, P.N., Lloyd, L.L. et al. (2001). Evaporative light-scattering detection coupled to microcolumn liquid chromatography for the analysis of underivatized amino acids: sensitivity, linearity of response and comparisons with UV absorbance detection. *Journal of Microcolumn Separations* 13 (4): 169–175.

23 Eom, H.Y., Park, S.Y., Kim, M.K. et al. (2010). Comparison between evaporative light scattering detection and charged aerosol detection for the analysis of saikosaponins. *Journal of Chromatography. A* 1217 (26): 4347–4354.

24 Gorecki, T., Lynen, F., Szucs, R., and Sandra, P. (2006). Universal response in liquid chromatography using charged aerosol detection. *Analytical Chemistry* 78 (9): 3186–3192.

25 Schilling, K., Pawellek, R., Lovejoy, K. et al. (2018). Influence of charged aerosol detector instrument settings on the ultra-high-performance liquid chromatography analysis of fatty acids in polysorbate 80. *Journal of Chromatography. A* 1576: 58–66.

26 Teutenberg, T., Tuerk, J., Holzhauser, M., and Kiffmeyer, T.K. (2006). Evaluation of column bleed by using an ultraviolet and a charged aerosol detector coupled to a high-temperature liquid chromatographic system. *Journal of Chromatography. A* 1119 (1): 197–201.

27 Zhang, Y.H., Yang, Y.G., Shi, L. et al. (2020). Fingerprinting analysis of non-saponins from water-soluble part and determination of dencichine from

Panax notoginseng, P. ginseng and *P. quinquefolium* based on liquid chromatography coupled with charged aerosol detection. *China Journal of Chinese Materia Medica* 45 (14): 3475–3480.

28 Ning, Z., Liu, Z., Song, Z. et al. (2016). A single marker choice strategy in simultaneous characterization and quantification of multiple components by rapid resolution liquid chromatography coupled with triple quadrupole tandem mass spectrometry (RRLC-QqQ-MS). *Journal of Pharmaceutical and Biomedical Analysis* 124: 174–188.

29 He, Y., Li, Q., and Bi, K. (2015). Simultaneous determination of six active components by a single standard to determine multicomponents combined with fingerprint analysis for the quality control of Rhizoma Chuanxiong. *Journal of Separation Science* 38 (7): 1090–1099.

30 Tan, P., Zhang, L., Zhao, Y.L. et al. (2016). A practical method for the simultaneous quantitative determination of twelve anthraquinone derivatives in rhubarb by a single-marker based on ultra-performance liquid chromatography and chemometric analysis. *Analytical Methods* 8 (19): 3927–3934.

31 Zhu, B., Hu, D., Zhao, J., and Li, S. (2024). Rapid identification and quantification of Pseudostellaria heterophylla with its adulterants by HPLC-CAD fingerprint combined with improved quantitative analysis of multi-components by single marker (QAMS). *Journal of Pharmaceutical and Biomedical Analysis* 247: 116205.

32 Yao, J., Sun, X.G., Dong, R. et al. (2021). Simultaneous quantitative analyses of six components in *Astragalus membranaceus* based on HPLC-CAD and quantitative analysis of multi-components with a single-marker. *Acta Pharmaceutica Sinica* 56 (2): 557–564.

33 Zhao, X., Yang, G., Zheng, G.S. et al. (2014). Simultaneous determination of four lignans in Magnoliae Flos extract by high performance liquid chromatography-electrospray ionization mass spectrometry. *Chinese Journal of Analytical Chemistry* 42 (12): 1804–1810.

34 Wu, L., Zhang, S., Zhou, L. et al. (2021). Establishment and validation of the quantitative analysis of multi-components by single marker for the quality control of Qishen Yiqi dripping pills by high-performance liquid chromatography with charged aerosol detection. *Phytochemical Analysis* 32 (6): 942–956.

35 Zheng, W., Zhou, M., Chai, R.P. et al. (2022). Quality analysis of hawthorn leaves (the leaves of Crataegus pinnatifida Bge. var major N.E.Br) in different harvest time. *Phytochemical Analysis* 33 (7): 1147–1155.

36 Shi, Y., Nan, Y., Zheng, W. et al. (2024). Qualitative and semiquantitative analyses of the chemical components of the seed coat and kernel of Ziziphi Spinosae Semen. *Chinese Journal of Chromatography* 42 (3): 234–244.

37 Lao, Y.Z., Su, J.M., Zhang, J. et al. (2023). Determination of stachydrine hydrochloride in compound *Leonurus japonicus* capsules by high performance liquid chromatography-charged aerosol detector. *Environment and Chemistry* 42 (2): 675–678.

38 Lin, C.H., Chai, R.P., Zhang, L. et al. (2022). Determination of six kinds of bile acids in Niuhuang Jiedu tablets using high performance liquid chromatography coupled with charged aerosol detector. *Chinese Journal of Analytical Chemistry* 50 (5): 764–771.

39 Shen, M., Chen, R., Qian, Y.F. et al. (2024). Simultaneous determination of eight components in Jiegeng Donghua tablets by high performance liquid chromatography-charged aerosol detection. *Chinese Journal of Analysis Laboratory* 43 (10): 1423–1429.

40 Du, L.L., Chen, C., and Ran, L.J. (2023). Simultaneous determination of five triterpenic acids in *Crataegi Fructus* by accelerated solvent extraction-charged aerosol detector. *Environment and Chemistry* 42 (6): 2132–2135.

41 Di, J.P., Zhao, Y.C., Peng, Q. et al. (2023). Determination multi-components contents of *Perilla frutenscens* seed and fried seed formulations based on charged aerosol detector. *Environment and Chemistry* 42 (1): 341–344.

42 Lu, P.X., Li, G., Zheng, W. et al. (2022). Determination of gypenoside XLVI and LVI in Gynostemma pentaphyllum from Fujian by ultra-high performance liquid chromatography-charged aerosol detector. *Chinese Journal of Chromatography* 40 (9): 833–842.

43 Zhang, S.H., Wang, J.H., Guo, H.Z. et al. (2018). Determination of fructose, glucose and sucrose in Imperatae Rhizoma by HPLC-CAD. *Chinese Journal of Pharmaceutical Analysis* 38 (6): 942–947.

44 Liu, Q., Wang, L., and Xiao, Y.N. (2021). Simultaneous determination of seven active ingredients in Chuanxiong Chatiao Granules by HPLC-CAD. *Chinese Journal of Hospital Pharmacy* 41 (14): 1417–1422.

45 Fang, L.Z., Wu, X.S., and Zhang, Y. (2019). Simultaneous determination of seven kinds of flavonoids components in Xiaoerjiegan tablets by HPLC-CAD. *Chinese Journal of Hospital Pharmacy* 39 (23): 2402–2406.

46 Bai, G.G., Yuan, F., Ye, H. et al. (2019). Determination of ginkgolides in four kinds of Ginkgo leaf preparations by HPLC-CAD. *Chinese Journal of Pharmaceutical Analysis* 39 (6): 1102–1107.

47 Zhang, H.B., Fu, Q.S., Yao, G.Z. et al. (2024). Establishment of HPLC-CAD characteristic chromatograms and content determination of six carbohydrates for Qingzhi Huayu Formula. *Chinese Traditional Patent Medicine* 46 (10): 3212–3217.

48 Li, T., Cheng, X.L., You, R.L. et al. (2022). Determination of saponins in Qizhitongluo capsules by HPLC-CAD. *Chinese Pharmaceutical Journal* 57 (14): 1193–1197.

49 Zou, Z., Min, J.G., Wu, M.J. et al. (2021). Determination of two isomers asiaticoside B and madecassoside in Sanjin tablets by HPLC-CAD. *Chinese Traditional and Herbal Drugs* 52 (14): 4217–4220.

50 Xu, X., Wang, S., Wang, H. et al. (2021). Simultaneous quantitative assays of 15 ginsenosides from 119 batches of ginseng samples representing 12 traditional Chinese medicines by ultra-high performance liquid chromatography coupled with charged aerosol detector. *Journal of Chromatography. A* 1655: 462504.

51 Ouyang, L.F., Wang, Z.L., Dai, J.G. et al. (2014). Determination of total ginsenosides in ginseng extracts using charged aerosol detection with post-column compensation of the gradient. *Chinese Journal of Natural Medicines* 12 (11): 857–868.

52 Baker, T.R. and Regg, B.T. (2018). A multi-detector chromatographic approach for characterization and quantitation of botanical constituents to enable in silico safety assessments. *Analytical and Bioanalytical Chemistry* 410 (21): 5143–5154.

53 Xu, X.Y., Jiang, M.T., Wang, Y. et al. (2023). Multiple heart-cutting two-dimensional liquid chromatography/charged aerosol detector assay of ginsenosides for quality evaluation of ginseng from diverse Chinese patent medicines. *Journal of Chromatography. A* 1708: 464344.

54 Song, J., Tu, G.Z., Liu, Y. et al. (2025). Hydroxyl groups introducing NMR strategy for structural elucidation of a heptasaccharide isolated from Trillium tschonoskii. *Carbohydrate Research* 549: 109359.

55 Huy, T.N., Kim, L.V.H., Hien, M.N. et al. (2021). Evaluation of the saponin content in *Panax vietnamensis* acclimatized to Lam Dong Province by HPLC-UV/CAD. *Molecules* 26: 5373.

56 Yang, Y.J., Sun, X.G., Yang, J. et al. (2017). Determination of three saponins in rhizoma and fibrous root of *Trillium tschonoskii* and *Trillium kamtschaticum. China Journal of Chinese Materia Medica* 42 (6): 1146–1151.

57 Wang, B., Gao, L., Wang, J. et al. (2018). Establishment of a liquid chromatographic method for determination of the content of steroidal saponin 25-epi-officinalisninII in *Asparagi Radix. Journal of International Pharmaceutical Research* 45 (10): 799–803.

58 Xu, S.S., Wu, S.J., Lu, W.X. et al. (2023). Simultaneous determination of monosaccharides and oligosaccharides in *Dendrobii Officinalis* Caulis by HPLC-CAD. *Chinese Journal of Pharmaceutical Analysis* 43 (8): 1302–1311.

59 Chen, R., Ma, D.Y., Guo, B., and Xue, M. (2021). Simultaneous quantification of saccharides in Glycyrrhizae Radix et Rhizoma by HPLC-CAD combined with chemometrics. *China Pharmacist* 24 (10): 1932–1937.

60 Xie, Q., Tian, H.M., Huan, X.H. et al. (2022). Quality evaluation of Codonopsis Radix and processed products based on the analysis of monosaccharides and oligosaccharides by liquid chromatography coupled with charged aerosol detector. *Phytochemical Analysis* 33: 262–271.

61 Wu, Z.N., Li, L., Li, N. et al. (2017). Optimization of ultrasonic-assisted extraction of fatty acids in seeds of *Brucea javanica* (L.) Merr. from different sources and simultaneous analysis using high-performance liquid chromatography with charged aerosol detection. *Molecules* 22: 931.

62 Yang, R., Wu, Z.N., Pu, Y.Q. et al. (2019). Fast and non-derivative method based on high-performance liquid chromatography-charged aerosol detection for the determination of fatty acids from *Agastache rugosa* (Fisch. et Mey.) O. Ktze. seeds. *Natural Product Research* 33 (13): 1969–1974.

63 Zhu, L. (2018). Quantitative analysis of alkaloids and nucleotides in medicinal herbs of *Fritiliaria*. Anhui University of Chinese Medicine. Anhui, China.

64 Deng, L., He, M.W., and Hou, L.B. (2019). Simultaneous determination of 6 alkaloids and coumarins in Yuanhu Zhitong capsules by HPLC-DAD. *Central South Pharmacy* 17 (1): 88–92.

65 Luo, C.M., Wu, X.S., Lei, M.Z. et al. (2019). Simultaneous determination of 8 ingredients in Fengshiding tablets by HPLC-CAD. *Central South Pharmacy* 17 (6): 863–867.

66 Xie, M.J., Yu, Y.T., Zhu, Z.Y. et al. (2021). Simultaneous determination of six main components in Bushen Huoxue prescription by HPLC-CAD. *Journal of Pharmaceutical and Biomedical Analysis* 201: 114087.

67 Liu, L.C., Fan, W.X., Zhang, H.Y. et al. (2024). A versatile economic strategy by HPLC-CAD for quantification of structurally diverse markers in quality control of Shengmai formula from raw materials to preparations. *Phytomedicine* 129: 155625.

68 Cao, J.L., Lei, T., Wu, S.J. et al. (2020). Development of a comprehensive method combining UHPLC-CAD fingerprint, multi-components quantitative analysis for quality evaluation of Zishen Yutai Pills: a step towards quality control of Chinese patent medicine. *Journal of Pharmaceutical and Biomedical Analysis* 191: 113570.

10

Analysis of Small Molecule Pharmaceuticals with Charged Aerosol Detection

Ulrike Holzgrabe

Institute of Pharmacy, University of Würzburg, Wuerzburg, Germany

CHAPTER MENU

10.1 Summary

The gold standard for quality control of active pharmaceutical ingredients (APIs) is currently high-performance liquid chromatography (HPLC) with UV detection. APIs without a chromophore are often derivatized to make them visible to UV; the alternative mass spectrometric (MS) detection is rarely used in pharmacopoeial methods. Charged aerosol detection (CAD) is an ideal solution for quality analysis of such APIs. Moreover, CAD can additionally detect and quantify counterions as well as unexpected impurities, because its response is independent of the physicochemical properties of the analyte. The hyphenation of CAD and UV detection can give a complete picture of the impurity profile of an API, especially after optimization of the response with regard to mobile phase composition during a gradient, and the power function value. This is exemplified here by the analysis of amino acids, dronates, gabapentin, and vigabatrin.

10.2 Introduction

The quality of drugs is well-regulated by national licensing authorities and international pharmacopoeias because drugs of low quality can be dangerous for the patients. There have been several cases where patients died or substantially suffered from low quality drugs, e.g., at the beginning of the century gentamicin contaminated with sisomicin and many other impurities and, more recently, contamination of histamine. Contamination for both examples originated from changes in the production process. Other examples include over-sulfated chondroitin sulfate (OSCS) from China—a contamination of unfractionated heparin which occurred 15 years ago, and more recently, issues with sartans and other drugs containing nitrosamines. All cases reported above have one thing in common—the toxic impurities did not possess a chromophore or fluorophore. Hence, the detection of unexpected impurities was challenging. While the impurities of gentamicin can be evaluated by HPLC with amperometric detection, this approach is both labor-intensive and not very robust. OSCS in heparin can be assessed by NMR spectroscopy and nitrosamine in different active pharmaceutical ingredients (APIs) by a hyphenation of HPLC and MS; these approaches tend to be expensive and complex. Thus, a universal detector such as the evaporative aerosol detectors can be advantageous.

The gold-standard of quality analysis used by all international pharmacopoeias, such as the European Pharmacopoeia (PhEur) [1], the United States Pharmacopeia (USP) [2] and the Japanese Pharmacopoeia (JP) [3], is HPLC with UV/Vis detection. In some rare cases, MS or electrochemical detection is used. Evaporate Light Scattering Detection (ELSD), which can be used for nonvolatile analytes, lacks sensitivity so is mostly used for the content determination of major natural products in mixtures where the largest peak is evaluated. When it comes to impurity profiling, peaks of very low intensity (often <0.1% of the API peak) have to be inspected. For this purpose, when using ELSD high amounts of the drug analyte have to be injected onto the column in order to quantify the low-level impurities. This results in non-reproducible spikes on the tail of the API peak and makes the evaluation of real impurities almost impossible [4]. Furthermore, ELSD does not provide a linear response and suffers from low sensitivity and a small and complex dynamic range. Another aerosol detector is the chemiluminescent nitrogen detector (CLND) but this is not compatible with nitrogen-containing mobile phases such as acetonitrile (ACN) and requires much more maintenance than CAD, ELSD, and UV.

CAD can be regarded as a kind of universal detector which can analyze all components that can be nebulized and dried into particles (for details see Chapter 1). CAD can measure all nonvolatile and many semi-volatile analytes over a wide enough dynamic range to allow simultaneous analysis of the APIs and low-level impurities. This is advantageous for the quality analysis of drugs and especially for impurity profiling, because CAD can see the APIs, their expected and unexpected impurities as well as counterions regardless of the physicochemical properties such as the presence of a chromophore. Hence, derivatization of the analytes is not necessary.

10.3 Driving the Sensitivity of CAD

The quality assessment of APIs requires the evaluation of so-called related substances originating from the synthesis and degradation of the drug, the assessment of content, in addition to the analysis of elemental impurities and residual solvents. While the international pharmacopoeias provide general methods for the latter two, the analytical methods needed to measure content and related substances must be developed for each drug individually. To establish a method for analysis of related substances, the analyst needs to know the production procedure on one hand and degradation pathways on the other. The methods for evaluation of expected impurities must be able to quantify them in a range between 0.05% up to 1% of the drug substance, in the presence of mostly >99% of the drug's substance, which can be a challenge with regard to separation and sensitivity.

Though CAD is a universal detection technique, the peak size (i.e., sensitivity), heavily depends on the composition of the mobile phase. A high portion of organic solvent in the mobile phase, e.g., ACN, decreases the size of the droplets of the primary aerosol and, thus, enhances the efficiency of the transport process, resulting in more particles arriving at the electrometer (for details see Chapter 1) and an enhanced response for the analytes. The impurity profiling, which needs high sensitivity, can take advantage of the solvent composition, namely highly organic mobile phases, but this is not compatible with every column chemistry and with all analytes. Hence, upon development and optimization of the chromatographic method, a three-dimensional problem consisting of the chemistry and polarity of the analytes to be separated, the stationary phase (reversed-phase (RP), ion-pair chromatography (IPC), mixed mode chromatography (MMC) and hydrophilic interaction chromatography (HILIC)), and the corresponding mobile phase must be solved. Additionally, the influence on the response combined with the background noise needs also to be considered. The general procedures laid down in Table 10.1 are especially worthwhile for the analysis of the impurities of drugs (APIs), which can be structurally diverse: besides structurally related substances, there can also be reagents used in synthesis as well as starting products of very different structure. Having said this, it becomes clear that the

Table 10.1 Semiqualitative relationship between the chromatographic parameters, response/sensitivity, and background noise for the application of ion-pair chromatography (IPC), mixed mode chromatography (MMC), and hydrophilic interaction chromatography (HILIC).

Parameter	IPC	MMC	HILIC
Stationary phase	Reversed phase	Bi–/trimodal functionalized columns (e.g., IEX/RP)	Normal phase (e.g., silica, amide, diol phases)
Percentage of organic content mobile phase	0–20%	0–50%	70–95%
Response	↑	↑↑	↑↑↑
Mobile phase additive	Volatile ion-pairing reagent	Volatile buffer	Volatile buffer
Background noise	↑↑↑	↑	↑
Sample dissolution solvent	Aqueous/organic	Aqueous/organic	Highly organic necessary

components of the analyte mixtures are of very different polarity, vapor pressure, volatility, molecular weight, and chargeability.

In order to demonstrate the pros and cons of different chromatographic systems with CAD, a study on underivatized amino acids was described for both HILIC and IPC [5]. For both cases a kind of generic method was established. For IPC, a C18 Acclaim™ Polar Advantage II column (Thermo Fisher Scientific) with a gradient of heptafluorobutyric acid (HFBA) and trifluoroacetic acid (TFA) in water and ACN was employed. For HILIC, an Accucore™ 150 Amide HILIC column (Thermo Fisher Scientific) with aqueous ammonium formate buffer and ACN (20 : 80) was used. The signal-to-noise ratio (*S/N*), the relative standard deviation (% RSD) for the content and the limit-of-quantification (LoQ) obtained for both methods are displayed in Table 10.2.

Some principal conclusions from the amino acid method development can be drawn with regard to detector response, which hold true for many HPLC-CAD methods:

- Both methods presented here can be used for impurity profiling; they are kind of generic methods
- Due to the high amount of ACN, HILIC shows a higher *S/N* attributed to higher analyte transport efficiency within the detector (in comparison to the IPC method). However, as detailed in Chapter 3, since the higher transport

Table 10.2 Performance comparison of S/N, % RSD and LoQ between IPC and HILIC for a series of underivatized amino acids according to ref. [5]; injection of 5 ng on column.

	S/N ($n = 6$)		% RSD ($n = 6$)		LoQ	
	IPC	**HILIC**	**IPC**	**HILIC**	**IPC**	**HILIC**
Ala	9.2	17.5	6.2	2.5	6	3
Cys	6.7	nd	9.7	nd	10	10
Ile	14.3	19.8	3.8	1.6	5	3
Leu	10.2	19.8	3.1	2.3	5	3
Met	9.8	19.7	5.5	2.9	5	3
Phe	5.8	26.9	5.0	1.8	10	3
Val	13.7	23.4	6.1	2.7	5	3

nd = due to a coelution of cysteine with the system peak, *S/N*, and RSD could not be assessed.
Source: [5] / with permission of Elsevier.

efficiency obtained with HILIC also applies to impurities, any gains in *S/N* ratio depend on the use of pure solvents, volatile additives and "low-bleed" columns. When applying IPC, background noise increases with increased length of the ion-pairing (IP) reagent. Hence, IP reagents of short chain length are recommended.

- High evaporation temperature (T_e) of the detector can, on one hand, decrease the background noise resulting in a better *S/N*, but on the other hand, it may be a problem for semi-volatile analytes because the detector response might be reduced. Hence, this parameter has to be chosen carefully.
- LoQs obtained with HILIC were lower than those with IPC.

Finally, the comparison between IPC and HILIC can be a kind of neck-and-neck race. Though the HILIC method was one-up on IPC in this amino acid race, it clearly shows that both methods, in addition to MMC, must be considered in order to find the best method with regard to separation and sensitivity.

10.4 Problem of the Nonlinear Response

The aerosol detectors are prone to nonlinear response, which can be described by a power law equation $A = a(m_{inj})^b$. One solution to this problem is a log–log transformation and the other is the application of a so-called power function value (PFV). The PFV can improve the method's linearity by optimizing the signal output so that the exponent "b" is closer to 1.0 (i.e., quasi-linear) over the method's range of quantitation. PFV is normally set to 1.0 but can be optimized

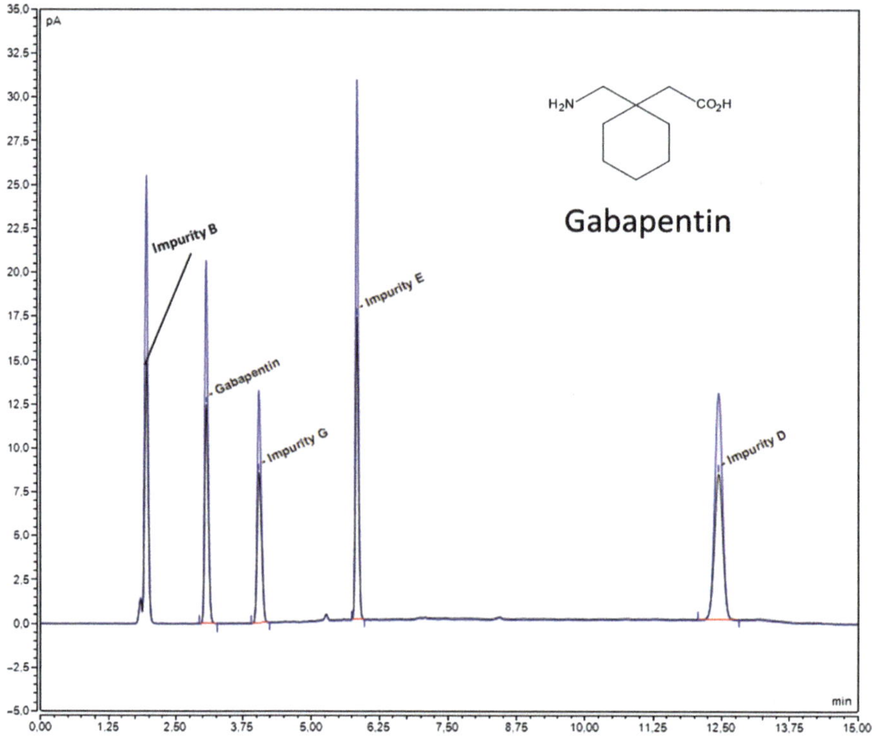

Figure 10.1 Overlay of the two chromatograms of gabapentin and its impurities with a PFV of 1.0 (black) and 1.5 (blue) according to ref. [8]. *Source:* [8] / with permission of Elsevier.

mostly by increasing the value. This can result in a broader linear dynamic range, better *S/N* and a better uniformity of the response [6–8]. However, care must be taken to avoid optimizing PFV solely to improve *S/N* since it is only intended to linearize response. This is especially true when low-level impurities of a drug have to be evaluated because the wrong, too high, PFV can create an exponential drop in response similar to that commonly observed with ELSD where their levels can be underestimated or completely missed [9]. But it can also enhance the response as has been shown for gabapentin and its impurities (cf. Figure 10.1) [8] where even the more volatile impurity B profits from a higher PFV. For additional details on the use of PFV see Chapter 3.

Log–log transformation is a manipulation of the measured data that can facilitate the use of linear regression. As discussed in Chapter 3, relying solely on a correlation coefficient (e.g., r^2) to assess linearity or goodness of fit is not sufficient even without transformation. Also, after log transformation, the calculated correlation coefficient by itself is unlikely to be mathematically valid. Use of a more robust approach that examines residual error especially for lower mass levels is recommended. This is especially important for log-transformed data where an r^2 very close to 1.0 is often observed even when significant absolute errors exist. Usage of the PFV has some similarities to log–log transformation [6] but use of an optimal PFV directly affects the signal output with respect to linearity. This approach also facilitates the use of linear regression and can also improve limits of detection, peak shapes and precision. However, before applying one of those options it should be checked whether a linearization is really necessary, because often CAD has a wider linear range than ELSD. The range of impurity profiling is typically 0.05–1% and of the content assessment 80–120%. In these small ranges, CAD response might be sufficiently linear although the additional benefits described above when using an optimal PFV might not be realized.

10.5 Advantage of a Universal Detector for Drug Quality Analysis

Whereas UV/Vis detectors are blind to many counterions, such as halides, phosphorus ions, sodium, or potassium, with CAD it is possible to identify and quantify them, the drug substance, API, ions and its organic impurities in one run!

Already in 2010, the K. Zhang group at Genentech laid the foundation for such analyses. They separated some 25 commonly used drug counterions of different charges using a mixed mode Acclaim Trinity™ P1 column (Thermo Fisher Scientific) [10] with a ternary gradient of ACN, water and 0.2 M ammonium formate (pH 4.0). Due to the importance of API analysis, scientists at the European Directorate for the Quality of Medicines and HealthCare (EDQM) validated the generic method and identified/verified some counterions in a couple of drug substances by means of HPLC-CAD-TOF/MS [11]. Besides the mapping of corresponding counterions in drugs and their impurities as well as non-expected counterions, such as traces of chloride in cloxacillin sodium, it is possible to check the stoichiometric ratio of drug/ion, which can disclose an ionic impurity.

A similar approach to the separation of ions, was applied to the analysis of the electrolyte supplement potassium aspartate/magnesium aspartate (PAMA). Depending on the production method, the mix from K^+- and Mg^{2+}- aspartate contains additional amounts of Na^+ ions. Li et al., applied the Acclaim Trinity P1 column and optimized the mobile phase T_e by using the response surface methodology (RSM) resulting in very similar chromatograms [12], which show varying amounts of aspartate and Na^+, depending on the manufacturer. The topic of ion analysis with CAD is further discussed in Chapter 5.

Bisphosphonates for the treatment of osteoporosis are chromophore-lacking compounds containing a basic nitrogen moiety in addition to two phosphonate

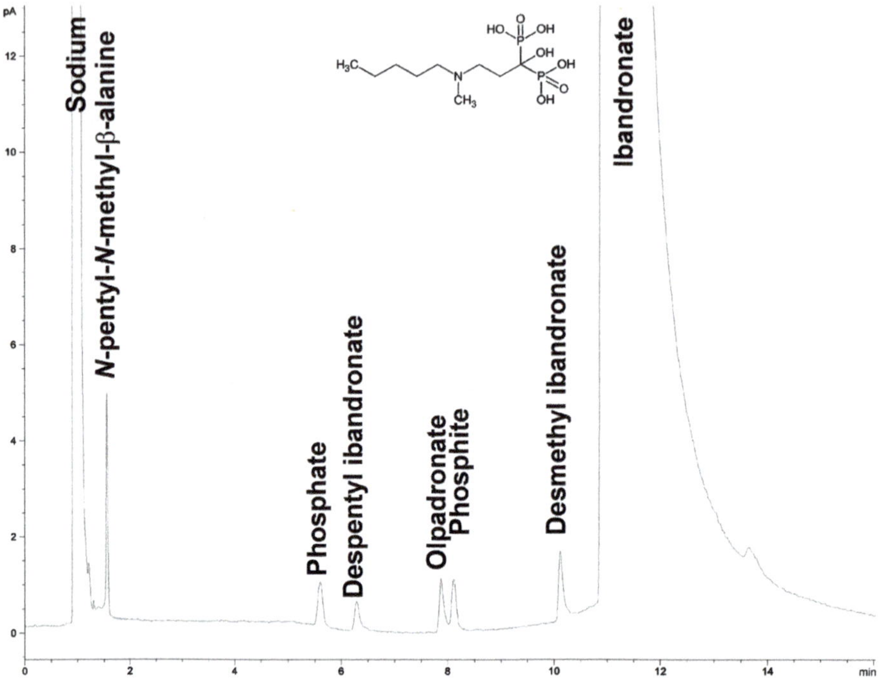

Figure 10.2 Chromatogram of ibandronate with potential impurities. *Source:* [13] / with permission of Elsevier.

groups presenting an analytical challenge with regard to both separation and detection. The separation of an important representative, ibandronate sodium, and its impurities can be achieved by using a mixed-mode column (SIELC Coresep® SB) and gradient elution by means of 4.5 mM TFA and 4.5% ACN to 15 mM TFA and 15% ACN (flow rate = 1.2 mL). Since the precursor compounds (*N*-methyl-*N*-pentyl-β-alanine, and desmethyl-ibandronate), as well as phosphate, phosphite, and the sodium counterion have to be expected as impurities, detection by CAD is the ideal solution [13]. As can be seen in Figure 10.2, all components can easily be separated, identified (by spiking), and quantified. However, batch studies revealed the dronate compound to be relatively pure with unspecified impurities below 0.10%. It is difficult to understand why the PhEur decided to apply a poor and not very robust RP chromatographic method (with a phenylhexylsilyl silica gel column) with an ACN/ethanol elution and detection by a differential refractometer, which is unable to separate any impurity besides acetone. With the pharmacopoeial method, it will be difficult to guarantee the quality of ibandronate from other batches or producers.

Due to the universal response of CAD, Ilko et al. [11] found unknown impurities in cloxacillin that were not visible in the HPLC-UV chromatogram. Because of the independence of the response from physicochemical properties, a quantification of unknown impurities can be performed without having a chemical reference substance (= CRS) at hand when using normalization. This is in stark contrast to UV detection.

Degradation studies can especially take advantage of CAD's universal response. Gao et al. [14] studied the photodegradation of lomefloxacin in eye drops. Using CAD, they found a very small range of the correction factor for the degradation products (0.94–1.07) in contrast to UV detection (ranging from 1.20 to 16.45) when applying CRS. This demonstrates the possibility of studying drug degradation and quantification of degradation products without a CRS. Furthermore, unknown degradants can be found as long as they are non or semi-volatile. This is supported by Liu et al. who studied the suitability of HPLC-CAD for impurity profiling of CRMs (Certified Reference Material) [15] whose impurities show similar responses.

10.6 Hyphenation of CAD and UV

As described above, CAD has the advantage of providing detection of compounds without chromophores, which can be beneficial for the impurity profiling of drugs. Conversely, API impurities might be volatile or semi-volatile but have a chromophore. In these cases, an in-line measurement with both CAD and UV can be of great value. E.g., amino acids which are mostly produced via fermentation are characterized by two kinds of impurities such as: (i) other amino acids which can be limited by the ninhydrin-positive substance tests and (ii) other dicarboxylic acids originating from the biosynthesis pathways and the citrate cycle; they must be quantified by an additional (HPLC) assay. Consequently, the PhEur describes two separate tests in these cases. For aspartic acid, the PhEur lists nine impurities: malic acid, fumaric acid, succinic acid, maleic acid in addition to glutamic acid, alanine, *L*-glutamine, asparagine, and *D*-aspartic acid. A simple ion-pair reversed phase chromatographic method using an Inertsil ODS three column with a mobile phase of perfluoroheptanoic acid (PFHA) in a water/methanol mix is appropriate to separate all impurities from each other and from the main peak. However, the organic diacids show a low intensity [16]. Slight changes of the column to an Acclaim Polar Advantage II and a mobile phase containing a nonafluoropentanoic acid/TFA ion pair reagent mix in water, with CAD-UV inline-measurement enables detection of all impurities in one run [17], as can be seen in Figure 10.3. In both studies, aspartic acids were found to be rather pure (content >99.7%) which is mirrored in the PhEur impurity limits of <0.10% and

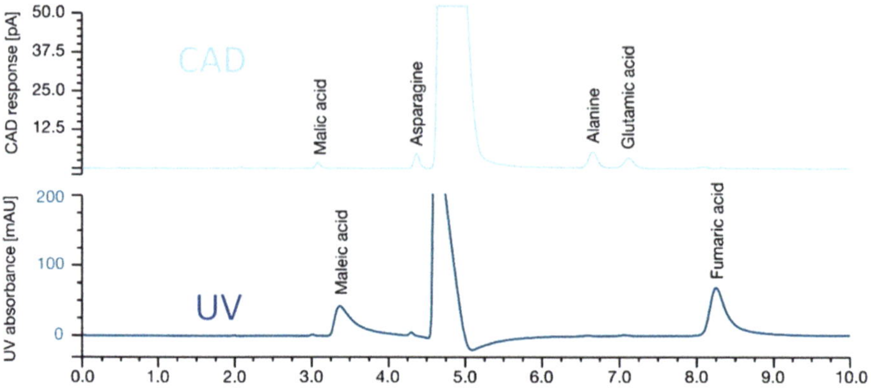

Figure 10.3 UV and CAD chromatograms of a sample solution of aspartic acid (5 mg/mL) spiked with possible impurities, in-line measurement according to ref. [17] JCA 2021, CAD: evaporation temperature 50 °C, PFV 1.0, UV detection $\lambda = 210$ nm. *Source:* [17] / with permission of Elsevier.

0.2%, respectively. The volatile maleic and fumaric acid can be quantified in the UV chromatogram and the others in the CAD chromatogram.

Similarly, impurity profiling of vigabatrin can also profit from hyphenated UV-CAD detection. Whereas the drug substance vigabatrin and two impurities (i.e., γ-aminobutanoic acid (GABA) and 2-[(2*RS*)-2-amino-3-en-1-yl]propanedioic acid are nonvolatile and can be assessed by CAD, the impurities (5*RS*)-5-ethenylpyrrolidin-2-one and (2*E*)-2-(2-aminoethyl)but-2-enoic acid are semi-volatile and volatile, respectively, but possess a sufficient chromophore to be measured by UV. The PhEur employs an ion-pairing chromatography test with a phenyl-hexyl column, and mobile phase composed of PFHA/MeOH at 45 °C and a flow rate of 1.0 mL. However, the high amount of the ion-pairing reagents needs to be diluted before detection because it increases the background noise of the CAD (see Table 10.2). Alternatively, a fully validated MMC method also provides a very good separation, higher sensitivity in a shorter run time and without the need for a second pump for dilution of the ion-pairing reagent. The Sielc Primesep® 100, a RP column with embedded strong cation exchange groups, in combination with mobile phase composed of water/ACN (85/15) and 0.1% TFA, at 25 °C and 1.2 mL/min flow rate, reduced the background noise substantially when applying a T_e of 70 °C. The UV detection was performed at $\lambda = 210$ nm. As can be seen in Figure 10.4, both detectors complement each other perfectly [18].

Likewise, the volatile and nonvolatile impurities of gabapentin can be analyzed by hyphenation of UV and CAD [8] (cf. above power function setting; see Figure 10.5). Also, the impurity proofing of simvastatin and lovastatin

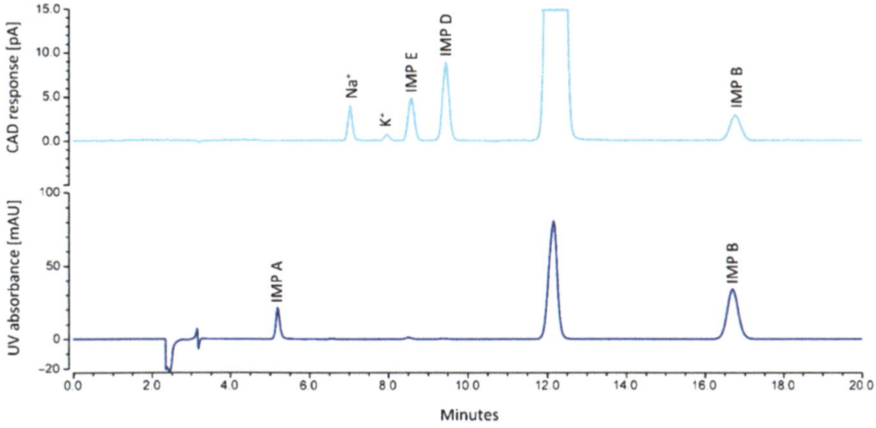

Figure 10.4 UV (below) and CAD (above) chromatograms of a 0.03% sample solution of vigabatrin spiked with possible impurities, in-line measurement according to ref. [18] JPBA 2021; CAD: evaporation temperature 70 °C; PFV 1.0 UV-detection $\lambda = 210$ nm. Looks like Figure 6 with a different sample description. *Source:* [18] / with permission of Elsevier.

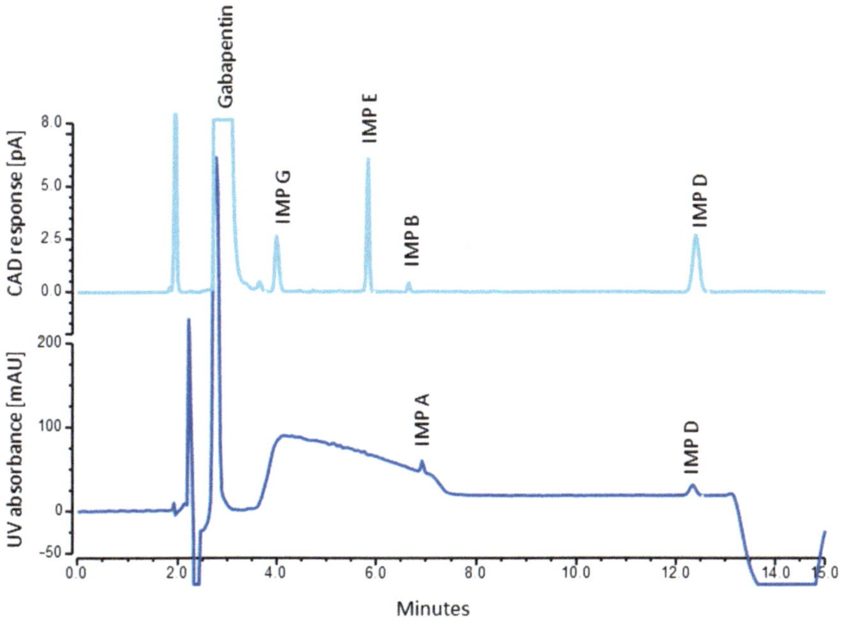

Figure 10.5 UV (below) and CAD (above) chromatograms of a gabapentin sample solution (8 mg/mL) spiked with 0.1% of possible impurities, in-line measurement according to ref. [8] JCA 2021; impurities A and B are semi volatile; CAD: evaporation temperature 30 °C, PFV 1.3, UV-detection $\lambda = 210$ nm. *Source:* [8] / with permission of Elsevier.

(both are contaminated with a corresponding dihydro-component) can take advantage of the UV/CAD hyphenation when the method of the PhEur is slightly adapted [19].

10.7 CAD in the International Pharmacopoeias

Even though CAD provides a lot of advantages for impurity profiling, i.e., detection of nonvolatile impurities with no chromophore, detection of counterions, detection of unexpected impurities etc., it is rarely used in the international pharmacopoeias. However, this is starting to change. The PhEur makes use of the CAD for the test for related substances of gadobutrol monohydrate, topiramate, and vigabatrin, and the USP applies the CAD for purity testing of deoxycholic acid, choline citrate, chloride, bitartrate, and gadobutrol, too.

For aminoglycosides which do not have any chromophore the CAD seems to be highly appropriate for impurity profiling and content determination and, thus, applications can be found in the literature, e.g., for gentamicin [20, 21], etimicin [22], amikacin [6, 23], and tobramycin [24]. Hence, it is not surprising that the USP now considers an RP18 HPLC method with a mobile phase composed of aqueous TFA and ACN (gradient elution) and a CAD for tobramycin sulfate and formulations, for content evaluation and impurity assessment [25], instead of the current method which employs a derivatization with 2,4-dinitrofluorobenzene. Time will tell if the CAD will be employed in all aminoglycoside monographs. In the case of metoprolol whose most prominent impurity is the starting product 3-isopropylamino-1,2-propanediol, the very valid HILIC-HPLC-CAD method developed by USP authors [26, 27] did not find its way into the USP until now. However, both pharmacopoeias mention the CAD in their chromatography general chapters (General Chapt. 621 Chromatography USP; Chapt. 2.2.29 PhEur) indicating that further applications are welcome.

Why are the pharmacopoeias so hesitant? On the one hand there is only one supplier for the CAD. If the supplier decides to stop production, then the monograph would become useless and hence, the quality control impossible. On the other hand, the CAD requires relatively pure solvents, in some cases LC-MS grade, and the ability to successfully use a monograph may be affected by a change in solvent supplier. The same holds true for chemically identical columns, the end results of which might differ from different providers. The same or similar challenges may also be seen with monographs that use HPLC with UV detection.

Nonetheless, the advantage of having universal, robust, and sensitive detection outweighs the above-mentioned disadvantages. Moreover, unexpectedly, maybe other potentially toxic impurities might also be discovered. Thus, there are more monographs using the CAD to come.

References

1 European Pharmacopoeia, Council of Europe (2023). 11th Ed., European Directorate for the Quality of Medincines and Healthcare, Strasbourg.

2 United States Pharmacopeia (USP) (2024). *USP-NF*, Maryland, USA. https://www.usp.org/.

3 Japanese Pharmacopoeia (2021). 18th Ed., Pharmaceutical and Medical Agency Tokyo, Japan. https://www.pmda.go.jp/english/rs-sb-std/standards-development/jp/0029.html.

4 Almeling, S. and Holzgrabe, U. (2010). Use of evaporative light scattering detection for the quality control of drug substances: influence of different liquid chromatographic and evaporative light scattering detector parameters on the appearance of spike peaks. *Journal of Chromatography. A* 1217 (14): 2163–2170.

5 Pawellek, R. and Holzgrabe, U. (2021). Performance of ion pairing chromatography and hydrophilic interaction liquid chromatography coupled to charged aerosol detection for the analysis of underivatized amino acids. *Journal of Chromatography. A* 1659: 462613.

6 Ahmad, I.A.H., Blasko, A., Wang, H. et al. (2021). Charged aerosol detection in early and late-stage pharmaceutical development: selection of regression models at optimum power function value. *Journal of Chromatography. A* 1641: 461997.

7 Barnhart, W.W., Farooq, M.Q., and Ahmad, I.A.H. (2024). A simplified tutorial on charged aerosol detection: understanding the basics, optimization, and troubleshooting. *Journal of Chromatography Open* 6: 100181.

8 Pawellek, R., Muellner, T., Gamache, P., and Holzgrabe, U. (2021). Power function setting in charged aerosol detection for the linearization of detector response–optimization strategies and their application. *Journal of Chromatography. A* 1637: 461844.

9 Handlovic, T.T., Roy, D., Barnhart, W.W., and Haidar Ahmad, I.A. (2024). Role of the power function value in linearity and universality for charged aerosol detectors: theoretical elucidations from a validated model. *Analytical Chemistry* 96 (40): 16045–16052.

10 Zhang, K., Dai, L., and Chetwyn, N.P. (2010). Simultaneous determinatieon of positive and negative pharmaceutical counterions using mixed-mode chromatography coupled with charged aerosol detector. *Journal of Chromatography. A* 1217 (37): 5776–5784.

11 Ilko, D., Nap, C., Holzgrabe, U., and Almeling, S. (2014). Validation and application of an HPLC-CAD-TOF/MS method for identification and quantification of pharmaceutical counterions. *Pharmeuropa Bio & Scientific Notes* 2014: 81–91.

12 Li, P., Sun, W., Zuo, L. et al. (2018). Fast simultaneous determination of main components and impurity sodium ion in PAMA injection by mixed-mode chromatography. *Journal of Pharmaceutical and Biomedical Analysis* 161: 407–413.

13 Wahl, O. and Holzgrabe, U. (2015). Impurity profiling of ibandronate sodium by HPLC–CAD. *Journal of Pharmaceutical and Biomedical Analysis* 114: 254–264.

14 Gao, J., Xu, B., Li, P. et al. (2023). Universal response method for the quantitative analysis of photodegradation impurities in lomefloxacin hydrochloride ear drops by liquid chromatography coupled with charged aerosol detector. *Journal of Pharmaceutical and Biomedical Analysis* 234: 115552.

15 Liu, S., Lu, B., Peng, Z. et al. (2023). HPLC-CAD as a supplementary method for the quantification of related structure impurities for the purity assessment of organic CRMs. *Analytical and Bioanalytical Chemistry* 415: 3375–3384.

16 Holzgrabe, U., Nap, C.-J., and Almeling, S. (2010). Control of impurities in l-aspartic acid and l-alanine by high-performance liquid chromatography coupled with a corona charged aerosol detector. *Journal of Chromatography. A* 1217 (3): 294–301.

17 Pawellek, R., Schilling, K., and Holzgrabe, U. (2020). Impurity profiling of l-aspartic acid and glycine using high-performance liquid chromatography coupled with charged aerosol and ultraviolet detection. *Journal of Pharmaceutical and Biomedical Analysis* 183: 113149.

18 Pawellek, R. and Holzgrabe, U. (2021). Influence of the mobile phase composition on hyphenated ultraviolet and charged aerosol detection for the impurity profiling of vigabatrin. *Journal of Pharmaceutical and Biomedical Analysis* 201: 114110.

19 Walther, R. and Holzgrabe, U. (2023). Simplification of pharmacopoeial liquid chromatography methods for related substances of statins by hyphenated ultraviolet and charged aerosol detection. *Journal of Pharmaceutical and Biomedical Analysis* 225: 115218.

20 Joseph, A. and Rustum, A. (2017). Paul H. Gamache, Thermo Fisher Scientific, Wiley, USA. Applying charged aerosol detection to aminoglycosides: development and validation of an RP-HPLC method for gentamicin and netilmicin. In: *Charged Aerosol Detection for Liquid Chromatography and Related Separation Techniques*, 393–423.

21 Stypulkowska, K., Blazewicz, A., Fijalek, Z., and Sarna, K. (2010). Determination of gentamicin sulphate composition and related substances in pharmaceutical preparations by LC with charged aerosol detection. *Chromatographia* 72 (11–12): 1225–1229.

22 Gao, J., Wang, Q., Zhu, B. et al. (2024). Quantitative analysis of the impurities in Etimicin using hydrophilic interaction liquid chromatography coupled with charged aerosol detector. *Journal of Pharmaceutical and Biomedical Analysis* 249: 116384.

23 Soliven, A., Haidar Ahmad, I.A., Tam, J. et al. (2017). A simplified guide for charged aerosol detection of non-chromophoric compounds—Analytical method development and validation for the HPLC assay of aerosol particle size distribution for amikacin. *Journal of Pharmaceutical and Biomedical Analysis* 143: 68–76.

24 Chopra, S., Vanderheyden, G., Hoogmartens, J. et al. (2010). Comparative study on the analytical performance of different detectors for the liquid chromatographic analysis of tobramycin. *Journal of Pharmaceutical and Biomedical Analysis* 53 (2): 151–157.

25 Zeine, C.B.J., Tran, D., Reddy, R., and Laranang, T. (2023). Use of impurities in the development and validation of a LC-method with CAD detection for tobramycin impurity analysis.

26 Xu, Q. and Tan, S. (2019). Quantitative analysis of 3-isopropylamino-1,2-propanediol as a degradation product of metoprolol in pharmaceutical dosage forms by HILIC-CAD. *Journal of Pharmaceutical Analysis* 9 (6): 431–436.

27 Xu, Q., Tan, S., and Petrova, K. (2016). Development and validation of a hydrophilic interaction chromatography method coupled with a charged aerosol detector for quantitative analysis of nonchromophoric α–hydroxyamines, organic impurities of metoprolol. *Journal of Pharmaceutical and Biomedical Analysis* 118: 242–250.

Section 3

Industrial Applications of Charged Aerosol Detection

11

Analysis of Lipid-Based Formulations with Charged Aerosol Detection

Claudia Seidl[1], Gabriel M. Leme[1], Charles R. Bupp[1], Andrei Blasko[2], and Imad A. Haidar Ahmad[1]

[1] *One Amgen Center Drive, AMGEN Inc., Thousand Oaks, CA, USA*
[2] *Clearview Pharma Solutions, LLC, Dallas, TX, USA*

CHAPTER MENU

11.1 Introduction

Lipid-based formulations (LBFs) have become an essential tool in modern pharmaceutical development, providing an effective strategy to overcome the common challenge of poor aqueous solubility in both established and emerging drug candidates [1, 2]. With an increasing proportion of drug candidates classified under the Biopharmaceutics Classification System (BCS) as Class II (low solubility, high permeability) and Class IV (low solubility, low permeability), formulation scientists are increasingly turning to LBFs to overcome absorption barriers, enhance oral bioavailability, and improve overall therapeutic performance [3].

LBFs are drug-delivery systems that utilize lipids as key excipients to enhance the solubility, stability, and absorption of poorly water-soluble drugs. These formulations cover a wide spectrum, ranging from simple oil solutions and lipid suspensions to more advanced emulsions, microemulsions, self-emulsifying drug delivery systems (SEDDS), and liposomes (when used primarily for solubilization or stabilization without additional delivery functionalities). Importantly, LBFs form the foundation of lipid-based drug delivery systems (LBDDS), a broader class that builds upon the principles of LBFs to achieve more sophisticated therapeutic objectives. While LBFs primarily focus on improving solubility and bioavailability, LBDDS are designed to precisely modulate pharmacokinetics, biodistribution, and site-specific delivery. More complex in composition, these systems often incorporate structural components such as cholesterol or block copolymers and are specifically designed to enable controlled or sustained release, enhance membrane permeability, and achieve targeted delivery to specific tissues or cells. In more advanced designs, functionalization with target ligands, such as antibodies, peptides, or nucleic acids like mRNA, further enhances site-specific delivery and therapeutic precision. Figure 11.1 illustrates the diversity of some types of LBFs, from basic emulsion-based formulations to sophisticated LBDDS comprised in vesicular-based and lipid-particulate systems [4–6].

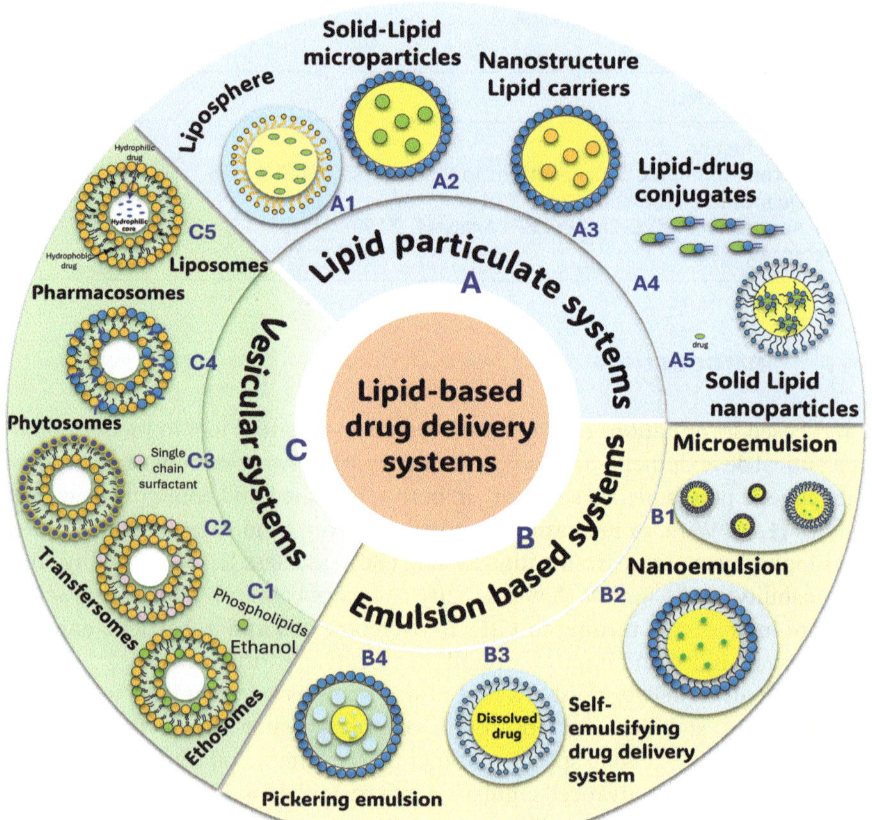

Figure 11.1 **Schematic representation of different types of LBDDS.** **A:** Lipid particle systems. **A1:** Liposphere is composed of a solid-lipid core, which contains the drug, surrounded by a single layer of phospholipid. **A2:** Solid-lipid microparticle composed of a solid-lipid core, which contains the drug, surrounded by a surfactant outer shell. **A3:** Nanostructure lipid carrier composed of a mixed solid-liquid lipid matrix, which encapsulates drug molecules, surrounded by a surfactant layer. **A4:** Lipid-drug conjugate composed of a drug molecule chemically bonded to a lipid. **A5:** Solid-lipid nanoparticles composed of a solid-lipid core containing an encapsulated drug surrounded by a surfactant layer. **B:** Emulsion-based systems. **B1:** Micro-emulsion which consists of an oil-water emulsion stabilized by a surfactant to form nanodroplets. These droplets are thermodynamically stable and are typically ~10–100 nm in size. **B2:** Nano-emulsions are kinetically stable dispersions formed of droplets up to 500 nm. **B3:** Self-emulsifying drug delivery systems are isotropic mixtures composed of a surfactant layer, lipid layer, and co-solvent or co-surfactant, carrying a drug. Upon dilution, these mixtures spontaneously form emulsions. **B4:** Pickering emulsion which consists of lipid and aqueous liquid layers, with one layer suspended into the other. The structure is stabilized by micro- or nanoparticles. **C: Vesicular** systems. **C1:** Ethosomes are made of phospholipids in the form of soft phospholipid vehicles containing a high concentration of ethanol and water. **C2:** Transferosomes are made of phospholipids, surfactants, and water, which are designed to carry drugs which could cross the skin due to their ultra deformability. **C3:** Phytosomes are made of phytochemicals which are chemically bound to phospholipid to improve their absorption and bioavailability. **C4:** Pharmacosomes are colloidal vesicles where the drug molecules are covalently bonded to phospholipids, which results in the formation of amphiphilic complexes. **C5:** Liposomes are spherical vesicles designed to deliver both hydrophilic and hydrophobic drugs. They are composed of phospholipid bilayers which are surrounded by aqueous layers.

The specific composition of an LBF is tailored to the physicochemical properties of the drug and the intended delivery route. Typically, these formulations combine one or more lipids, such as medium or long-chain triglycerides, phospholipids, cholesterol, fatty acids, sterols, or PEGylated lipids, which act as the primary carriers and structural matrix. To promote emulsification and ensure colloidal stability, surfactants (nonionic or ionic, such as polysorbates) are incorporated, often alongside co-solvents or co-surfactants like ethanol, propylene glycol, or polyethylene glycol 400 (PEG400) to improve miscibility and maintain homogeneity. Depending on the formulation goal, additional water-soluble or water-insoluble excipients and an aqueous phase may be included to achieve the desired physicochemical characteristics and facilitate dispersion upon administration [7].

The intricate composition of LBFs inherently contributes to significant analytical and manufacturing challenges. Their complexity arises from the coexistence of components that vary widely in polarity, molecular weight, and chemical functionalities, factors that can lead to incomplete solubilization, broad particle size distribution, and physical instability caused by the self-assembly properties of certain lipid formulations [8, 9]. Consequently, LBFs are commonly analyzed as dispersions rather than true solutions [6]. Moreover, lipids are particularly prone to aggregation and degradation during storage and transportation, leading to drug leakage, reduced potency, and diminished safety [10]. Therefore, maintaining strict control over the identity and concentration of individual lipid components during the production and release stages is essential to ensure product consistency and therapeutic reliability. To address these challenges, regulatory agencies have established specific expectations for the design, control, and analysis of LBF. The US Food and Drug Administration (FDA) provides detailed guidance on lipid-containing products, including both traditional liposomes and modern lipid nanoparticle (LNP) systems developed for RNA and other biotherapeutic applications. According to the FDA Guidance for Industry: Liposome Drug products, manufacturers must implement comprehensive chemistry, manufacturing and controls (CMC) documentation describing each lipid component and its critical quality attributes. These requirements include defining the identity, purity, and source of all lipids; establishing lipid-to-drug ratios; and validating analytical methods capable of detecting impurities or degradation products [11].

Among modern detection techniques, charged aerosol detection (CAD) has emerged as a powerful and versatile solution for addressing the inherent analytical challenges of LBFs. As a near-universal, mass-sensitive detector, CAD operates independently of a compound's chromophoric properties and provides consistent, reproducible responses across a broad range of nonvolatile analytes. Unlike ultraviolet (UV) or evaporative light scattering detection (ELSD), CAD relies on a mass-dependent detection principle generating a direct, proportional signal that more accurately reflects true component ratios in complex multicomponent samples. This capability makes CAD particularly valuable for the quantitative analysis of non-chromophoric compounds surpassing the sensitivity of ELSD. When coupled with high-performance liquid chromatography (HPLC), CAD combines good sensitivity and robustness, making HPLC-CAD an indispensable analytical tool for applications ranging from early-stage formulation screening to advanced, regulatory-compliant method validation [12, 13].

This chapter explores the application of CAD as a powerful analytical technique for the characterization of LBF. It highlights how CAD effectively addresses key analytical challenges associated with the complex, multicomponent nature of these systems by discussing applications across diverse formulation types. The discussion also includes key chromatographic considerations for the analysis of LBFs such as sample preparation, dilution strategies, selection of chromatographic mode, stationary-phase chemistry, pore sizes, and mobile phase optimization. Emphasis is placed on the use of CAD for quantitative analysis of individual formulation components, including phospholipids, PEGylated lipids, cholesterol, surfactants in liposomes, and LNPs, as well as to its use as a tool for in-process control, stability assessment, and formulation development.

11.2 Chromatographic Considerations in the Analysis of LBFs

Effective chromatographic separation is fundamental to the characterization of LBFs, which often contain chemically diverse components that must often be quantified within a single analytical run. Even when using a universal detector such as CAD, the quality of the chromatographic data ultimately depends on appropriate sample preparation and optimization of separation and detection conditions. Inadequate extraction, dilution, or solvent selection can alter sample composition and compromise quantitation, while poor chromatographic resolution can obscure critical information, leading to inaccurate quantitation or misinterpretation of formulation behavior. Achieving accurate, reproducible results therefore require careful optimization of sample preparation and handling, chromatographic mode selection, column chemistry and pore size, as well as mobile phase composition and optimization, discussed in the following sections.

11.2.1 Sample Preparation and Dilution Strategies

Accurate HPLC-CAD analysis of LBFs requires rigorous control of sample and standard preparations. Insoluble or poorly dispersed components can lead to precipitation and inconsistent sampling, while variation in solvent properties and matrix composition can disrupt aerosol formation and compromise detector response [14].

Solvent selection is critical. Lipid in LBF samples should be dissolved or diluted in organic solvent, such as 100% methanol, ethanol, or isopropanol because they are compatible with both lipid and amphiphilic components and with CAD nebulization [15]. Prior to injection, samples may need to be further diluted with a suitable solvent system, often methanol/water (90:10, v/v) with a small percentage of an additive, such as trifluoroacetic acid (TFA), to improve peak shape and retention consistency. This is crucial to prevent sample aggregation by optimizing sample solvent and minimizing sample precipitation and phase separation, through organic solvent blending, gradual dilution and temperature control [16]. Overly aqueous or high-boiling solvents can suppress signal or increase baseline noise, whereas high-volatile systems may cause unstable aerosol formation and erratic detector response. Care must also be taken to prepare standards and samples using the same solvent matrix to maintain consistent aerosol formation and signal response. Filtration through low-binding, hydrophobic-resistant filters (preferably PTFE or glass fiber) is advised to remove particulates without lipid loss.

Another potential issue is sample loss due to adherence to glassware and other labware, which can result in low recovery and poor reproducibility. Given the hydrophobic nature and solvent sensitivity of lipids, it is strongly recommended to use silanized glassware exclusively, including pipettes, vial inserts, and storage bottles, to minimize adsorption losses and prevent the leaching of plasticizers that may interfere with detection [17].

These practices collectively promote stable aerosol formation, minimize matrix-related variability and enhance the accuracy and precision of HPLC-CAD quantitation across diverse LBF components.

11.2.2 Choosing the Chromatographic Mode for Analyzing LBFs

For the analysis of LBFs, reversed-phase high-performance liquid chromatography (RP-HPLC) is the most widely used approach due to its high reproducibility, broad accessibility, and compatibility with analytes spanning across a wide range of polarities. The variety of available column chemistries and mobile phases options provides flexibility to tailor selectivity for specific formulation components.

A key aspect of method development for LBFs lies in the careful selection of the chromatographic column, as it directly influences separation efficiency, resolution, and overall method performance. In reversed-phase separations, the retention of lipids is governed by a combination of head group polarity, degree of tail unsaturation, and alkyl chain length. Less polar head groups, longer saturated tails, and fewer double bonds increase hydrophobic interactions and increase retention times. Conversely, highly polar head groups, shorter chains, or greater unsaturation reduce hydrophobic interactions, leading to earlier elution. Understanding how these structural factors influence hydrophobicity is essential for designing chromatographic conditions that can resolve diverse lipid species in complex LBFs [18].

Structural variations among PEGylated lipids (Figure 11.2) can markedly influence chromatographic behavior. DMG-PEG (1,2-dimyristoyl-*rac*-glycero-3-methoxy-polyethylene glycol) and DMPE-PEG (1,2-dimyristoyl-sn-glycero-3-phosphoethanolamine-N-PEG) contain short, fully saturated C14 alkyl chains that limit hydrophobic interactions relative to longer-chain analogs. However, differences in head-group chemistry significantly affect polarity: the methoxy end group at the polyethylene glycol moiety in DMG-PEG is less polar than the phosphoethanolamine group in DMPE-PEG, resulting in stronger hydrophobicity and stronger retention for DMG-PEG.

Cationic surfactants such as DOTAP (1,2-dioleoyl-3-trimethylammonium propane) and DOBAQ (*N*-(4-carboxybenzyl)-*N*,*N*-dimethyl-2,3-bis(oleoyloxy) propan-1-aminium) (Figure 11.2) feature long, unsaturated C18 hydrocarbon chains that confer substantial hydrophobic character to their molecular structure. However, DOTAP is more hydrophobic than DOBAQ because it has a larger uninterrupted hydrophobic segment and lacks the polar benzoic acid moiety present in DOBAQ. Phospholipids like DSPC (distearoylphosphatidylcholine),

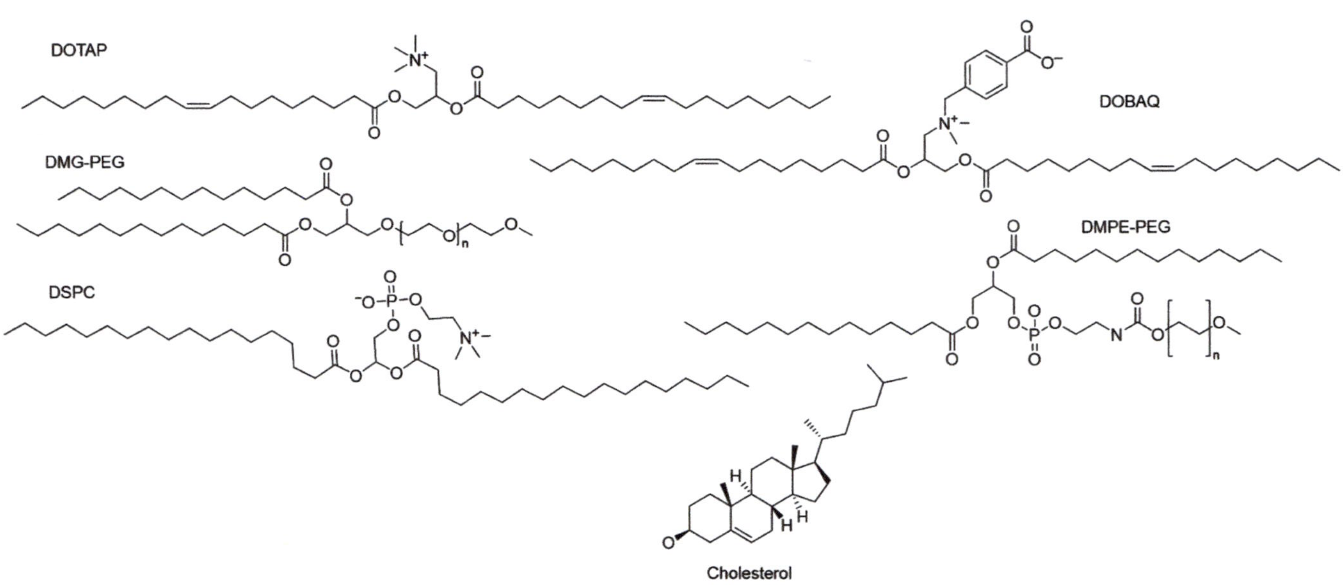

Figure 11.2 Chemical structures of selected lipid species relevant to LBFs. PEGylated lipids (DMG-PEG, DMPE-PEG), cationic lipids (DOTAP, DOBAQ), the phospholipid DSPC, and cholesterol. Structural variations in head-group polarity and hydrocarbon chain composition govern hydrophobicity and retention in RP-HPLC.

with two fully saturated C18 chains, exhibit high hydrophobicity despite their polar phosphatidylcholine head group. Full saturation promotes dense chain packing and a larger hydrophobic surface area, often resulting in longer retention times than less saturated or shorter-chain phospholipids.

Cholesterol (Figure 11.2) is a common steroid used in LBFs for its role in modulating membrane packing, reducing permeability, and enhancing structural stability. It is largely hydrophobic due to its rigid polycyclic hydrocarbon structure. The presence of a single hydroxyl group induces localized polarity, which can slightly reduce its retention in RP-HPLC compared to lipids with similar hydrophobicity but with more flexible alkyl chains, leading to earlier elution despite its high overall hydrophobicity.

These examples highlight that even small structural variations, such as differences in chain length, saturation, or head group chemistry, can significantly influence hydrophobic interaction with the stationary phase and, consequently, chromatographic behavior.

11.2.3 Stationary-Phase Chemistry and Pore Size

Reversed-phase chromatographic columns with C18 bonded stationary phase are widely regarded as the standard choice for LBFs analysis, offering strong retention and high resolution for a broad spectrum of lipid species, including triglycerides, phospholipids, and PEGylated lipids. However, when the goal is to analyze less hydrophobic components or to improve peak symmetry, alternative stationary phases such as C8 and C4 can be employed to moderate retention strength. Phenyl columns represent another versatile option, combining hydrophobic interactions with π–π interactions to enhance selectivity for chemically diverse lipids and lipid–PEG conjugates. This dual-mode of retention can be particularly advantageous for separating mixtures that include phospholipids, sterols, cationic lipids, and PEGylated lipids [19]. During column screening, Tam et al. [20] applied the hydrophobic subtraction model (HSM) to evaluate 15 reversed-phase columns with varied selectivity parameters for the separation of the hydrolytic products of DSPC, 1-SLPC (2-stearoyl-sn-glycero-3-phosphocholine) and 2-SLPC (1-stearoyl-sn-glycero-3-phosphocholine), which are stearoyl-lysophosphatidylcholine (S-LPC) positional isomers formed from partially hydrolyzed DSPC in DSPC containing formulations. The goal was to achieve baseline separation of the S-LPC isomers from each other and from the API degradants and process impurities. Many columns failed to meet these criteria due to co-elution of one or both S-LPC isomers with late-eluting API-related impurities (e.g., Phenomenex Kinetex, Biphenyl, Halo Phenyl-Hexyl, and multiple C18 phases), insufficient retention or weak interactions with the target analyte (e.g., Agilent Poroshell EC-CN), or excessive retention that reduced method efficiency (e.g., Agilent Zorbax Extend

C18, ACE Excel Super C18). The Poroshell Bonus-RP column, a solid-core phase with a polar-embedded amide group, demonstrated the most suitable selectivity for this application. It successfully placed both S-LPC peaks in regions of the chromatogram free from potential interference, achieving clean resolution from the API peak. Although the conclusion about the choice of column in the above-mentioned example applies specifically to the sample analyzed in this study, evaluating multiple selectivities by screening multiple combinations of chromatographic columns and mobile phases (including variations in mobile phase pH) remains the most effective strategy to find conditions that achieve the desired resolution between peaks of interest.

Column pore size selection is as critical as stationary phase and mobile phase choice, particularly for analytes of large molecular volume present in LBFs. For bulky lipids such as PEGylated lipids (e.g., DMG-PEG 2000, which is ~2500 Da) or long-chain phospholipids, wider pores facilitate unhindered diffusion into the stationary phase, improve mass transfer kinetics, and reduce peak broadening from restricted diffusion. Kim et al. [15] applied an analytical quality by design (AQbD) framework to compare C18 columns with pore sizes of 100, 130, and 300 Å. They concluded that the 300 Å Waters Inc. XBridge Peptide BEH C18 column provided superior resolution and efficiency because its wide pores accommodated larger lipids without steric hindrance, whereas smaller pore columns risked reduced accessibility and broader peaks. In contrast, Beckert et al. [19] successfully used a phenyl column with 130 Å pore size for six lipids, demonstrating that moderate pore sizes can still deliver baseline separation when analytes are within the pore access range. Su et al. [21] targeted simultaneous separation of lipids and large nucleic acids (mRNA, sgRNA), favored 1000 Å pore C4 and C18 columns to maximize pore accessibility for these very large biomolecules.

Collectively, these examples highlight that both column chemistry and pore size are critical parameters involved in chromatographic performance in HPLC-CAD analysis of LBFs. Stationary phase chemistry dictates the dominant interaction mechanism while pore size selection must align with the molecular dimensions and diffusivity of the analytes. Ultimately, optimal method performance arises from the combined tuning of stationary phase chemistry, pore size and mobile phase composition, which together balance retention, selectivity, and diffusion.

11.2.4 Mobile Phase Considerations

Mobile phase selection in RP-HPLC-CAD analysis of LBFs is a critical parameter for balancing solubility, CAD detection compatibility, and resolution across chemically diverse lipid classes and, in some cases, co-analysis with other formulation components. Typical mobile phase systems for LBFs analysis combine an aqueous buffer containing volatile additives with a strong organic solvent to achieve elution of highly hydrophobic species. Acetonitrile (ACN) and methanol (MeOH) are the most commonly used organic solvents, chosen for their polarity, viscosity and elution strength, while volatile additives are incorporated to control peak shape and retention, particularly for ionizable lipids.

Beckert et al. [19] optimized a mobile phase composition for the simultaneous quantification of six lipids in LNP formulations. Their final method used 0.1% (v/v) TFA in both aqueous (water) and organic (acetonitrile) phases, providing mild ion pairing that improved retention and peak shape for cationic lipids while avoiding the baseline noise and peak broadening often associated with stronger alkyl ammonium ion-pairing reagents. Su et al. [21] developed an ion-pair reversed-phase platform method for the simultaneous separation of lipids and large therapeutic nucleic acids [mRNA, sgRNA, antisense oligonucleotides (ASO)] within complex LBFs. Optimization focused on identifying an ion-pair reagent that would provide sufficient retention of highly polar, negatively charged nucleic acids while maintaining resolution for co-formulated lipids. After screening alkylammonium salts of varying chain length and concentration, 100 mM dibutylammonium acetate (DBAA) was selected for its balanced hydrophobicity and volatility. The final gradient system used DBAA in water as mobile phase A and isopropanol as mobile phase B, achieving baseline separation of nucleic acids from PEGylated and phospholipid species in a single run, an important advance towards unified characterization of lipid-nucleic acid LBFs. Kim et al. [15] applied an analytical quality-by-design (AQbD) approach to systematically optimize the mobile phase for RP-HPLC analysis of four lipid classes relevant to LBFs. Multiple solvent types (acetonitrile, methanol and their mixtures), pH levels, and buffer concentrations were evaluated to balance resolution, analysis time, and CAD compatibility. The final method used methanol containing 7.5 mM ammonium formate as the organic phase and an aqueous phase with the same additive, providing volatile CAD-compatible buffering while maintaining sufficient ionic strength for peak shape control. Methanol was ultimately preferred over acetonitrile for its superior selectivity in resolving closely eluting lipid species in the test panel.

Yu et al. [16] developed an RP-HPLC-CAD method for quantifying lipid content in LNP-mRNA vaccines, with mobile phase composition tailored to maintain lipid stability and consistent detector response during analysis. The optimized system used 0.01 M triethylamine acetate (TEAA) as a volatile buffer in both aqueous and ethanol phases. TEAA stabilized charged lipid species and improved peak shape while remaining fully compatible with CAD detection. Ethanol was chosen as organic modifier instead of acetonitrile or methanol because of its favorable impact on lipid solubility and elution behavior, particularly for highly hydrophobic lipid components. Tam et al. [20] investigated the influence of TFA concentrations on RP-HPLC-CAD separation of S-LPC degradants derived from DSPC, focusing on its impact on separation quality and CAD performance. The selected mobile phase consisted of 0.1% (v/v) TFA in both water (A) and (B) methanol, providing volatile ion-pairing that improved retention and peak symmetry for zwitterionic LPCs. Comparative testing using 0.2% (v/v) TFA achieved similar separations but produced higher baseline noise in CAD due to additive impurities. The authors recommended using minimum percentage of mobile phase modifier to reduce baseline noise arising from potential impurities in the additive. In addition, the use of LC-MS grade chemicals is strongly recommended to minimize baseline drift. Li et al. [22], developed a single ion-pair RP-UHPLC method capable of simultaneously separating siRNA duplexes and functional lipids in LNPs using dual detection (UV for siRNA and CAD for lipids). The main challenge was the large retention time gap between highly polar siRNA and hydrophobic phospholipids under conventional C18/TEAA conditions. While siRNAs eluted at relatively low organic content (~15–22% acetonitrile), most lipids required stronger elution conditions (~83–100% acetonitrile). In addition, siRNAs showed poor peak shapes due to the stereochemical complexity of their phosphorothioate linkages [23, 24]. The authors addressed this by optimizing selectivity using a two-tiered approach. At the stationary phase level, they screened multiple chemistries and identified the ethylene-bridged hybrid (BEH) phenyl column as optimal, leveraging π–π interactions to enhance siRNA retention while narrowing the lipid retention range. At the ion-pair level, they systematically varied alkylammonium homologs (C4–C10) and established a first-order linear correlation between the siRNA logarithmic retention factor ($\ln k$) and alkyl-chain length, enabling predictable, tunable siRNA retention without disturbing phospholipid elution order. From this screening, DBAA was selected over other structural analogs such as TEAA, DPAA, and DAAA because it produced cleaner baselines while maintaining comparable selectivity. The choice of ion-pair reagent had minimal influence on phospholipid separation, consistent with the hydrophobic-dominated retention mechanism of lipids. The optimized short gradient was achieved on a BEH phenyl column using 0.1 M DBAA in a 25 minutes gradient method, resolving multiple siRNAs, zwitterionic and cationic lipids, and degradation products in real LNP samples. This workflow established a practical, unified RP-HPLC-CAD approach for the comprehensive siRNA-lipid characterization in LNP formulations.

Collectively, these studies underscore the importance of balancing solvent strength, additive volatility, and ionic character when designing CAD-compatible mobile phases for LBFs analysis. Methanol and acetonitrile remain the primary organic modifiers, selected based on the desired elution range and selectivity, while ethanol is emerging as a practical alternative for enhancing lipid solubility in highly hydrophobic systems. Volatile ion-pair reagents such as TFA, TEAA, and DBAA, allow modulation of retention for charged analytes without compromising CAD detection. Ultimately, the most successful methods couple volatile, CAD-compatible additives with optimized organic strength, enabling simultaneous quantification of chemically diverse components, from polar nucleic acids to hydrophobic lipids within a single run.

11.3 Quantitation and Validation

The performance of liquid chromatography-charged aerosol detection (LC-CAD) methods can be on par with LC–UV when proper precautions are followed during method development and implementation (Figure 11.3). Furthermore, LC-CAD methods can also be fully validated to meet stringent modern analytical requirements for accuracy, precision, and robustness according to International Council

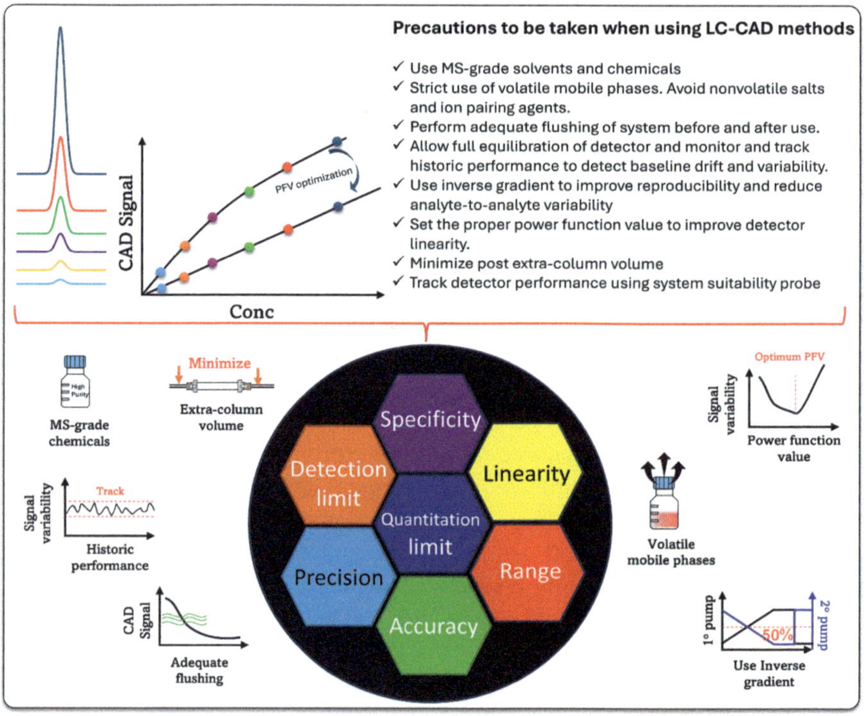

Figure 11.3 **Key operational precautions and validation criteria for LC-CAD methods.** The upper panel summarizes essential practices for achieving stable detector performance and reproducible quantitative data, including solvent selection, mobile phase volatility, system equilibration, and power function value (PFV) optimization. The lower panel highlights core validation parameters—specificity, linearity, range, accuracy, precision, and quantitation and detection limits—that define method reliability and compliance with analytical standards [10, 25].

for Harmonization Guideline Q2(R2): Validation of Analytical Procedures [ICH Q2(R2)] guidelines [10, 13, 26].

Table 11.1 summarizes representative LC-CAD methods developed for the quantitation of lipids commonly used in LBFs. These examples encompass a broad range of lipid classes, including phospholipids, cholesterol, PEGylated lipids, and ionizable or cationic lipids, and report key analytical performance parameters such as linearity, range, precision, accuracy or recovery, and limits of quantitation (LOQ). Method-specific annotations highlight differences in calibration strategy, detector sensitivity, and chromatographic conditions. Collectively, the consistent and reproducible performance achieved across these studies underscores the versatility, quantitative reliability, and growing analytical relevance of CAD for the analysis of LBFs.

A review of published CAD-based LC methods reveals a broad consensus on the detector's suitability for quantifying diverse components in LBFs. Collectively, these studies demonstrate that, when properly optimized, CAD delivers reliable, reproducible, and robust quantitative performance, making it well suited for routine analytical applications in the characterization and quality control of LBFs.

Across the methods surveyed, linearity consistently met regulatory requirements, with correlation coefficients (R^2) typically exceeding 0.995 and often approaching 0.999. To accommodate CAD's characteristic nonlinear response at higher analyte loads, many studies implemented quadratic calibration models or weighted regressions (e.g., $1/x^2$). The validated concentration ranges varied by lipid class: phospholipids and cholesterol were generally quantified from low μg/mL level to ~1000 μg/mL, whereas PEGylated lipids were frequently detected at sub-μg or ng levels, underscoring CAD's high sensitivity for amphiphilic and low-mass lipid species.

Precision was likewise strong. In most studies, repeatability and intermediate precision expressed as relative standard deviation (RSD) were below 4%, even in complex formulation matrices. Multi-analyte or platform methods occasionally reported higher variability (up to ~14%) at concentration extremes, a reflection of the broader dynamic range required for such assays.

Accuracy and recovery results were equally robust, typically within 90–110%, and often tighter (97–103%), aligning with regulatory expectations under ICH Q2 guidelines. Several reports extended validation to mixture-level performance, an increasingly relevant practice for LNP and liposomal formulations that contain multiple co-formulated lipids.

LOQ is governed by analyte chemistry, the mobile phase and additives used, and the chromatographic method conditions.

From a method development perspective, several CAD-specific strategies were used to enhance analytical performance. These included diverting high-abundance components to protect detector stability, optimizing

Table 11.1 Representative LC-CAD methods for the quantitative analysis of key lipid classes in LBFs.

Lipid(s)	Linearity (range; R^2)	Precision	Accuracy/ recovery (%)	LOQ (μg/mL)	Notes	Mobile phase	Column	References
1- & 2-stearoyl-lysophosphatidyl-choline (LPCs)—degradants of DSPC	0.5–12.5 μg/mL; $R^2 = 0.999(8)$ and 0.999(9) (2 instruments)	Repeatability 0.3–2.8% RSD; injection RSD at LOQ 1.7–2.4%	97.6–103.6% ($n = 5$, average result from spiked placebo)	0.5 (0.1% of target sample concentration)	Weighted ($1/x^2$) linear regression; flow-divert around DSPC; stability-indicating	A: 0.1% (v/v) TFA in water B: 0.1% (v/v) TFA in methanol	Agilent Poroshell Bonus-RP, 2.7 μm, 4.6 × 100 mm	[20]
DPPC, DOPC, POPC (and additional natural/synthetic PLs)	0.5–100 μg/mL; $R^2 \geq 0.975–0.999$	Reproducible, but RSD is not shown.	Deviation from nominal concentration 1.4–5.7% (selected PLs at different concentrations)	Analyte-dependent. < 0.02–25	RPLC-CAD; analyte-specific sensitivity.	A: 0.2% TFA in ACN B: 0.2% TFA in MeOH C: 0.2% (v/v) TFA in water	Thermo Hypersil Gold (C18), 1.9 μm, 2.1 × 150 mm	[27]
DSPC	10.3–30.9 μg/mL; $R^2 = 0.996$	RSD < 5.0%	97.2–100.5% (low, medium and high levels)	1.6	AQbD; ICH Q2; MODR robustness	MeOH: Water (96 : 4, v/v) with 7.5 mM AF	Waters Xbridge Peptide BEH C18 column 3.5 μm, 4.6 × 250 mm	[15]
DSPC, DOPE (plus other non-PEG lipids)	9.375–1000 μg/mL (quadratic); $R^2 \geq 0.999$	Intra-laboratory Repeatability RSD < 2%; inter-day RSD < 2.5%	100 ± 10%	0.43–2.02 (non-PEGylated) 3.8–6.2 (PEGylated)	S/N (3 : 1, 10 : 1) for LOD/LOQ; forced-degradation specificity	A: 0.15% TFA in H_2O (v/v) B: 0.1% TFA in MeOH (v/v)	Phenomenex Luna C18(2), 3 μm, 4.6 × 150 mm	[28]

DSPC	10–400 ng on-column; R^2 0.995–0.999	RSD < 4% (repeatability & IP)	DSPC spike 96–101%	NR	Mixture-level validation; range restriction to linearize CAD	A: 0.1% TFA in H_2O (v/v) B: 0.1% TFA in ACN (v/v)	Waters Acquity BEH Phenyl 1.7 µm, 2.1 × 150 mm	[19]
DSPC	10–40 µg/mL; $R^2 = 0.998$ (DSPC)	Overall RSD 2.41% (DSPC)	97–103% (DSPC)	NR	Two-lab validation; quadratic fit at method level	A: 0.01 M TEAA in H_2O B: 0.01 M TEAA in EtOH	Waters Xbridge Peptide BEH C18, 3.5 µm, 4.6 × 150 mm	[16]
DSPC, DOPE	LOQ–200 µg/mL; $R^2 \geq 0.990$	Overall RSD 0.3–3.5%	94.6–107.3% (across levels)	20 & 40 (DOPE & DSPC)	Platform IPRP-UV-CAD method. CAD diverted early to avoid matrix; single-run UV + CAD	A: 100 mM DBAA, pH 6.8 B: IPA	AMT Halo C4 (400A, 3.4 µm, 2.1 × 150 mm)	[21]
Cholesterol	0.5–100 ug/mL, $R^2 = 0.990$	NR (robustness shown)	Deviation from nominal <5%	2.5	Analyte-specific calibration	A: 0.2% TFA in ACN B: 0.2% TFA in MeOH C: 0.2% (v/v) TFA in water	Thermo Hypersil Gold (C18), 1.9 µm, 2.1 × 150 mm	[27]
		RSD ≤1.3%	98.39–99.07% (across levels)	3.2	ICH Q2; MODR robustness	MeOH: Water (96 : 4, v/v) with 7.5 m AF	Waters Xbridge Peptide BEH C18 column (3.5 µm, 4.6 × 250 mm)	[15]

Table 11.1 (Continued)

Lipid(s)	Linearity (range; R^2)	Precision	Accuracy/ recovery (%)	LOQ (µg/mL)	Notes	Mobile phase	Column	References
		Intra-laboratory Repeatability RSD < 2%; inter-day RSD < 2.5%	100 ± 10%	0.79	S/N criteria; forced degradation	A: 0.15% TFA in H2O (v/v)spienter spiops xmltex140B: 0.1% TFA in MeOH (v/v)	Phenomenex Luna C18(2), 3 µm, 4.6 × 150 mm	[28]
		Intra 1–3% RSD; inter 1–2% RSD	99–103%	Lowest calibration point taken as LOQ	–	Condition 1: A: 0.5% Formic/TEA complex (5:2) in 50% MeOH (v/v); B: 0.5% Formic/TEA complex (5:2) in MeOH Condition 2: A:19.2 mM GAA, 16.2 mM TEA in 50% MeOH (v/v); B:19.2 mM GAA, 16.2 mM TEA in MeOH (v/v)	Waters BEH C18, 1.7 µm, 2.1 × 150 mm	[29]

		Overalls RSD 5.15%	98–108%	NR	Two-lab validation	A: 0.01 M TEA in H2O; B: 0.01 M TEA in EtOH	Waters Xbridge Peptide BEH C18, 3.5 μm, 4.6 × 150 mm	[16]
		3.4–5.2% RSD	97.6–109.1%	20	Platform method	A: 100 mM DBAA, pH 6.8; B: IPA	AMT Halo C4 (400A, 3.4 μm, 150 × 2.1 mm)	[21]
DMG-PEG2000	4.9–14.7 μg/mL; $R^2 = 0.997$	RSD ≤ 4.4%	97.79–100.48%	0.818	ICH Q2; MODR robustness	MeOH:Water (96 : 4, v/v) containing 7.5 mM AF	Waters Xbridge Peptide BEH C18 column (3.5 μm, 4.6 × 250 mm)	[15]
DMG-PEG2000, DSPE-PEG2000	9.375–1000 μg/mL (quadratic); $R^2 \geq 0.999$	Intra-laboratory Repeatability (ILR) RSD < 2%; inter-day RSD < 2.5%	100 ± 10%	DMG-PEG2000: 6.16 DSPE-PEG2000: 3.81	ICH Q2; several formulations tested	A: 0.15% TFA in H_2O (v/v) B: 0.1% TFA in MeOH (v/v)	Phenomenex Luna C18(2), 3 μm, 4.6 × 150 mm	[28]
DMG-PEG2000, DMPE-PEG	10–400 ng on-column; $R^2 = 0.995$–0.999	RSD < 4% (repeatability & IP)	Mixture-level acceptance met	NR	Mixture-level validation	A: 0.1% TFA in H_2O (v/v) B: 0.1% TFA in ACN (v/v)	Waters Acquity BEH Phenyl 1.7 μm, 2.1 × 150 mm	[19]

Table 11.1 (Continued)

Lipid(s)	Linearity (range; R^2)	Precision	Accuracy/ recovery (%)	LOQ (µg/mL)	Notes	Mobile phase	Column	References
PEG-DMG, PEG-lipid-2	PEG_DMG: 7–19 µg/mL PEG-lipid-2: 17–27 µg/mL $R^2 = 0.993$–1.000	Intra-assay: 1–7% RSD; Inter-assay: 1–6% RSD	93–102% (PEG Lipid Specific)	7; 17	Lowest calibration points taken as LOQs	Condition 1: A: 0.5% Formic/TEA complex (5 : 2), in 50% MeOH (v/v) B: 0.5% Formic/TEA complex (5 : 2) in MeOH Condition 2: A:19.2 mM GAA, 16.2 mM TEA in 50% MeOH (v/v) B:19.2 mM GAA, 16.2 mM TEA in MeOH (v/v)	Waters BEH C18, 1.7 µm, 2.1 × 150 mm	[29]
DMG-PEG2000	Working range 5–20 µg/mL; $R^2 = 0.993$	Overall RSD 4.38%	101–111%	NR	Two-lab validation; quadratic at method level	A: 0.01 M TEA in H_2O B: 0.01 M TEA in EtOH	Waters Xbridge Peptide BEH C18, 3.5 µm, 4.6 × 150 mm	[16]

DSPE-PEG2000 DSPE-PEG0.55k DSPE-2 arm-PEG2000 DMG-PEG2000	LOQ–200 μg/mL; $R^2 \geq 0.990$	0.7–13.1% RSD	89.1–109.4%	100	Platform method; early CAD diversion	A: 100 mM DBAA, pH 6.8 B: IPA	AMT Halo C4 (400A, 3.4 μm, 150 × 2.1 mm)	[21]
ALC-0315 (ionizable)	50–150 μg/mL; $R^2 = 0.997$	RSD ≤ 0.9%	97.15–100.05%	9.971	ICH Q2; MODR robustness	MeOH:Water (96 : 4, v/v) with 7.5 mM AF	Waters XBridge Peptide BEH C18 column (3.5 μm, 4.6 × 250 mm)	[15]
SM-102; DOTAP (cationic)	9.375–1000 μg/mL (quadratic); $R^2 \geq 0.999$	Intra-laboratory Repeatability (ILR) RSD < 2%; inter-day RSD < 2.5%	100 ± 10%	DOTAP: 0.69 SM-102: 1.30	S/N criteria; forced degradation	A: 0.15% TFA in H_2O (v/v) B: 0.1% TFA in MeOH (v/v)	Phenomenex Luna C18(2), 3 μm, 4.6 × 150 mm	[28]
DOTAP, DOBAQ	10–400 ng on-column; $R^2 = 0.996–0.999$	RSD < 4% (repeatability & IP)	Mixture-level acceptance met	NR	Mixture-level validation	A: 0.1% TFA in H_2O (v/v) B: 0.1% TFA in ACN (v/v)	Waters Acquity BEH Phenyl 1.7 μm, 2.1 × 150 mm	[19]

Table 11.1 (Continued)

Lipid(s)	Linearity (range; R^2)	Precision	Accuracy/ recovery (%)	LOQ (µg/mL)	Notes	Mobile phase	Column	References
Two cationic lipids (unspecified, proprietary)	0.067–0.163; & 0.086–0.130 $R^2 = 0.993$–1.000	Intra/inter-assay: 1–6% RSD	95–102%	67; 86	Lowest calibration points taken as LOQs	Condition 1: A: 0.5% Formic/TEA complex (5 : 2), in 50% MeOH (v/v) B: 0.5% Formic/TEA complex (5 : 2) in MeOH Condition 2: A:19.2 mM GAA, 16.2 mM TEA in 50% MeOH (v/v) B:19.2 mM GAA, 16.2 mM TEA in MeOH (v/v)	Waters BEH C18, 1.7 µm, 2.1 × 150 mm	[29]
Cationic lipid "9001"	Working range 60–240 µg/mL; $R^2 = 0.997$	Overall RSD 3.01%	98–103%	NR	Two-lab validation	A: 0.01 M TEA in H_2O B: 0.01 M TEA in EtOH	Waters Xbridge Peptide BEH C18 column (3.5 µm, 4.6 × 250 mm)	[16]
Multiple ionizable/ cationic lipids DLin-KC2-DMA DLin-MC3-DMA SM-102 ALC-0315; DOTAP	From LOQ to 200 µg/mL; $R^2 > 0.990$	1.4–14.4% RSD	91.8–110.2%	20–50	Platform method; early CAD diversion	A: 100 mM DBAA, pH 6.8 B: IPA	AMT Halo C4 (400A, 3.4 µm, 150 × 2.1 mm)	[21]

The table summarizes method ranges, precision, accuracy and detection limits, and chromatographic conditions across representative studies.
NR = not reported.

chromatographic separation to reduce signal suppression, and coupling CAD with UV detection to expand dynamic range and meet orthogonal detection requirements [21].

Taken together, these findings demonstrate that LC-CAD provides a robust, precise, and sensitive platform for quantitative analysis of LBFs. When coupled with appropriate calibration models and matrix-specific optimization, CAD enables reliable quantitation across a wide range of LBFs components and concentrations. The growing adoption of platform methods and mixture-level validation approaches reflects the field's shift towards comprehensive, real-world testing strategies that support both development and quality control of increasingly complex LBFs.

11.4 Applications of LC-CAD in LBFs Analysis

Building on the analytical principles and method-development strategies described in the preceding sections, the following case studies highlight the practical implementation of LC-CAD in lipid-based formulations (LBFs) analysis. These examples, drawn from recent peer-reviewed publications, demonstrate how CAD-based methods have been applied to evaluate stability and during formulation development, providing quantitative insight into lipid composition, degradation pathways, and formulation consistency during LBFs manufacturing.

11.4.1 Stability Assessment

Stability assessment of LBFs is essential, as physicochemical changes, such as lipid degradation, aggregation, or loss of encapsulated material can compromise delivery efficiency and therapeutic performance. Because lipids are particularly susceptible to degradation and aggregation during storage and transportation, accurate quantitation of individual lipid components is critical to ensure product quality and consistency [16, 30]. The US FDA also recognizes lipid degradation as a critical quality attribute for drug products formulated with LNP underscoring the need for robust analytical methods to monitor these changes throughout product development and shelf life [11].

RP-HPLC-CAD methods have been developed to resolve both intact lipids and their degradation products proving to be highly effective in stability-indicating assays for LBFs [20, 22, 27–30].

Bender et al. [28] developed an RP-HPLC-CAD method capable of resolving both intact lipids and their degradation products, enabling specific testing for a panel of LNP lipids that included the cationic species DOTAP and SM-102, the zwitterionic phospholipids DOPE (1,2-dioleoyl-sn-glycero-3-phosphoethanolamine) and DSPC, cholesterol and the PEGylated lipids DMG-PEG2000 and DSPE-PEG2000 (1,2-distearoyl-sn-glycero-3-phosphoethanolamine-PEG2000).

Specificity was evaluated through forced-degradation studies, in which each lipid was subjected to alkaline hydrolysis (0.1 M NaOH, 2 hours) and, in separate experiments, acidic conditions to promote lyso-lipid formation. Under alkaline conditions, DOTAP yielded additional oleic acid peaks while retaining its undegraded parent peak; DOPE, DSPC and DSPE-PEG2000 degraded completely, producing stearic- or oleic-acid products; SM-102 (8-[(2-hydroxyethyl)[6-oxo-6-(undecyloxy)hexyl]amino]-octanoic acid, 1-octylnonyl ester) showed partial degradation, resulting in two additional peaks; DMG-PEG2000 degraded completely without generating secondary products, and cholesterol remained stable. CAD detection provided clear, quantitative resolution of these degradation products with no interference with the non-degraded lipids, confirming the method's specificity and suitability as stability indicating assay for complex LNP formulations.

Li et al. [22] reported an ion-pair RP-HPLC-UV/CAD method which enabled the simultaneous separation and detection of small interfering RNA (siRNA) and functional lipids, from experimental LNP formulations (consisting of two types of phospholipids, cholesterol and PEGylated short-chain lipid) in a single analytical run. When applied to LNP stability samples stressed with base (0.01 M NaOH) and 0.3% hydrogen peroxide, the method could monitor siRNA, lipids and their potential degradation products.

The method exhibited good selectivity and resolution across all component types, demonstrating its capability to distinguish intact lipids from emerging degradants. Although not designed as a full stability study, this application highlights the method's value for detecting lipid degradation and assessing formulation integrity of siRNA-LNP systems.

Kinsey et al. [29] optimized an RP-UHPLC-CAD method to evaluate the stability of LNP by monitoring degradation behavior of individual lipid components. Forced-degradation experiments were conducted with representative LNP lipids (cationic Lipid 1 and 2) exposed to basic (0.1 M NaOH) and oxidative stress (0.03% H_2O_2) conditions to generate degradants and assess chromatographic selectivity. To further confirm the method's specificity, an impurity reference standard comprising myristic acid, Lyso-PC1, Lyso-PC2, stearic acid and a known degradant of cationic lipid 1 was also analyzed, representing typical hydrolysis products of PEG-lipids and DSPC. The method achieved baseline separation for all degradants from their corresponding parent lipids. Subsequently the method was applied for stability studies of LNP formulations stored at −70, −20, 4 and 25 °C for 10 months, demonstrating suitability for stability evaluation and tracking of lipid integrity in LNP.

Weber et al. [27] developed and validated a rapid RP-UHPLC-CAD method for simultaneous assessment of composition and stability in liposomal formulations.

Applied directly to bupivacaine-loaded liposomes subjected to controlled photodegradation stress to model light-induced degradation relevant to long-term storage, the method sensitively captured time-dependent composition shifts (notably loss of cholesterol followed by moderate reductions in phospholipid content). By combining broad lipid coverage with single-run quantitation, the CAD method provides a practical, stability-indicating platform for detecting lipid degradation and assessing formulation integrity.

11.4.2 Formulation Development

Ensuring that lipid composition and nucleic acid content remain within target specifications during LNP manufacturing is critical for achieving consistent performance and meeting quality standards. In one example, an ion-pair reversed-phase (IPRP) HPLC method was developed to enable the simultaneous separation and quantification of both lipid species and nucleic acids in LNP formulations containing ASO, single-guide RNA (sgRNA), or messenger RNA (mRNA). The assay was applied to study the impact of downstream purification steps such as tangential flow filtration (TFF) and dialysis on formulation composition. Quantitative results revealed whether purification altered lipid molar ratios, removed specific lipid components preferentially, or affected nucleic acid loading and integrity. This integrated approach provided a process monitoring tool that could identify subtle but potentially critical changes in composition, supporting both early-stage formulation optimization and later process validation activities [21].

Another complementary application involved the use of a reversed-phase high-performance liquid chromatography-charged aerosol detection (RP-HPLC-CAD) method designed for high-resolution quantification of individual lipid species in mRNA-loaded LNPs [28]. This method targeted seven lipids spanning four broad lipid categories, representative of clinically relevant LNP formulations: the cationic lipids DOTAP and SM-102, the phospholipids DOPE and DSPC, the sterol cholesterol, and the PEGylated lipids DMG-PEG2000 and DSPE-PEG2000. In formulation development studies, the assay enabled tracking of lipid molar ratios across manufacturing batches, detection of composition shifts arising from changes in process parameters, and identification of lipid losses during purification steps such as dialysis. By quantifying each lipid class in a single run, the method streamlined analytical workflows and provided formulation scientists with actionable data to guide both composition targeting and process optimization.

Kim et al. [15] applied their optimized RP-HPLC-CAD method to real LBF samples to evaluate its suitability for supporting formulation development activities. The method was used to quantify four key lipid classes within complex formulation matrices, confirming its capability to resolve target analytes from excipients and other formulation components without interference. By accurately measuring lipid content in these samples, the assay enabled verification that formulation batches met their intended lipid ratios and provided a means to detect any compositional deviations that could arise from process variations. This application demonstrated the method's value as a reliable tool for routine composition checks during formulation optimization and scale-up, ensuring that critical lipid attributes were maintained throughout development.

Overall, charged aerosol detection (CAD) has proven to be a valuable and versatile technique for studying LBFs. Its broad sensitivity, robustness and ability to simultaneously quantify non-chromophoric components make it especially useful for analyzing complex mixtures that challenge traditional detection methods. When paired with well-designed chromatographic separations, CAD enables reliable, high-quality data that can guide formulation development, stability assessment and quality control. As LBDDS continue to advance and diversify, CAD will remain an important analytical tool for understanding and ensuring consistency of these innovative drug products.

References

1 Mu, H., Holm, R., and Mullertz, A. (2013). Lipid-based formulations for oral administration of poorly water-soluble drugs. *International Journal of Pharmaceutics* 453 (1): 215–224.

2 Yan, S., Cheng, Y., Li, L. et al. (2023). Lipid-based formulations: a promising approach for poorly soluble drug delivery via the intestinal lymphatic system. *Journal of Drug Delivery Science and Technology* 87: 104770.

3 Goke, K., Lorenz, T., Repanas, A. et al. (2018). Novel strategies for the formulation and processing of poorly water-soluble drugs. *European Journal of Pharmaceutics and Biopharmaceutics* 126: 40–56.

4 Karmaker, S., Rosales, P.D., Tirumuruhan, B. et al. (2025). More than a delivery system: the evolving role of lipid-based nanoparticles. *Nanoscale* 17 (19): 11864–11893.

5 Yadav, P. (2025). Lipid-based drug delivery systems: formulation and applications. In: *Next-Generation Drug Delivery Systems* (ed. A. Pathak and S.P. Singh), 89–110. New York, NY: Springer US.

6 Pouton, C.W. and Porter, C.J. (2008). Formulation of lipid-based delivery systems for oral administration: materials, methods and strategies. *Advanced Drug Delivery Reviews* 60 (6): 625–637.

7 Nakmode, D., Bhavana, V., Thakor, P. et al. (2022). Fundamental aspects of lipid-based excipients in lipid-based product development. *Pharmaceutics* 14 (4): 831.

8 Fox, C.B., Sivananthan, S.J., Mikasa, T.J.T. et al. (2013). Charged aerosol detection to characterize components of dispersed-phase formulations. *Advances in Colloid and Interface Science* 199-200: 59–65.

9 Rehman, M., Tahir, N., Sohail, M.F. et al. (2024). Lipid-based nanoformulations for drug delivery: an ongoing perspective. *Pharmaceutics* 16 (11): 1376.

10 Anderson, M. and Omri, A. (2004). The effect of different lipid components on the in vitro stability and release kinetics of liposome formulations. *Drug Delivery* 11 (1): 33–39.

11 U.S Food and Drug Administration (2018). Liposome drug products: chemistry, manufacturing, and controls; human pharmacokinetics and bioavailability; and labeling documentation. Available from: https://www.fda.gov/regulatory-information/search-fda-guidance-documents/liposome-drug-products-chemistry-manufacturing-and-controls-human-pharmacokinetics-and (accessed 10 September 2025).

12 Haidar Ahmad, I.A., Blasko, A., Wang, H. et al. (2021). Charged aerosol detection in early and late-stage pharmaceutical development: selection of regression models at optimum power function value. *Journal of Chromatography. A* 1641: 461997.

13 Soliven, A., Haidar Ahmad, I.A., Tam, J. et al. (2017). A simplified guide for charged aerosol detection of non-chromophoric compounds: Analytical method development and validation for the HPLC assay of aerosol particle size distribution for amikacin. *Journal of Pharmaceutical and Biomedical Analysis* 143: 68–76.

14 Lynnerup, J.T., Lillotte, T., Feldmueller, M. et al. (2025). Untangling "dissolved" drug species from various formulations of a poorly soluble drug: sampling methods, mechanistic insights, and IVIVC. *Journal of Pharmaceutical Sciences* 114 (8): 103884.

15 Kim, K.H., Lee, J.E., Lee, J.C. et al. (2023). Optimization of HPLC-CAD method for simultaneous analysis of different lipids in lipid nanoparticles with analytical QbD. *Journal of Chromatography. A* 1709: 464375.

16 Yu, X., Yu, C., Wu, X. et al. (2023). Validation of an HPLC-CAD method for determination of lipid content in LNP-encapsulated COVID-19 mRNA vaccines. *Vaccines (Basel)* 11 (5).

17 Batishchev, O.V. and Indenbom, A.V. (2008). Alkylated glass partition allows formation of solvent-free lipid bilayer by Montal-Mueller technique. *Bioelectrochemistry* 74 (1): 22–25.

18 Cajka, T. and Fiehn, O. (2014). Comprehensive analysis of lipids in biological systems by liquid chromatography-mass spectrometry. *Trends in Analytical Chemistry* 61: 192–206.

19 Beckert, N., Dietrich, A., and Hubbuch, J. (2024). RP-CAD for lipid quantification: systematic method development and intensified LNP process characterization. *Pharmaceuticals (Basel)* 17 (9): 1217.

20 Tam, J., Ahmad, I.A.H., and Blasko, A. (2018). A four parameter optimization and troubleshooting of a RPLC - charged aerosol detection stability indicating method for determination of S-lysophosphatidylcholines in a phospholipid formulation. *Journal of Pharmaceutical and Biomedical Analysis* 155: 288–297.

21 Su, W.C., Lieu, R., Fu, Y. et al. (2025). A platform method for simultaneous quantification of lipid and nucleic acid components in lipid nanoparticles. *Journal of Chromatography. A* 1746: 465788.

22 Li, L., Foley, J.P., and Helmy, R. (2019). Simultaneous separation of small interfering RNA and lipids using ion-pair reversed-phase liquid chromatography. *Journal of Chromatography. A* 1601: 145–154.

23 Chen, T., Tang, S., Fu, Y. et al. (2022). Analytical techniques for characterizing diastereomers of phosphorothioated oligonucleotides. *Journal of Chromatography. A* 1678: 463349.

24 Kadlecova, Z., Kalíková, K., Tesařová, E., and Gilar, M. (2022). Phosphorothioate oligonucleotides separation in ion-pairing reversed-phase liquid chromatography: effect of temperature. *Journal of Chromatography. A* 1681: 463473.

25 Barnhart, W.W., Farooq, M.Q., and Ahmad, I.A.H. (2024). A simplified tutorial on charged aerosol detection: understanding the basics, optimization, and troubleshooting. *Journal of Chromatography Open* 6: 100181.

26 (ICH), International Council for Harmonisation of Technical Requirements for Pharmaceuticals for Human Use (2025). Validation of analytical procedures: text and methodology Q2(R2). Available from: https://www.ich.org/page/quality-guidelines (accessed 10 October 2025).

27 Weber, F., Rahnfeld, L., and Luciani, P. (2020). Analytical profiling and stability evaluation of liposomal drug delivery systems: a rapid UHPLC-CAD-based approach for phospholipids in research and quality control. *Talanta* 220: 121320.

28 Bender, V., Fuchs, L., and Suss, R. (2024). RP-HPLC-CAD method for the rapid analysis of lipids used in lipid nanoparticles derived from dual centrifugation. *International Journal of Pharmaceutics: X* 7: 100255.

29 Kinsey, C., Lu, T., Deiss, A. et al. (2022). Determination of lipid content and stability in lipid nanoparticles using ultra high-performance liquid chromatography in combination with a corona charged aerosol detector. *Electrophoresis* 43 (9–10): 1091–1100.

30 Wu, J., Crist, R.M., McNeil, S.E., and Clogston, J.D. (2019). Ion quantification in liposomal drug products using high performance liquid chromatography. *Journal of Pharmaceutical and Biomedical Analysis* 165: 41–46.

12

Application of Charged Aerosol Detection in Polysorbate Analysis

Jessica Lin, Juan Bian, Frank Hrovat, and Kelly Zhang

Synthetic Molecule Analytical Chemistry, Genentech Inc., South San Francisco, CA, USA

CHAPTER MENU

12.1 Introduction

12.1.1 Overview of Polysorbates in Biopharmaceuticals

Polysorbates, particularly polysorbate 20 (PS20) and polysorbate 80 (PS80), are nonionic surfactants that are used widely in biopharmaceutical formulations. Their primary function is to stabilize protein-based drugs by preventing protein aggregation, denaturation, and surface adsorption during manufacturing, transportation, storage, and administration [1, 2], especially under stress conditions such as agitation, pumping, and freeze-thaw cycles, where proteins tend to lose their native structure and form aggregates [3].

Structurally, polysorbates are complex mixtures, mainly composed of fatty-acid esters of polyethoxylated (POE) sorbitan, with each mole of sorbitan esterified with roughly 20 moles of ethylene oxide (EO). PS20 and PS80 differ in the fatty acids and the degree of esterification, where PS20 uses lauric acid and PS80 uses oleic acid [4]. Their amphiphilic nature, combining hydrophilic polyethoxylated head groups and hydrophobic fatty-acid tails, enables them to interface effectively between aqueous environments and hydrophobic surfaces, thus stabilizing protein molecules [5, 6].

Polysorbates are highly heterogeneous due to the complex reactions involved in their manufacturing process, comprising mono-, di-, tri-esters, and unesterified POE sorbitans, as well as free fatty acids (FFAs) [7]. This chemical diversity complicates their analysis, necessitating advanced and robust analytical techniques.

12.1.2 Challenges in Polysorbate Analysis

Polysorbates pose significant analytical challenges due to their complex nature of heterogeneous chemical composition and their susceptibility to degradation. These components may vary not only among different suppliers but also between different batches from the same supplier. This compositional complexity hampers the development of universal analytical standards and complicates both qualitative and quantitative characterization. Polysorbates are prone to degradation through two primary pathways: hydrolysis and oxidation. The detection of degradation products such as FFAs and aldehydes adds another layer of complexity.

Moreover, polysorbates lack strong chromophores, making them inherently difficult to detect with conventional ultraviolet (UV)-based detection methods. Alternative assays, such as fluorescence micelle assays (FMA) take advantage of the ability of polysorbates to form micelles above their critical micelle concentrations (CMCs). With the incorporation of fluorescent dyes such as *N*-phenyl-1-naphthylamine into the hydrophobic core of the micelles, the fluorescence emission can be measured to determine the concentration of polysorbates. However, FMA is an indirect measurement of micelles, and compounds other than polysorbates that form micelles can also contribute to the signals, leading to overestimation of the polysorbate concentration. Additionally, the polysorbate concentration in the product may be below the CMC needed for micelle formation. Furthermore, the measurement is non-specific, and cannot differentiate between individual components or degradants of polysorbates [2, 8, 9]. Studies [8] have demonstrated that critical polysorbate components such as POE sorbitan monolaurate may escape detection with FMA, highlighting the need for more specific and reliable analytical methods.

12.2 Analytical Technology of Polysorbates

12.2.1 Separation of Polysorbates

Advanced analytical techniques, such as high-performance liquid chromatography (HPLC) have become indispensable for distinguishing intact polysorbates from their degradation products, and separating different impurities and degradants.

Reversed-phase HPLC (RPLC) has been a standard method for polysorbate separation. In general, non-esterified POE species elute first due to higher hydrophilicity, followed by mono-esters, di-esters, and higher-order esters. The fatty acid composition also contributes to the retention time, where the clusters with lower fatty acid carbon numbers elute earlier than the ones with higher fatty acid carbon numbers. A representative RPLC chromatogram of PS80 is shown in Figure 12.1a [1]. Hydrophilic interaction chromatography (HILIC) was also explored to separate polysorbates but the application is less common. In HILIC, the separation is primarily governed by the number of terminal hydroxyl groups [10]. The elution order shown in Figure 12.1b illustrates this principle where the tri-esters elute first, followed by di-esters and mono-esters. The non-esterified PEG, ethoxylated isosorbide and ethoxylated sorbitan elute toward the end of the chromatogram [10, 12]. Mixed-mode chromatography, which incorporates both anion exchange and hydrophobic interaction, was also reported. The elution order is similar to that of RPLC, where mono-esters elute earlier than di-esters and tri-esters, and fatty acids with higher carbon numbers elute later than those with lower carbon numbers. An example utilizing an Oasis MAX column is shown in Figure 12.1c [8]. A key advantage of mixed-mode chromatography lies in its ability to retain polysorbates by hydrophobic retention, while excluding proteins through anion exchange [5, 13]. This dual retention mechanism enables polysorbate analysis in the presence of proteins, streamlining the analysis workflow. Furthermore, it eliminates the need for a separate protein-removal step by either protein precipitation [14] or solid-phase extraction [15]. Typically, the PS components are partially resolved, yielding characteristic fingerprints.

Notably, two-dimensional HPLC (2DLC) has been successfully applied to characterize polysorbates in the presence of protein drug substances. The first dimension typically employs hydrophobic interaction chromatography (HIC) [16], mixed-mode chromatography, or ion exchange chromatography (IEC) [17, 18], to separate PS from the proteins. The second dimension offers further separation of polysorbate components using RPLC. Polyols such as POE sorbitan, POE isosorbide and POE are hydrophilic species that coelute with proteins in HIC or mixed-mode columns, but can be separated from proteins using IEC. Therefore, IEC is a preferred option for the first dimension when the analysis includes the unesterified polyol species. Zegota et al. [11] combined anion and cation exchange columns in a single inline 2D-LC system, and this dual-exchange methodology efficiently removed monoclonal antibody (mAb) from the formulations with protein concentrations as high as 180 mg/mL in the first dimension, followed by polyphenyl reversed phase separation in the second dimension. A representative chromatogram of PS20 on the second dimension is shown in Figure 12.1d. In contrast to a typical run time over 1 hour for other 2D-LC analyses, this method achieves separation of PS20 and PS80 within 8 minutes, enabling high-throughput

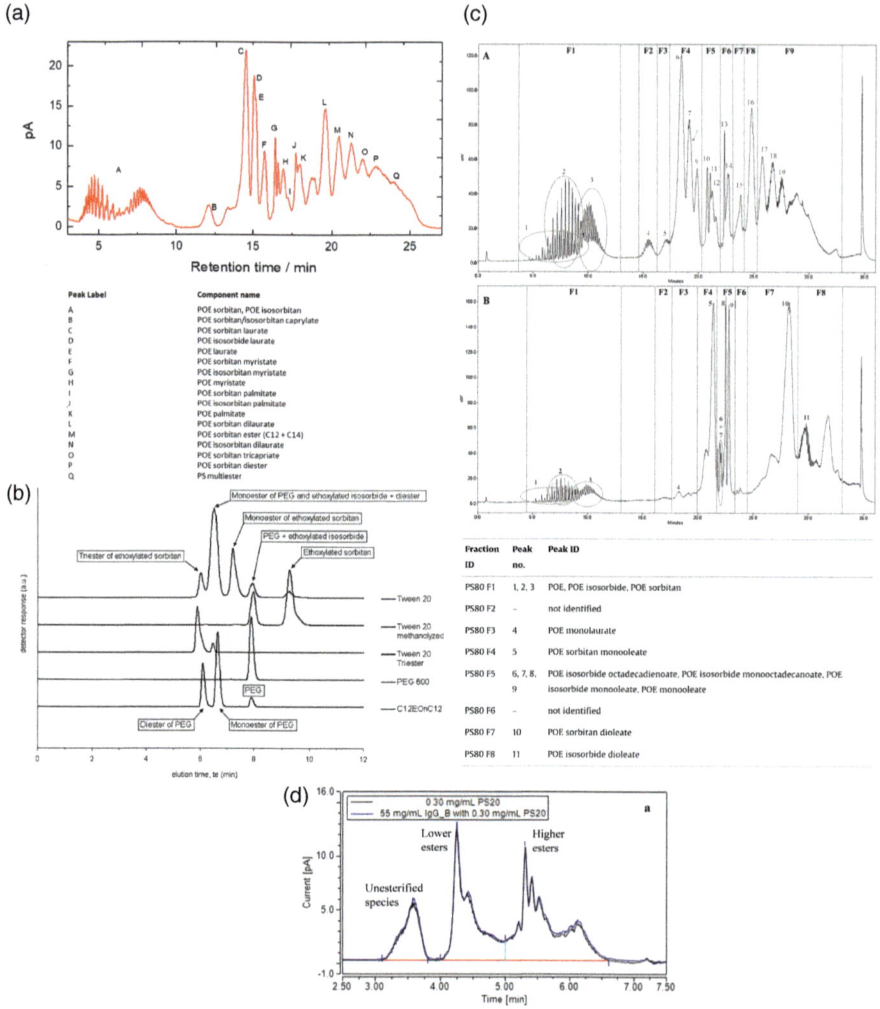

Peak Label	Component name
A	POE sorbitan, POE isosorbitan
B	POE sorbitan/isosorbitan caprylate
C	POE sorbitan laurate
D	POE isosorbide laurate
E	POE laurate
F	POE sorbitan myristate
G	POE isosorbitan myristate
H	POE myristate
I	POE sorbitan palmitate
J	POE isosorbitan palmitate
K	POE palmitate
L	POE sorbitan dilaurate
M	POE sorbitan ester (C12 + C14)
N	POE isosorbitan dilaurate
O	POE sorbitan tricapriate
P	POE sorbitan diester
Q	PS multiester

Fraction ID	Peak no.	Peak ID
PS80 F1	1, 2, 3	POE, POE isosorbide, POE sorbitan
PS80 F2	-	not identified
PS80 F3	4	POE monolaurate
PS80 F4	5	POE sorbitan monooleate
PS80 F5	6, 7, 8, 9	POE isosorbide octadecadienoate, POE isosorbide monooctadecanoate, POE isosorbide monooleate, POE monooleate
PS80 F6	-	not identified
PS80 F7	10	POE sorbitan dioleate
PS80 F8	11	POE isosorbide dioleate

Figure 12.1 Representative HPLC chromatograms for polysorbates. (a) RPLC-CAD chromatogram of PS80 on Zorbax 300 SB-C8 column [1], (b) HILIC-ELSD chromatogram of PS20 on LUNA HILIC column [10], (c) mixed-mode HPLC-CAD chromatogram of PS20 on Oasis MAX column [8], (d) second dimension chromatogram of PS20 on 2DLC-CAD using Waters BioResolve RP mAb polyphenyl column [11]. *Source:* (a) [1] / with permission of Elsevier. (b) [10] / with permission of Elsevier. (c) [8] / with permission of Elsevier. (d) [11] / with permission of Elsevier.

analysis. The method reported excellent quantitation accuracy of 96–109% and precision of less than 1% relative standard deviation (RSD).

12.2.2 Detection of Polysorbates

Various detectors have been employed in HPLC analysis of polysorbates, including mass spectrometry (MS), evaporative light scattering detectors (ELSD), and charged aerosol detector (CAD). While MS provides unparalleled capabilities for identifying individual polysorbate components, its quantitative application is considerably limited due to its susceptibility to instrument and sample variations, high cost, and the requirement for specialized operator expertise. Nevertheless, a few quantitative MS applications have been reported, such as single quadrupole MS based on the extracted ion intensity of dioxolanylium diagnostic ions for determining total polysorbate content in the presence of high-concentration proteins [5]. Evers et al. reported another application using single quadrupole MS for quantifying individual degradation products of PS20 in a stability-indicating HPLC method, demonstrating excellent selectivity and sensitivity [9].

Despite these MS applications, ELSD and CAD remain the mainstream detection methods for HPLC analysis of polysorbates. CAD functions by nebulizing the HPLC eluent into a fine aerosol. The solvent is then evaporated, and the resulting dry analyte particles are charged using an ionized nitrogen stream. These charged particles subsequently pass through an electrometer, generating a signal proportional to the analyte concentration [19]. Similarly, ELSD operates by nebulizing the liquid chromatography eluent with an inert gas, forming an aerosol mist. This mist then travels through a heated drift tube, where the volatile mobile phase evaporates. The primary difference from CAD lies in the detection technique: with ELSD, nonvolatile analyte particles scatter a directed light beam, and the intensity of the scattered light increases exponentially with the analyte mass. In contrast, the charge generated in CAD follows a parabolic relationship with the analyte mass. Consequently, CAD is generally expected to provide a more uniform response across diverse chemical classes and over a wider range of concentrations compared to ELSD [8, 20]. Several studies have compared the quantitative performance of CAD and ELSD. For instance, Lippold et al. [9] reported that CAD exhibits higher sensitivity and stability, and offers better coverage of PS components, with a performance comparison summarized in Table 12.1.

While CAD is relatively powerful, it also comes with limitations. It is generally considered to be more complex and less robust compared to UV, or even ELSD. It requires meticulous optimization in detector parameters, such as power function value (PFV) and evaporation temperature to maximize the performance. Similar to ELSD, a significant drawback of CAD is its unsuitability for volatile compounds, including fatty acid components with short chains. These volatile

Table 12.1 Comparison of CAD and ELSD performance in quantitative analysis for polysorbates.

Criteria	CAD	ELSD
Sensitivity	Higher sensitivity across polysorbate subspecies	Lower sensitivity especially for low-mass and semi volatile species
Quantitation robustness	Stable even with degradation products	More variability for degradation products
Linearity and calibration	Better linearity and reproducibility	More challenging to maintain consistent calibration
Response uniformity	More uniform across mono- and polyesters	Response varies with analyte size and volatility
Effect of analyte volatility	Less loss of signal for semi-volatile components	More loss of signal for semi-volatile components

compounds typically evaporate along with the mobile phase before reaching the detection stage, resulting in little or no response. For semi volatile or nonvolatile compounds, the detector response is inherently nonlinear, particularly at higher concentrations, necessitating the application of a PFV to linearize the signal [21, 22]. Furthermore, CAD is sensitive to mobile phase compositions, with its response increasing with a higher organic solvent content due to enhanced transport efficiency and volatility. Consequently, the response can deviate significantly from universality during gradient elution. A common approach to mitigate this issue involves employing an inverse gradient, which introduces a second flow after the analytical column and before the detector that is the inverse of the separation gradient [23, 24]. However, this necessitates considerable effort to install an additional pump, rendering the analytical method challenging in a quality control (QC) environment due to the required qualification and validation of the new hardware setup for good manufacturing practice (GMP) compliance. Another limitation in the context of polysorbate analysis is the potential peak fronting. Paschen et al. [25] observed more pronounced peak fronting with CAD compared to ELSD at higher PS20 concentrations, attributed to overloading of the aerosol formation process. At elevated analyte concentrations, an excessive number of particles can be generated too rapidly, saturating the charging field and causing the earlier-eluting particles at the front of the peak to be charged more efficiently than the later ones at the back. Fortunately, this potential peak shape issue can be resolved by diluting the samples to a concentration range suitable for CAD. And it could be argued that the peak fronting is less prominent with ELSD due to its limited sensitivity so that the analyte particles in the peak front did not generate sufficient response.

12.3 Applications of CAD in Polysorbate Analysis

12.3.1 Characterization of Polysorbate Components

PS20 and PS80 are both POE derivatives of sorbitan monoesters, containing approximately 20 moles of POE per mole of sorbitan. Polysorbate synthesis typically begins with the cyclization of sorbitol, yielding sorbitan alongside the undesired isosorbide. The subsequent step involves either esterification of the resulting sorbitans and isosorbides with lauric acid (for PS20) or oleic acid (for PS80), followed by polyethoxylation, or reverse sequence [27]. Given the nonspecific nature of the synthesis and the quality of the raw materials, commercial polysorbate batches comprise a heterogeneous mixture of mono-, di-, and tri-esters, POE chains with varying lengths, unesterified polyoxyethylene sorbitans, and other byproducts such as isosorbide derivatives as shown in Figure 12.2 [26]. A comprehensive compositional analysis of PS80 by Keire et al. using high-resolution mass spectrometry with rapid hydrogen/deuterium (H/D) exchange revealed a remarkable number of 17,000 potential species [28].

Each component—monoesters, diesters, triesters, and unesterified POE sorbitans—contributes uniquely to surfactant properties such as micelle formation, hydrophobic interaction, and protein stabilization [29]. For instance, longer and unsaturated fatty acid chains enhance hydrophobicity and lower the CMC. Polyunsaturated fatty acids, such as linoleic and linolenic acid esters, may increase interfacial fluidity but are more susceptible to oxidation, potentially compromising the long-term stability of the surfactant. This inherent chemical diversity leads to significant batch-to-batch variability, affecting both the surfactant's functional performance and its susceptibility to degradation.

To ensure its functional consistency in biopharmaceutical formulations and mitigate degradation risks, health authorities expect a thorough characterization of the polysorbate components during the development and manufacturing of biopharmaceuticals. The European Medicines Agency (EMA) Q&A document (September 2023) explicitly advises that if polysorbate is present in the formulation and is critical for preventing protein aggregation, or if its degradation poses potential quality risks, then a polysorbate assay should be included in the release and stability specifications of the product [30]. The International Council for Harmonisation (ICH) Q6B guideline requires that critical excipients in biologic drug products, such as PS20 and PS80, meet pharmacopeial standards. The United States Pharmacopeia (USP) [31, 32] outlines the acceptable percentages of different fatty acid esters present in formulation solutions, as detailed in Table 12.2. For PS20, the laurate ester content is typically 40–60%, with the remaining esters containing fatty acids ranging from C8 to C18. In PS80, the oleate ester content constitutes approximately 60%, with the remaining esters containing fatty acids

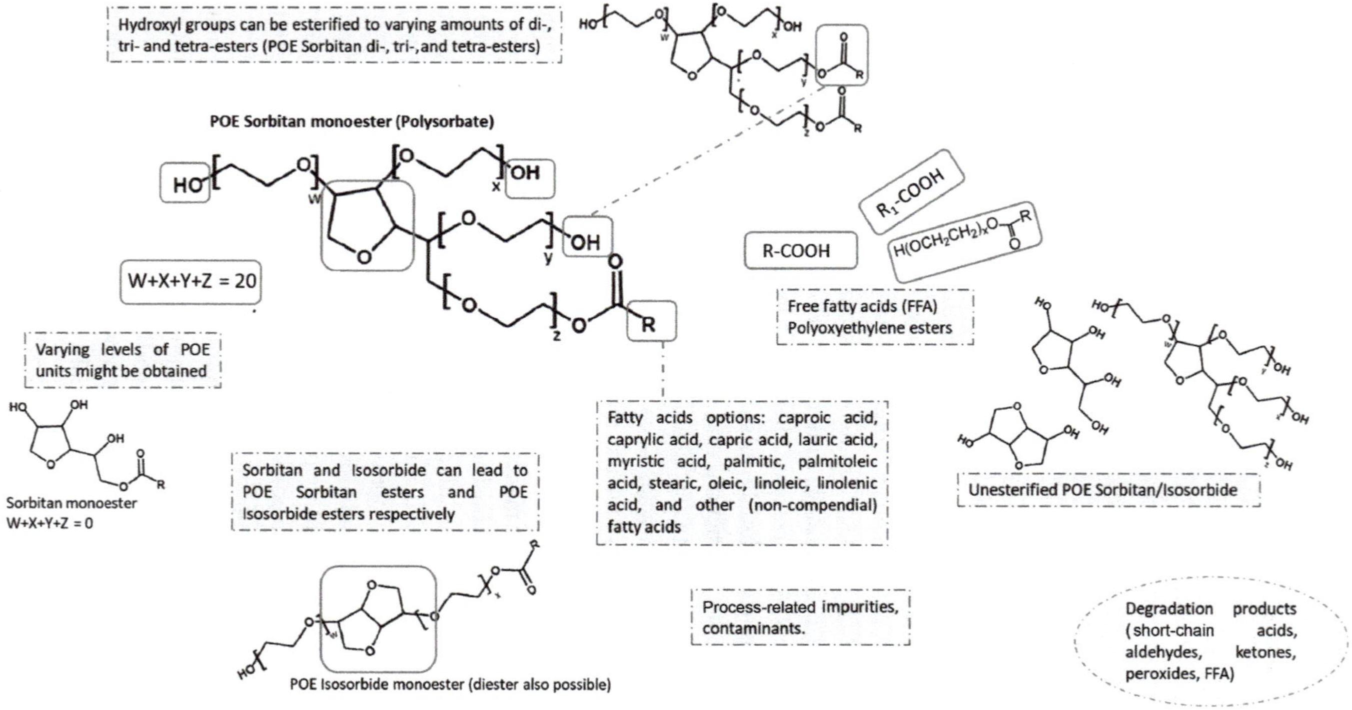

Figure 12.2 Representation of molecular heterogeneity of polysorbates [26]. *Source:* [26] / with permission of Elsevier.

Table 12.2 US Pharmacopeia composition profiles of fatty esters in PS80 and PS20.

Acid	Carbon number	PS80 percentage (%)	PS20 percentage (%)
Caproic	6		≤1.0
Caprylic	8		≤10.0
Capric	10		≤10.0
Lauric	12		40.0–60.0
Myristic	14	≤5.0	14.0–25.0
Palmitic	16	≤16.0	7.0–15.0
Palmitoleic	16 (with a double bond)	≤8.0	
Stearic	18	≤6.0	≤11.0
Oleic	18 (with a double bond)	≥58.0	≤11.0
Linoleic	18 (with two double bonds)	≤18.0	≤3.0
Linolenic	18 (with three double bonds)	≤4.0	

ranging from C14 to C18 [8]. It was reported that [18] by tightening the control of long-chain fatty acid esters in PS20, hydrolytic degradation could be reduced, thereby lowering the risk of FFA particle formation and improving quality profile for the drug product.

12.3.2 Polysorbate Stability in Biopharmaceutical Formulations

Polysorbates primarily degrade via hydrolysis and oxidation pathways. These degradation processes can compromise the integrity of biopharmaceuticals by altering surfactant functionality, inducing protein instability and generating potentially toxic degradants, thus raising significant quality and safety concerns.

Hydrolysis of the fatty acid ester bonds can occur both chemically or enzymatically. The chemical hydrolysis usually occurs under acidic or basic pH conditions, yielding fatty acids and POE sorbitans. An RPLC-CAD method has been utilized to study the kinetics and mechanism of chemical hydrolysis [1]. As shown in Figure 12.3, under acidic conditions (pH 1), the ester peaks gradually decreased, giving way to the non-esterified POE sorbitan peaks. Conversely, under basic conditions (pH 11), the degradation of esters with lower carbon numbers was much more significant than the ones with higher carbon numbers. The observed difference in the extent of degradation at acidic and basic pH aligns with the general principle that ester hydrolysis is more efficient under basic conditions due to the irreversible nature of base-catalyzed hydrolysis, in contrast to the equilibrium-driven acid-catalyzed process. Another difference between acidic

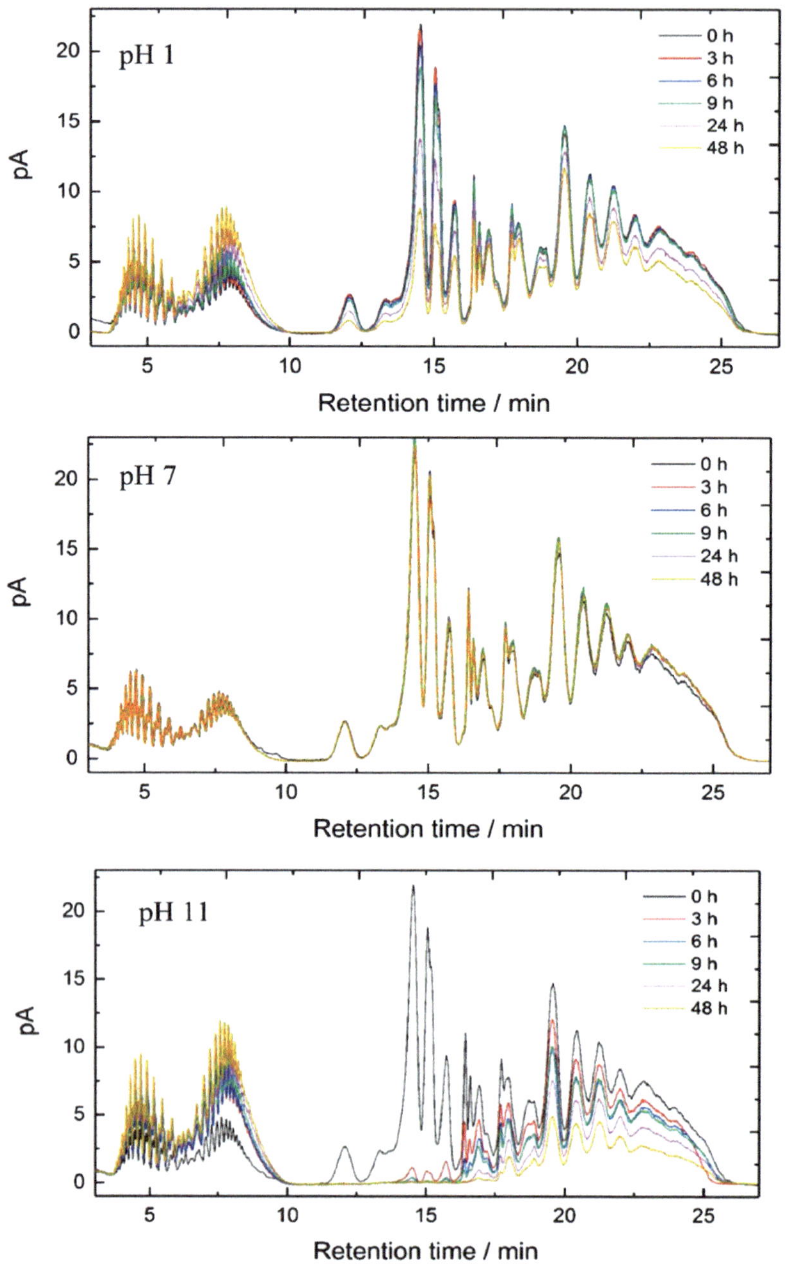

Figure 12.3 HPLC-CAD chromatogram of PS20 in 50 mM phosphate buffer incubated at pH 1, pH 7 and pH 11 at 40 °C with samples pulled and analyzed at 0, 3, 6, 9, 24, and 48 hours [1]. *Source:* [1] / with permission of Elsevier.

and basic hydrolysis is that the esters of fatty acids with lower carbon numbers hydrolyze much more significantly under basic conditions (pH 11), as evident from the peak reduction in the 13–16 minutes elution window. These peaks are POE sorbitan laureate, POE isosorbide laureate, and POE laureate. The variation in degradation extent based on fatty acid carbon numbers can be rationalized by the distinct mechanisms of acidic and basic hydrolysis. In acid-catalyzed hydrolysis, the rate-limiting step is protonation, and the subsequent water attack is less sensitive to the fatty acid chain length. On the contrary, in base-catalyzed hydrolysis, the rate-limiting step is the nucleophilic attack of the hydroxide ion on the carbonyl carbon of the ester, where shorter fatty acid chains offer less steric hindrance around the reaction center. Because the ester bond is intrinsically stable at neutral pH, minimal degradation has been observed at pH 7 as evident from the superimposable traces over time in Figure 12.3.

In recent years, enzymatic degradation has been increasingly recognized as the primary cause of hydrolysis [26]. This enzymatic degradation is mediated by residual esterases or lipases from the host cell proteins [29] present in the drug product. Tomlinson et al. [33] examined 20 enzymes for their ability to degrade polysorbates, and identified 13 of them with polysorbate hydrolysis activity. The study utilized HPLC-CAD to monitor the polysorbate profile changes upon incubation with studied enzymes as shown in Figure 12.4. The study reported substrate specificity of hydrolysis enzymes. For example, PPT1 hydrolyzes more hydrophobic mono-esters and higher order esters, while rhLPL hydrolyzes all esters concurrently. Both chemical and enzymatic hydrolysis reduces the PS concentration, and leads to particulate formation due to the low solubility of the FFA degradants.

Oxidative degradation is a radical-driven chain reaction that primarily affects the POE chains and, to a lesser extent, unsaturated fatty acid esters. This degradation pathway can be initiated by oxygen, light, heat, and trace metal ions [34]. The degradation path of the POE chains is shown in Figure 12.5. The process starts with an initiation phase where free radicals are generated through thermal energy, light exposure, or metal-catalyzed redox reactions. During the subsequent propagation phase, an alkyl radical reacts with molecular oxygen to form a peroxyl radical, which then abstracts a hydrogen atom from another POE chain, generating a hydroperoxide (ROOH) and a new alkyl radical, thus perpetuating the chain reaction. The final stage is termination, where the hydroperoxides decompose. The oxidation process yields peroxides, aldehydes, ketones, epoxides, and short-chain POE esters. Similar to hydrolysis, oxidation compromises polysorbate activity, which is detrimental to protein formulation stability. Furthermore, the reactive aldehydes and peroxides can oxidize proteins in the formulation, posing significant quality concerns [35]. Oxidation of unsaturated fatty acids, such as oleic acid, the predominant fatty acid in PS80, has also been reported [8, 36, 37], yielding

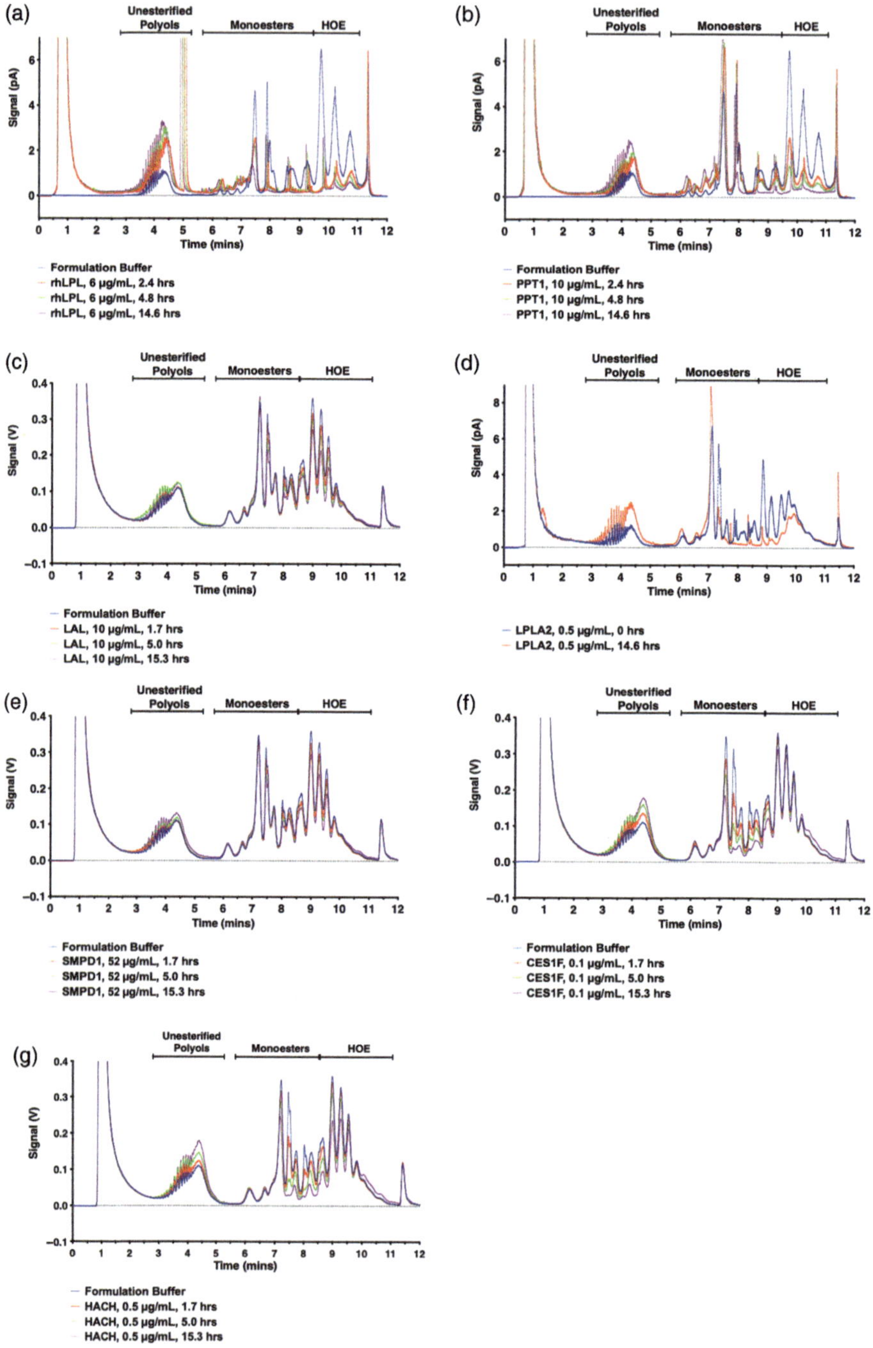

Figure 12.4 HPLC-CAD chromatograms for PS20 during incubation study with different recombinant enzymes: (a) rhLPL, (b) PPT1, (c) LAL, (d) LPLA2, (e) SMPD1, (f) CES1F, and (g) HACH [33]. *Source:* [33] / with permission of Elsevier.

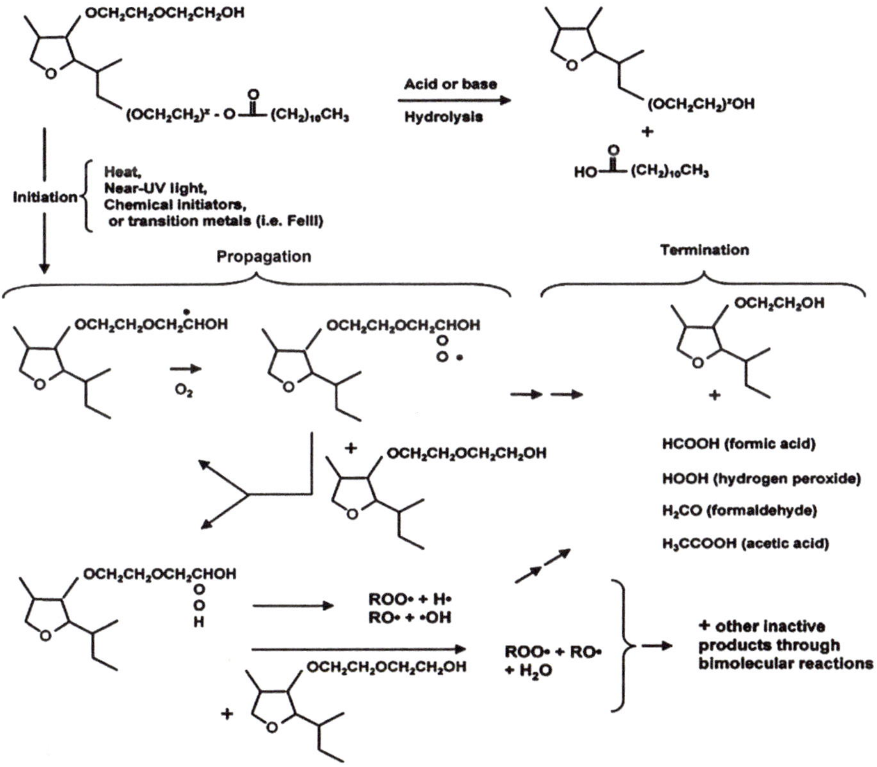

Figure 12.5 Hydrolysis and oxidation degradation pathways of polysorbates [34]. *Source:* [34] / with permission of Elsevier.

degradants such as hydroxyl oleate and keto oleate. PS80 tends to be more susceptible to oxidation than PS20 due to the presence of esters with unsaturated fatty acids like oleic acid. Notably, HPLC-CAD has been utilized to elucidate the degradation patterns of polysorbates. For instance, Lippold et al. [8] reported, using a CAD assay on PS20 fractions, that more hydrophobic subspecies, such as poly-esters, oxidized more rapidly than more hydrophilic subspecies, such as mono-esters. Guo et al. employed an HPLC-CAD assay to investigate the oxidation of PS80 in different buffer systems and attributed oxidation to metal ions originating from the excipients, demonstrating that different buffers can provide varying degrees of protection against metal ion-induced oxidation [38].

12.3.3 Quantification of Polysorbates

Typical polysorbate concentrations in drug formulations range from 0.01% to 0.05% weight/volume (w/v), as reported by 60% of surveyed biopharmaceutical companies in an industry-wide survey study conducted in 2022 [26], underscoring the need for methods with high sensitivity. According to this survey, 88% of participating companies have implemented PS content assay in GMP release testing, 75% in developmental studies, and 69% in GMP stability studies. Among the methodologies for PS content assay, 94% used HPLC-based methods, 44% used FMA and 19% used thiocyanate complexation. HPLC-based methods offer superior sensitivity and versatility in profiling and quantification simultaneously.

HPLC-CAD has been the standard method for quantification of PS content. While earlier studies utilized mixed-mode HPLC with rapid gradients to separate polysorbates from proteins as a single peak for quantification [39], current state-of-the-art HPLC methods employ shallow gradients to resolve polysorbates into subspecies for simultaneous profiling and quantification.

Hawe et al. [11] reported a fully validated 2DLC-CAD method that demonstrated linearity from 0.1 to 0.5 mg/mL, accuracy between 96% and 109%, sensitivity down to 2 μg/mL, and precision lower than 1% RSD for both PS20 and PS80. Beyond content assay, LC-CAD has been utilized to quantify polysorbate subspecies to monitor product quality and stability, control lot-to-lot variability, and elucidate the degradation mechanisms. Lippold et al. employed LC-CAD for the quantification of subspecies such as mono- and poly-esters [8].

However, when it comes to quantitation of PS components, CAD exhibits limitations in sensitivity especially for short-chain FFAs. A significant reduction in sensitivity has been reported for fatty acids with a carbon chain length of 12 or less [21, 40]. This limitation restricts the applicability of LC-CAD in quantitative composition analysis primarily to PS80 and less so to PS20, where shorter chain fatty acids are more prevalent. FFAs are direct indicators of polysorbate degradation and pose significant product quality concerns, necessitating their thorough investigations. LC can be coupled with MS for the quantification of FFAs to improve method sensitivity. For instance, liquid chromatography-multiple reaction monitoring (LC-MRM) has demonstrated superior sensitivity for FFAs compared to 2DLC-CAD [41]. LC-MS in negative-ion mode has also been successfully applied to the analysis of FFAs, achieving sensitivities down to 100 ng/mL for lauric acid [6, 38].

12.4 Emerging Trends and Future Directions

12.4.1 Advancements in CAD Technology

While CAD parameters such as evaporation temperature, filter constant, and the PFV can be optimized to enhance its quantitative performance, modern CAD instruments could benefit from improvement of the hardware to further enhance its sensitivity, compatibility, and robustness. Recent years have witnessed hardware advancements, including the replacement of cross-flow nebulizer designs with concentric nebulizers, which has been shown to enhance sensitivity to the ng/mL range [42]. Another example is the introduction of an adjustable evaporation temperature over a wider range of ambient +5 to 100 °C, allowing users more flexibility to fine-tune the temperature for the analytes, especially the semi volatiles [43]. A recent study reported a novel charging mechanism based on atmospheric pressure chemical ionization (APCI) charging principle, termed direct corona charging mode (DCCM) in CAD. Compared to the traditional plasma collision charging mode (PCCM), DCCM was demonstrated to improve sensitivity through dilution-free charging, enhance repeatability by reducing airflow disturbance, and lower operational costs through gas conservation [44].

Significant enhancements have also been observed in streamlined CAD control and data processing capabilities integrated within instrument software. For example, the new Chromeleon software allows for dynamic linking of data channels, instantaneous recalculation of results using various fitting mechanisms such as log–log or quadratic curves, and the application of post-processing "power law" adjustments. Recent work has explored the application of machine learning to model CAD signals of FFAs, developing a quantitative structure–property relationship (QSPR) model using Gradient Boosted Trees (GBT). This model predicts CAD response based on analyte molecular properties, mobile phase flow rate, evaporation temperature, filter constants, and PFV [45]. Upon further development and validation, such models could be integrated into instrument control and data analysis software to significantly enhance the efficiency of LC-CAD method development.

12.4.2 Automated and High-Throughput Applications

High-throughput screening and analysis (HTS and HTA), enabling the simultaneous analysis of a large number of samples, are revolutionizing drug discovery and development. This approach unlocks the potential to rapidly screen extensive compound libraries and process conditions, leading to reduced costs, improved efficiency, and enhanced data quality. In the era of Artificial Intelligence (AI), HTS and HTA serve as powerful data engines, where the resulting large, standardized, and diverse datasets provide the necessary fuel to train sophisticated models. These models can significantly accelerate the identification of drug candidates, the optimization of their properties, and the overall development cycle.

Recent years have witnessed the application of CAD for quantitative analysis in high-throughput HPLC for compound assay and quality control [46, 47]. While the traditional HPLC analysis of PS usually takes longer time in the range of 20–30 minutes instead of sub-5 minute windows, a recent study was able to decrease the PS analysis time to 8 minutes by leveraging 2D-LC [11]. This approach holds significant potential for high-throughput screening of polysorbate batches, analysis of polysorbate stability samples, and formulation development.

12.4.3 Multi-Detector Systems

Although CAD has become a widely accepted method for PS quantification due to its near-universal response to nonvolatile analytes and its superior sensitivity over ELSD, its use as a standalone detector presents notable limitations. Primarily, CAD lacks molecular specificity, thus failing to differentiate between structurally similar polysorbate subspecies or to identify degradation products. This non-specificity can pose challenges in the analysis of complex biopharmaceutical formulations where co-eluting impurities or degraded excipients are present. The integration of orthogonal detectors, such as MS or UV, effectively addresses these shortcomings by providing impurity identification and structural confirmation alongside quantitative data. For instance, combining CAD with MS enables the simultaneous quantitation and identification of polysorbate components and their hydrolytic or oxidative degradation products, thereby enhancing the analytical resolution, reliability, and interpretability of polysorbate stability studies [17]. Another limitation of CAD is its diminished sensitivity toward semi volatile compounds, such as short-chain fatty acids. To accurately quantify these FFAs, the integration of complementary detectors like MS is highly recommended. The topic of multi-detector configurations is further discussed in Chapter 13.

12.5 Conclusion

Polysorbates are essential excipients in biopharmaceutical formulations, yet their inherent chemical complexity and susceptibility to degradation pose significant analytical challenges. HPLC-CAD has emerged as a staple technology for polysorbate characterization and quantification, offering broad applicability, near-universal response, and operational robustness. However, inherent limitations, such as reduced sensitivity to volatile compounds like short-chain fatty acids and the lack of molecular specificity, must be acknowledged.

The implementation of multi-detector systems that integrate CAD with MS emerges as a crucial strategy for comprehensive polysorbate characterization. This approach enables both accurate quantification and detailed structural elucidation of polysorbate components and their degradation products. Further improvements in CAD instrumentation, such as enhanced nebulization and charging mechanisms, coupled with advanced data processing, will further enhance the efficacy and reliability of polysorbate analysis in biopharmaceutical research and development, as well as quality control.

References

1 Dwivedi, M., Buske, J., Haemmerling, F. et al. (2020). Acidic and alkaline hydrolysis of polysorbates under aqueous conditions: towards understanding polysorbate degradation in biopharmaceutical formulations. *European Journal of Pharmaceutical Sciences* 144: 105211.

2 Dwivedi, M., Blech, M., Presser, I., and Garidel, P. (2018). Polysorbate degradation in biotherapeutic formulations: identification and discussion of current root causes. *International Journal of Pharmaceutics* 552 (1): 422–436.

3 Bai, L., Zhang, Y., Zhang, C. et al. (2023). Investigation of excipients impact on polysorbate 80 degradation in biopharmaceutical formulation buffers. *Journal of Pharmaceutical and Biomedical Analysis* 233: 115496.

4 Borisov, O.V., Ji, J.A., Wang, Y.J. et al. (2011). Toward understanding molecular heterogeneity of Polysorbates by application of liquid chromatography–mass spectrometry with computer-aided data analysis. *Analytical Chemistry* 83 (10): 3934–3942.

5 Kopf, R., Paschen, C., Müller, L. et al. (2023). Leveraging mass detection to simultaneously quantify surfactant content and degradation mode for highly concentrated biopharmaceuticals. *Journal of Pharmaceutical and Biomedical Analysis* 236: 115651.

6 Honemann, M.N., Wendler, J., Graf, T. et al. (2019). Monitoring polysorbate hydrolysis in biopharmaceuticals using a QC-ready free fatty acid quantification method. *Journal of Chromatography B* 1116: 1–8.

7 Shi, L., Shen, G., Chai, R. et al. (2025). A review of polysorbate quantification and its degradation analysis by liquid chromatography coupled with charged aerosol detection. *Journal of Chromatography. A* 1742: 465651.

8 Lippold, S., Koshari, S.H.S., Kopf, R. et al. (2017). Impact of mono- and poly-ester fractions on polysorbate quantitation using mixed-mode HPLC-CAD/ELSD and the fluorescence micelle assay. *Journal of Pharmaceutical and Biomedical Analysis* 132: 24–34.

9 Evers, D.H., Schultz-Fademrecht, T., Garidel, P., and Buske, J. (2020). Development and validation of a selective marker-based quantification of polysorbate 20 in biopharmaceutical formulations using UPLC QDa detection. *Journal of Chromatography B* 1157: 122287.

10 Abrar, S. and Trathnigg, B. (2010). Separation of nonionic surfactants according to functionality by hydrophilic interaction chromatography and comprehensive two-dimensional liquid chromatography. *Journal of Chromatography. A* 1217 (52): 8222–8229.

11 Zegota, M.M., Schuster, G., De Pra, M. et al. (2024). High throughput multidimensional liquid chromatography approach for online protein removal and characterization of polysorbates and poloxamer in monoclonal antibody formulations. *Journal of Chromatography. A* 1720: 464777.

12 Abrar, S. and Trathnigg, B. (2011). Separation of polysorbates by liquid chromatography on a HILIC column and identification of peaks by MALDI-TOF MS. *Analytical and Bioanalytical Chemistry* 400 (7): 2119–2130.

13 Bottger, R., Chao, P.H., Al Fayez, N. et al. (2022). Simultaneous chromatographic quantitation of drug substance and excipients in nanoformulations using a combination of evaporative light scattering and absorbance detectors. *Molecular Pharmaceutics* 19 (6): 1882–1891.

14 Dixit, N., Salamat-Miller, N., Salinas, P.A. et al. (2016). Residual host cell protein promotes Polysorbate 20 degradation in a sulfatase drug product leading to free fatty acid particles. *Journal of Pharmaceutical Sciences* 105 (5): 1657–1666.

15 Kim, J. and Qiu, J. (2014). Quantitation of low concentrations of polysorbates in high protein concentration formulations by solid phase extraction and cobalt-thiocyanate derivatization. *Analytica Chimica Acta* 806: 144–151.

16 He, Y., Brown, P., Bailey Piatchek, M.R. et al. (2019). On-line coupling of hydrophobic interaction column with reverse phase column -charged aerosol detector/mass spectrometer to characterize polysorbates in therapeutic protein formulations. *Journal of Chromatography. A* 1586: 72–81.

17 Li, Y., Hewitt, D., Lentz, Y.K. et al. (2014). Characterization and stability study of Polysorbate 20 in therapeutic monoclonal antibody formulation by multidimensional ultrahigh-performance liquid chromatography–charged aerosol detection–mass spectrometry. *Analytical Chemistry* 86 (10): 5150–5157.

18 Doshi, N., Ritchie, K., Shobha, T. et al. (2021). Evaluating a modified high purity Polysorbate 20 designed to reduce the risk of free fatty acid particle formation. *Pharmaceutical Research* 38 (9): 1563–1583.

19 Magnusson, L.E., Risley, D.S., and Koropchak, J.A. (2015). Aerosol-based detectors for liquid chromatography. *Journal of Chromatography A* 1421: 68–81.

20 Martos, A., Koch, W., Jiskoot, W. et al. (2017). Trends on analytical characterization of Polysorbates and their degradation products in biopharmaceutical formulations. *Journal of Pharmaceutical Sciences* 106 (7): 1722–1735.

21 Schilling, K., Pawellek, R., Lovejoy, K. et al. (2018). Influence of charged aerosol detector instrument settings on the ultra-high-performance liquid chromatography analysis of fatty acids in polysorbate 80. *Journal of Chromatography. A* 1576: 58–66.

22 Pawellek, R., Muellner, T., Gamache, P., and Holzgrabe, U. (2021). Power function setting in charged aerosol detection for the linearization of detector response – optimization strategies and their application. *Journal of Chromatography. A* 1637: 461844.

23 Behrens, B., Baune, M., Jungkeit, J. et al. (2016). High performance liquid chromatography-charged aerosol detection applying an inverse gradient for quantification of rhamnolipid biosurfactants. *Journal of Chromatography. A* 1455: 125–132.

24 Górecki, T., Lynen, F., Szucs, R., and Sandra, P. (2006). Universal response in liquid chromatography using charged aerosol detection. *Analytical Chemistry* 78 (9): 3186–3192.

25 Paschen, C.A., Klemm, D., Graf, T. et al. (2021). Simultaneous quantification of polysorbate 20 and poloxamer 188 in biopharmaceutical formulations using evaporative light scattering detection. *Journal of Pharmaceutical and Biomedical Analysis* 192: 113640.

26 Wuchner, K., Yi, L., Chery, C. et al. (2022). Industry perspective on the use and characterization of polysorbates for biopharmaceutical products part 1: survey report on current state and common practices for handling and control of polysorbates. *Journal of Pharmaceutical Sciences* 111 (5): 1280–1291.

27 Zhang, R., Wang, Y., Tan, L. et al. (2012). Analysis of polysorbate 80 and its related compounds by RP-HPLC with ELSD and MS detection. *Journal of Chromatographic Science* 50 (7): 598–607.

28 Yang, K., Hewarathna, A., Geerlof-Vidavsky, I. et al. (2019). Screening of polysorbate-80 composition by high resolution mass spectrometry with rapid H/D exchange. *Analytical Chemistry* 91 (22): 14649–14656.

29 Kranz, W., Wuchner, K., Corradini, E. et al. (2019). Factors influencing polysorbate's sensitivity against enzymatic hydrolysis and oxidative degradation. *Journal of Pharmaceutical Sciences* 108 (6): 2022–2032.

30 EMEA. Available from: https://www.ema.europa.eu/en/human-regulatory-overview/research-and-development/scientific-guidelines/biological-guidelines/questions-answers-biological-medicinal-products.

31 USP. Available from: https://doi.org/10.31003/USPNF_M66830_04_01.

32 USP. Available from: https://doi.org/10.31003/USPNF_M66800_05_01.

33 Kovner, D., Yuk, I.H., Shen, A. et al. (2023). Characterization of recombinantly-expressed hydrolytic enzymes from Chinese hamster ovary cells: identification of host cell proteins that degrade polysorbate. *Journal of Pharmaceutical Sciences* 112 (5): 1351–1363.

34 Kerwin, B.A. (2008). Polysorbates 20 and 80 used in the formulation of protein biotherapeutics: structure and degradation pathways. *Journal of Pharmaceutical Sciences* 97 (8): 2924–2935.

35 Ha, E., Wang, W., and John, W.Y. (2002). Peroxide formation in polysorbate 80 and protein stability. *Journal of Pharmaceutical Sciences* 91 (10): 2252–2264.

36 Dahotre, S., Tomlinson, A., Lin, B., and Yadav, S. (2018). Novel markers to track oxidative polysorbate degradation in pharmaceutical formulations. *Journal of Pharmaceutical and Biomedical Analysis* 157: 201–207.

37 Li, X., Wang, Z., Zheng, B. et al. (2023). Novel strategy to rapidly profile and identify oxidized species of polysorbate 80 using ultra-high-performance liquid chromatography coupled with high-resolution mass spectrometry. *Analytical Chemistry* 95 (24): 9156–9163.

38 Cheng, Y., Hu, M., Zamiri, C. et al. (2019). A rapid high-sensitivity reversed–phase ultra high performance liquid chromatography mass spectrometry method for assessing polysorbate 20 degradation in protein therapeutics. *Journal of Pharmaceutical Sciences* 108 (9): 2880–2886.

39 Hewitt, D., Zhang, T., and Kao, Y.H. (2008). Quantitation of polysorbate 20 in protein solutions using mixed-mode chromatography and evaporative light scattering detection. *Journal of Chromatography. A* 1215 (1): 156–160.

40 Ilko, D., Braun, A., Germershaus, O. et al. (2015). Fatty acid composition analysis in polysorbate 80 with high performance liquid chromatography coupled to charged aerosol detection. *European Journal of Pharmaceutics and Biopharmaceutics* 94: 569–574.

41 Zhang, S., Riccardi, C., Kamen, D. et al. (2022). Monitoring polysorbate hydrolysis in therapeutic proteins using an ultrasensitive extraction-free fatty acid quantitation method. *Analytical Biochemistry* 637: 114472.

42 Staff A (2021). Charged aerosol detection (CAD): inspiring next-generation separations [Internet]. AnalyteGuru. Available from: https://www.thermofisher.com/blog/analyteguru/charged-aerosol-detection-inspiring-next-generation-separations/ (accessed 14 May 2025).

43 ThermoFisher. Available from: https://assets.thermofisher.com/TFS-Assets/CMD/Specification-Sheets/ps-70661-lc-corona-veo-cad-ps70661-en.pdf.

44 Feng, S., Liu, H., Zhou, C. et al. (2025). Direct corona charging: a new strategy for enhancing sensitivity and stability in charged aerosol detection. *Analytical Chemistry* 97 (13): 7079–7088.

45 Pawellek, R., Krmar, J., Leistner, A. et al. (2021). Charged aerosol detector response modeling for fatty acids based on experimental settings and molecular features: a machine learning approach. *Journal of Cheminformatics* 13 (1): 53.

46 Jones, M. and Goodyear, R.L. (2023). High-throughput purification in drug discovery: scaling new heights of productivity. *ACS Medicinal Chemistry Letters* 14 (7): 916–919.

47 Wu, N.P., Wang, W., Gadiagellan, D. et al. (2024). An automated robotic interface for assays: facilitating machine learning in drug discovery by the automation of physicochemical property assays. *ACS Omega* 9 (23): 24948–24958.

13

Use of Orthogonal Detectors in Comprehensive Analyses: A Focus on Safety Assessment of Pharmaceuticals, Including Botanicals

Katherine S. Lovejoy and Frank Steiner

Dionex Softron GmbH, a Part of Thermo Fisher Scientific, Product Applications, Germering, Germany

CHAPTER MENU

13.1 Summary

This chapter will discuss comprehensive untargeted and partially targeted analysis of complex samples. An explanation of comprehensive analysis (CA) and an introduction to charged aerosol detection (CAD) in comprehensive analysis, including multidimensional liquid chromatography, will reveal the extensive use of CAD with complex samples. A look at the fundamental considerations associated with the use of CAD in multidetector systems and with multidimensional chromatography will provide practical implementation guidelines and insight into technical challenges and approaches to data analysis. The chapter will close with a focus on two consequential and noteworthy examples of comprehensive analysis using orthogonal detection and CAD: the characterization and quantitation of components in botanical medicines (i.e., dietary supplements and traditional medicines) to enable in silico safety assessments prior to their market introduction, and a consumer product safety assessment scheme for evaluating menstrual cups, including extractables analysis.

13.2 Introduction to Comprehensive Analysis

Comprehensive analyses performed by high-performance liquid chromatography (HPLC) involve separating, identifying, and quantifying multiple components within a sample. Comprehensive analysis of complex samples differs from comprehensive 2D-HPLC, although sometimes 2D-HPLC is used for comprehensive analysis. Comprehensive analysis involves accurately quantifying and identifying unknown samples, whereas 2D-HPLC aims to improve chromatographic resolution of complex samples with closely related components. Orthogonal detection techniques, including ultraviolet-diode array detection (UV-DAD) and mass spectrometry (MS) are commonly paired with CAD. Techniques including refractive index (RI) detection, chemiluminescent nitrogen detection (CLND) [1], nuclear magnetic resonance (NMR) spectroscopy [2], radio detection, and infrared spectroscopy are occasionally paired with CAD. The NMR pairing example has the disadvantage that the system was run in deuterium oxide and that the analysis of one peak required up to 10 hours. After analysis, the resulting data from *all* sources is integrated and analyzed to form an overall picture. The chemical data can be chemometrically surveyed to find meaningful relationships to the hypothesis being tested. Examples of common comprehensive analyses of pharmaceuticals include detection of leachables (i.e., from plastics used in production and packaging) in finished products; analysis of degradation products formed during stability testing to determine shelf life [3]; metabolite profiling [4]; evaluation of sample purification protocols [5]; and safety analysis of botanical medicines.

13.2.1 Advantages of Untargeted Analysis by CAD

Targeted analysis aims to characterize specific known compounds within a sample, whereas untargeted analysis aims to detect, identify, and quantify as many compounds as possible in a sample. CAD is highly effective for untargeted analysis due to its ability to accurately quantify peaks of nonvolatile unknown substances and to detect nonvolatile and semi volatile substances without dependencies on chemical structure, such as the presence of a chromophore or the potential to ionize in a mass spectrometer. The accurate quantification of unknown substances, based on the uniform response property, allows multiple unknown species to be quantified with the calibration curve of one well-characterized standard. Uniform response is explored in greater detail in Chapter 4 and practical quantification using uniform response is described below. The unique detection abilities independent of the presence of a chromophore or the ease of ionization are termed universal detection. Chapters 1 and 4 contain a thorough examination of universal detection.

There are many applications using CAD for quantification of unknowns in combination with one or more complementary detection techniques for identification of unknowns. The use of UV detection upstream, in series, with CAD is now commonplace. The ever-increasing need to rapidly identify and quantify many diverse species in complex samples has furthermore led to significant growth in use of multi-detector systems with CAD for CA. A very common configuration involves liquid splitting after a UV DAD flow cell to parallel CAD and MS devices.

13.2.2 LC × LC with CAD

Two-dimensional liquid chromatography (2D LC) systems use different separation mechanisms, usually two or more columns with different chemistries, to maximize the peak capacity of a system, which is the maximum possible number of compounds that can be separated. By adding extra separation dimensions to the system, the maximal theoretical peak capacity that can be achieved is the product of the peak capacity of each individual separation dimension, providing much greater resolving power than each individual separation on its own. Unlike MS, charged aerosol detectors do not provide peak deconvolution, structural information and identification. However, their capability to provide a uniform response for nonvolatile species warrants its inclusion in 2D systems to ensure that all of these analytes are detected and quantified. Band broadening in online 2D LC can be severe when the transfer volume into the second dimension depends on the flow rate from the first dimension and this can be worsened by the fluidics of an aerosol detector or MS [6]. Because band broadening can be an issue for CAD in some instrument setups due to the 0.15 mm inner diameter of the nebulizer capillary, the volume of in-line UV flow cells, additional valves, or capillaries, etc., analytical methods should be developed to promote analyte focusing on the column head in the second dimension of the LC. The fluidics and nebulizer in MS sources can also pose the same problem.

Examples of CA with CAD in 2D systems include metabolomic studies on lyophilized urine [7], screening of food packaging [8], analysis of heavy oil [9], analysis of phospholipids [10] (further explored in Chapter 5, Section 4) and characterization of hydroxypropylmethylcellulose [11]. Additionally, routine comprehensive impurity analysis of pharmaceuticals using 2D LC with CAD is the standard practice in some labs. For example, in an automated online multi-heartcutting system for pharmaceutical impurity analysis with UV and CAD, users can select between six orthogonal columns in the second dimension [12]. Chapters 11 and 12 treat comprehensive lipid and polysorbate analysis with multidimensional systems in detail.

13.3 Principles of the Implementation of Orthogonal Detectors

The advantages of a comprehensive analysis using CAD in a multidetector setup include sensitivity, orthogonal detection to MS and UV, and quantitation of unknowns with greater accuracy than MS or UV. The details of the implementation affect all these advantages, and the current state of the art is described below.

13.3.1 Sensitivity in Multidetector Setups

Sensitivity is a key factor in the choice of CAD over evaporative light scattering detection (ELSD) for multidetector setups. Both CAD and ELSD can be used as an orthogonal detection technique to MS and UV but, practically speaking, CAD provides 10-fold or better limits of detection (LOD) than ELSD. The greater sensitivity of CAD renders it more valuable for CA because CAD is more likely than ELSD to uncover substances present at very low levels [10, 13–15]. The other advantages of CAD over ELSD, including response curve shape and quality of uniform response, are addressed in Chapter 4.

Sensitivity, along with flow rate requirements, also governs the split ratio between CAD and MS. The 4 : 6 split used for the system in Figure 13.1 [16] is useful for simultaneous implementation of CAD and single quadrupole MS because the techniques have similar sensitivity. A split ratio of 1 : 10 [mass spectrometry-charged aerosol detection (MS:CAD)] is often implemented for high resolution MS, for example for the Orbitrap Exploris 120 [17] and the microTOF MS [18], because CAD has roughly 10× lower sensitivity, depending on the application. A split of 1 : 3 (MS:CAD) has been used for the quadrupole time-of-flight mass spectrometry (qTOF MS) [19]. The technical implementation of the split is discussed below, in Section 13.3.3.

13.3.2 Quantification and Uniform Response of CAD

CAD is often assigned the task of quantifying unknown peaks when combined with orthogonal detection techniques. One example is its use for in silico safety assessment of botanical products where peak areas greater than a certain threshold are flagged for identification by MS and in-depth toxicological evaluation [20]. Peaks under a certain threshold, based on the educated assumption that exposure to a miniscule dose of any toxic substance is acceptable, are of low toxicological concern and are disregarded.

A detector with a uniform response is one for which any analyte at a given concentration yields the same detector response as any other analyte. Relative

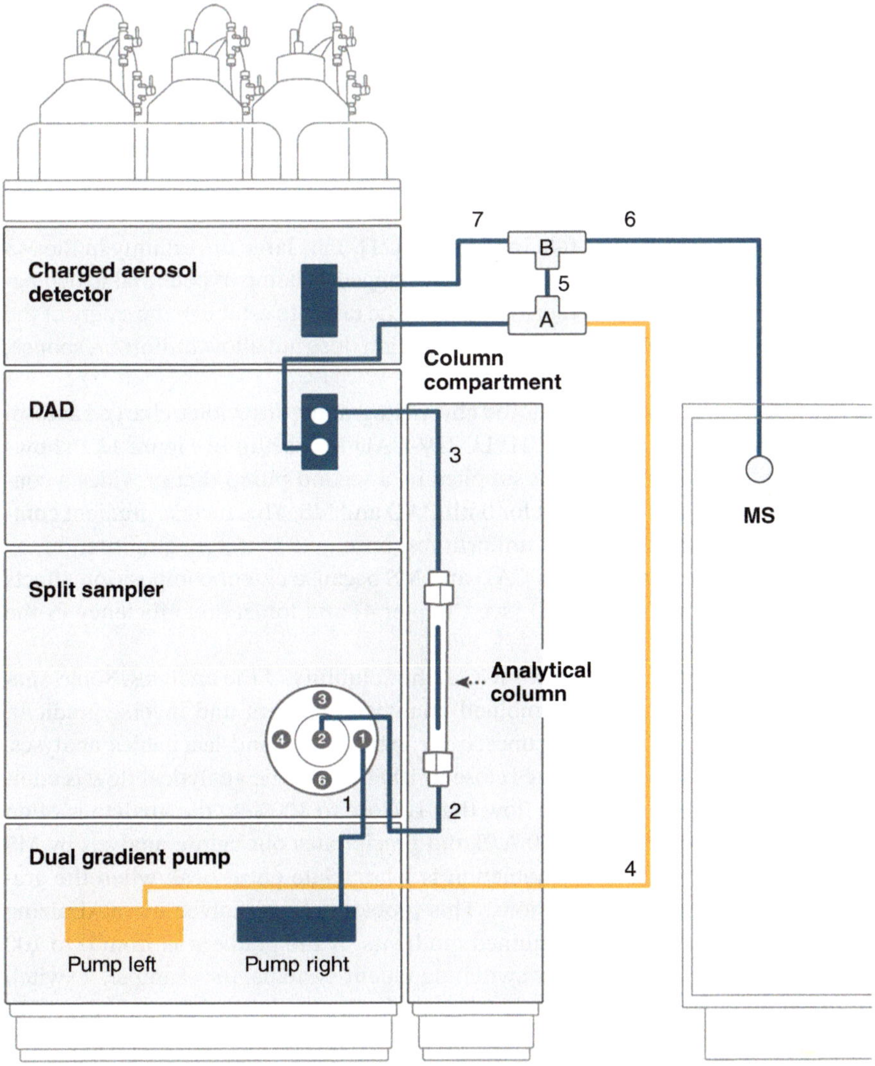

Figure 13.1 Schematic view of the UHPLC multidetector system used in [16]. Flow capillaries from analytical gradient are (1) 0.1 mm × 350 mm, (2) mobile phase preheater 0.1 × 650, and (3) 0.1 mm × 350 mm. The inverse gradient: (4) 0.1 mm × 950 mm) merges with the mobile phase downstream from the column at mixing point (A) (inv. gradient T-piece: 0.5 mm ID); flow is split toward the detectors at point (B) (T-vici: 0.15 mm ID); capillary between mixing and split point (5) is either a standard capillary (0.1 mm × 50 mm) or a linear capillary mixer (25 μL) for radial mixing. Split ratio between CAD (7) 0.1 mm × 550 mm and MS detector (6) 0.1 mm × 550 mm was 4 : 6 (MS:CAD). *Source:* [16] / Elsevier / CC BY 4.0.

response factors obtained from compound libraries capture the range of responses for a particular detector. From the relative response factors, an uncertainty is determined that accounts for response factor variation. The threshold of toxicological concern is adjusted lower to account for the uncertainty in the response factor variation and to ensure that no peaks are disregarded just because the instrument response is relatively low for that substance. For example, one study for LC/Q-TOF MS showed relative responses that ranged from 0.2% to 438% of the true value, or range of 2190-fold in response [21]. This large uncertainty in the MS response means that many more peaks are flagged as being of potential toxicological concern and that much more effort must be taken to establish the safety of the pharmaceutical or medical device. MS detection does not allow uniform response, which is why CAD is necessary.

The ultra-high-performance liquid chromatography-ultraviolet-charged aerosol detection-mass spectrometry (UHPLC-UV-CAD-MS) setup in Figure 13.1 shows an inverse, or makeup, gradient supplied by a second pump that provides a constant mobile phase composition for both CAD and MS. This inverse gradient composition (IGC) is important for uniform response, and therefore quantitation, by CAD. IGC is beneficial for both CAD and MS because eluent composition affects both particle formation in CAD (see Chapter 4) and ionization efficiency in the MS [22, 23].

An important consideration for IGC is the solubility of the analytes. Some analytes are not soluble in the combined analytical gradient and inverse gradient. This problem is especially of concern for extractables and leachables analyses. For example, if a substance elutes close to 100% B, but the analytical flow is combined with an inverse gradient flow that is close to 100% A, the analyte is often not sufficiently soluble in 50/50 A/B and precipitates out before analysis by MS or CAD. Evidence of this phenomenon is a large late ghost peak when the gradient returns to starting conditions. This problem can be solved by maximizing the organic content of the combined gradients. If the gradient is from 0 to 100 in both A and B, then consider switching eluent systems, for example, a switch from methanol to acetonitrile enabled quantitative analysis of the polymer additives TOTM (tris[2-ethylhexyl]trimellitate, a plasticizer) (results not published) and Irganox 1076 (an antioxidant) [24].

The uniform response property of CAD applies to nonvolatile substances. Semivolatile substances can have calibration curves that greatly differ from the calibration curve of a nonvolatile surrogate standard. The volatility, and therefore the uniformity of the response factor, by CAD of completely unknown substances can be determined by analyzing a sample at a given evaporation tube temperature (T_E) and then reanalyzing at a higher T_E. Substances that are lost at the higher T_E are likely semivolatiles and the detector response may not match that of the surrogate nonvolatile standard. One example of thresholds defined for an analytical determination of nonvolatility, also applicable for unknown analyte substances, determined that the peak area at $T_E = 50\,°C$ should be at least 85% that of the peak area at 35 °C. Using this threshold, originated by Eckardt et al. and called $Q_{50/35}$ [25], as well as IGC, relative response ranged from 82.4% to 127.3% of the true amount in the worst case, a range of 1.5-fold [25]. Relative response in a screening of extractables from menstrual cups [20] has been shown to be no worse than two for any analyte.

Another study showed relative responses by CAD *without* IGC and *without* screening for nonvolatiles that ranged from 0.149% to 2.680% of the true value, or a range of 18-fold, which was somewhat worse than the $UV_{230\,nm}$ relative response, which ranged from 0.334 to 1.976, or 6-fold. The response by CAD *without* IGC but *with* screening for nonvolatiles using the $Q_{50/35}$ showed relative responses that ranged from 0.44% to 1.95% of the true value, or a range of 4.5-fold [26]. Additional typical relative response ranges are listed in Section 13.4.

Although this screening approach at $T_E = 35$ and 50 °C marks some known nonvolatiles, such as bisphenol A, as semivolatiles, it is a conservative method of ensuring the accuracy of the quantitation of all unknowns in a sample. This simple calculation is a valuable and unique tool for determining whether unknowns can reliably be quantified by CAD with a universal standard. There is no comparable simple determination that can be made to improve quantification reliability of unknowns by MS or UV detection.

There are two key points to keep in mind when trying to optimize response uniformity. The first is that the uncertainty factor can be minimized and quantification is improved through three modifications: (i) classifying unknowns as nonvolatile or semivolatile, (ii) using IGC, and (iii) mixing the inverse and analytical gradients with an inline capillary mixer after the merge point and before the flow split, an idea that is further developed in the next section. The second point is that the purpose of IGC is to ensure uniform particle formation in the detector over the gradient, not to address baseline changes over the gradient. Baseline changes over the gradient observed by CAD are largely attributable to old or improperly degassed eluents, eluents with unacceptably high levels of impurities ("residue after evaporation" being more commonly associated with a solvent's specification), column bleed, or sample matrix components and can be almost always resolved with proper troubleshooting.

13.3.3 Control of the Split to Destructive Detectors

Destructive detectors render a sample unusable for further analysis or collection. CAD, ELSD, and MS are all destructive detector techniques. Stable control of the split conditions when multiple destructive detectors are present is critical to peak area reproducibility. CAD and MS technology are both nebulizer-based and therefore must be configured in parallel at the end of separate flow paths after a flow split. The actual split ratio is dependent on the back pressures of the detectors used.

When using multiple destructive detectors, the total sample mass is divided among the detectors. A common multidetector setup for CAD, UV, and MS with IGC and a split ratio of 4 : 6 (MS:CAD) is shown in Figure 13.1. The advantage of using CAD in such a split is that the detector is a mass flow-dependent device. For a mass flow-dependent detector, the flow split reduces not only the sample mass seen by the detector, but also the mass of the noise-causing impurities. The signal-to-noise ratio (SNR) for each peak should, in most cases, roughly match the SNR from the unsplit sample. Conversely, the MS is not a mass flow dependent device, so splitting the sample may lead to poorer MS detection limits, especially with single quadrupole MS instruments. The reason why ESI-MS detectors behave as concentration-sensitive detectors is because lower flow rates allow for more efficient electrospray ionization processes and better ion sampling into the MS. As discussed above in Section 13.3.1, split ratios of 1 : 10 MS:CAD are common for high resolution MS and 1 : 1 or 4 : 6 are common for single quadrupole MS detection. Most high-resolution MS instruments still easily detect split samples and even benefit from the lower flow rate.

When splitting with a simple T-connector, which is a commonly used approach with single quadrupole mass spectrometers, the split ratio depends on the relative flow restriction (backpressure) between the two paths. Even if both paths have identical tubing lengths and inner diameters after the splits, the fluidics inside the detectors will not usually be identical. For example, an electrospray MS source will have unique tubing leading to the needle and ion transfer tube that is different from the fluidics in the charged aerosol detector, namely the diverter valve and the nebulizer tube. For most applications, it is not critical to establish an exact split ratio, but it is important to ensure that the flow is sufficiently above the minimum required for each detector.

A variable flow splitter [24] will support 1 : 20 or even greater split ratios and improves split precision. Viscosity changes associated with gradient flow runs do not affect the split delivered by modern flow splitters with compact fluid resistors. In particular, the variable flow splitter ensures a stable flow when the flow rate to the MS nears the microflow range. Keep in mind that a flow of 0.4 mL/min split 1 : 10 yields a microflow of less than 40 μL/min to the MS. With such low flows, users often implement tubing of smaller inner diameter, so that acceptably narrow peaks and acceptable resolution are seen in the MS.

If a flow splitter is added to a system running the inverse gradient, the analytical and inverse flows must be well mixed before splitting the flow into two or more detectors. As previously published [16] for polysorbate 80 analysis, the

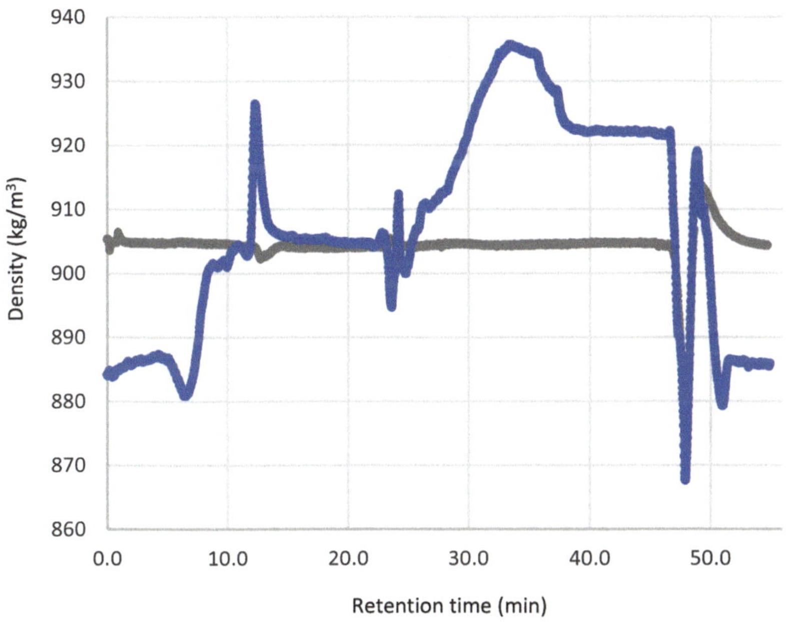

Figure 13.2 Density values measured between point B and MS detector of Figure 13.1. Blue trace: points A and B connected by standard capillary; black trace: points A and B connected by a 25 μL capillary mixer. Data collection rate was set to 1 Hz. *Source:* [16] / Elsevier / CC BY 4.0.

analytical and inverse flows can be combined with a t-piece with relatively high dead volume or an inline capillary mixer can be added after the merge point of the analytical and inverse gradients, and before the flow split (position 5 between points A and B in Figure 13.1). A mixer inserted at this point, when compared to a standard 0.1 mm × 50 mm capillary (Figure 13.2), yielded an eluent density that varied by ±0.5% whereas the eluent density after the standard capillary varied by ±5.5%. The density measurement was made only to evaluate mixing completeness and was not part of the polysorbate analysis. The relative standard deviations for the peak area of the monoesters and diesters (Figure 13.3, peaks G2 and G3) were 0.59% and 0.88%, respectively with the mixer and 2.20% and 1.18%, respectively with the standard capillary. Because CAD behaves as a mass-dependent, not a concentration-dependent detector, additional dilution of the sample due to the added volume of the mixer does not lead to lower response, as discussed in Chapter 4.

In another example, the inter-analyte %RSD for the area of six nonvolatile extractable and leachable standard substances as measured by CAD was 15% with the mixer and 26% with the standard capillary [26].

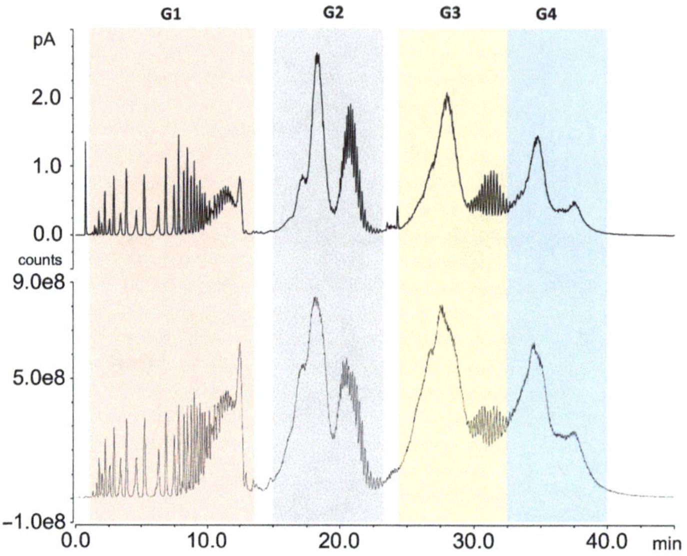

Figure 13.3 CAD chromatogram (top) and MS total ion chromatogram (bottom) of injection of a 20 μg polysorbate 80 sample. G1 retention time window: polyoxyethylene (POE) sorbitan, POE isosorbide, and POE; G2 retention time window: POE oleic acid monoesters, and POE oleic acid monoesters and isosorbide; G3 retention time window: POE oleic acid diesters of sorbitan and isosorbide; G4 retention time window: POE oleic acid triesters and tetraesters of sorbitan. *Source:* [16] / Elsevier / CC BY 4.0.

13.3.4 Eluent Compatibility, System Back Pressure and System Dead Volume

Other critical technical concerns when combining CAD with other detection techniques include eluent compatibility, system back pressure, and system dead volume.

All the detectors in a multidetector system should have the same eluent requirements. CAD and MS both have stringent but very similar eluent requirements. In some instances, analyte-specific eluent requirements for MS may influence CAD response. For example, the use of an acidic modifier to promote positive electrospray ionization may improve CAD response for volatile bases because, during droplet evaporation, the acid forms a less volatile salt with the base. In most cases, eluent requirements for MS are more stringent than for CAD and can be accommodated within the guidelines described in Chapter 3.

The backpressure and dead volume added by the extra tubing, splitters, and flow cells needed in a multidetector system can adversely impact the quality of the separation. The disadvantages of increased backpressure may include damaging the column by exceeding the pressure rating, exceeding the specifications of the pump [27], or cracking the upstream flow cell. To reduce backpressure, a reduced flow rate, an eluent system with low viscosity, an increased temperature or a backpressure relief valve can be used. A backpressure relief valve is also commonly used for MS detectors in the case of an upstream backpressure-sensitive flow cell, such as a light pipe flow cell commonly used with DAD and photodiode arrays (PDAs). The disadvantages of extra dead volume include peak broadening, reduced peak resolution, and poorer chromatogram quality. Dead volume can be reduced by choosing smaller volume flow cells.

The critical message is that flow cells upstream of a charged aerosol detector, commonly UV, should be low volume and must have a high back pressure rating. Fluorescence and RI flow cells contribute significant dead volume to the system and often have a low pressure rating. For these reasons, they are not often placed in line before other detectors in a multidetector system.

13.3.5 Strategies for Data Analysis

CA generates complicated datasets of CAD data. Well-designed strategies and computational power are required for meaningful interpretation of the data.

Principal component analysis (PCA), a powerful computational technique, has been used in over 40 peer-reviewed papers to interpret CAD chromatograms of complex samples. For example, it has been used to evaluate how much of an adulterant is present in an olive oil sample [28, 29], to distinguish among edible oils [30], to trace oxidation in krill oil [31], for assessing authenticity of coffee beans [32], to investigate metabolomics in turmeric dietary supplements [33], and to identify adulteration in black cohosh (a supplement for menopausal symptoms) [34]. Studies on oils are further detailed in Chapter 5, Section 4.

PCA is a technique that highlights differences and similarities between various samples. For example, it can highlight anomalous samples that do not fit the general pattern of the data. PCA simplifies variables associated with each peak, such as retention times and peak areas, into a set of linearly uncorrelated principal components that capture most of the variance in the data. By projecting the data onto these principal components, PCA allows identification of samples with different chromatographic profiles based on their relative positions in a three-dimensional space. This means that samples with distinct patterns will be separated from each other, making it easier to analyze and interpret the data. For a full PCA with HPLC data, raw chromatograms can be used to compare samples. This kind of analysis, however, requires careful alignment of analyte retention times and a consistent baseline, which is difficult to control even under the best of conditions. In a targeted analysis of HPLC data, integrated peak areas or sample amounts should be used. This minimizes the influence of baseline and retention shifts on PCA results [35].

Use of integrated peak areas also allows use of combined information from multiple detectors. For example, the carbohydrates of herbs, specifically the glycomes of milk vetch samples, a traditional Chinese medicine with scientific name *Astragalus membranaceus*, were analyzed by GC-MS and CAD [36]. LOD for the carbohydrate standards was less than 128 ng by HPLC-CAD and less than 51 ng by GC-MS. PCA was used on the combined gas chromatography-mass spectrometry (GC-MS) and CAD data to classify samples by region of origin.

Sometimes multicomponent samples cannot be baseline-resolved, meaning that multiple substances coelute. The components of these samples can be quantified by group analysis. For example, the degradation of polysorbate 80 (PS80), a component of most parenteral drug formulations of protein therapeutics, can be quantitatively characterized using four peak groups (Figure 13.3) in a multidetector setup with CAD and MS (Figure 13.1) [16]. Each group consists of coeluting components with the same degree of esterification. Both MS and CAD show the same peaks and peak groups, but relative intensity varies. In particular, the relative abundance is strongly influenced by the injected volume by MS, but relative abundance is similar regardless of injected amount by CAD. When 10 μL injections of PS80 containing 2.5, 5.0, 12, and 40 μg were evaluated, the relative standard deviation across all concentrations for groups 2 and 3 was 1.3% and 1.1%, respectively by CAD, but 6.0% and 2.5%, respectively by MS.

13.4 Application Examples for Comprehensive Analysis

Comprehensive analysis (CA) is performed so frequently with CAD as part of a multidetector system that the reader is directed to the CAD bibliography, maintained by Thermo Fisher Scientific, to find examples for the analysis of interest [37] Recent comprehensive analyses using CAD include studies on traditional Chinese medicine [38], oils, honey, polysorbates, and polyphenols [39], to name a few. Chapter 9 has additional examples of analyses of traditional Chinese medicine and natural products. This section highlights two examples of CA of complex samples that have been highly effectively reduced to practice: safety assessment of botanical medicines and safety assessment of extractables and leachables.

13.4.1 Comprehensive Analysis by CAD: Safety Assessment of Botanicals

Safety analysis of botanical medicines and dietary supplements is challenging because it requires estimation of the amounts of each of hundreds or thousands of components. Moreover, the determination of safety depends on the form of the product (e.g., tea, capsule, gummy), part of the plant (e.g., roots or leaves), growing conditions, post-harvest conditions, plant variety, processing, and purification [40]. Because testing in mammalian models is resource-intensive and regulatory agencies advocate predictive approaches instead of animal testing [41], the botanical medicine industry sought a reliable, regulation-compliant in silico approach.

Because toxicity always depends on the dose, a general limit of 10 μg botanical/person per day for ***any*** unconcentrated, non-purified botanical preparation or simple extract has been proposed [42]. CAD is used to determine which substances are present at concentrations exceeding this threshold. Accurate quantification of unknown substances with unknown toxicity in this way is crucial for demonstrating the safety of botanical products to regulatory bodies. Analytical quantification of components in a botanical product is also performed for authentication of the plant species and to protect against fraud. Although this section focuses on safety assessments, examples of authenticity and fraud testing can be found in Section 13.3.5, above, and in the CAD bibliography [37].

The use of the multidetector system liquid chromatography-ultraviolet-charged aerosol detection-high-resolution mass spectrometry (LC-UV-CAD-HRMS) for authentication and component characterization was the joint recommendation of a paper by an industry consortium [40] because the multidetector approach allows simultaneous identification by MS and estimation of substance concentration by CAD. The multidetector analytical approach for botanicals is termed "chemical constituent identification" (CCID) [43]. An early study verified the preference for CAD over ELSD for such botanical matrices due to better linearity over a wider range, higher sensitivity and better reproducibility [44].

A 2018 publication by Baker and Regg [20] laid the groundwork for the widespread use of CAD for safety and quality evaluations of botanical mixtures on a chemical constituent basis. The system contained a UV detector directly after a C18 2 mm × 150 mm column. An additional pump delivering an inverse gradient was placed after the UV detector and before an equal split to CAD and MS. The second pump ensured a constant mobile phase composition for CAD and MS. Figure 13.4 compares chromatograms from the three detectors for the 125 minute method. The peak numbers are labeled on the chromatogram acquired by CAD because it was used as the basis for quantitation. With IGC, the individual response factors for the nonvolatile substances were all two or better, meaning that CAD accuracy was always better than a factor of 2. This assessment was made comparing quantitation of each peak as a percentage of the overall CAD signal to quantitation using 34 well-characterized reference standards. Because of this accuracy, unknown substances could be sufficiently well quantified to allow a safety assessment. The information gained from such careful comprehensive analyses enables in silico safety assessments and minimizes the need for *in vitro*

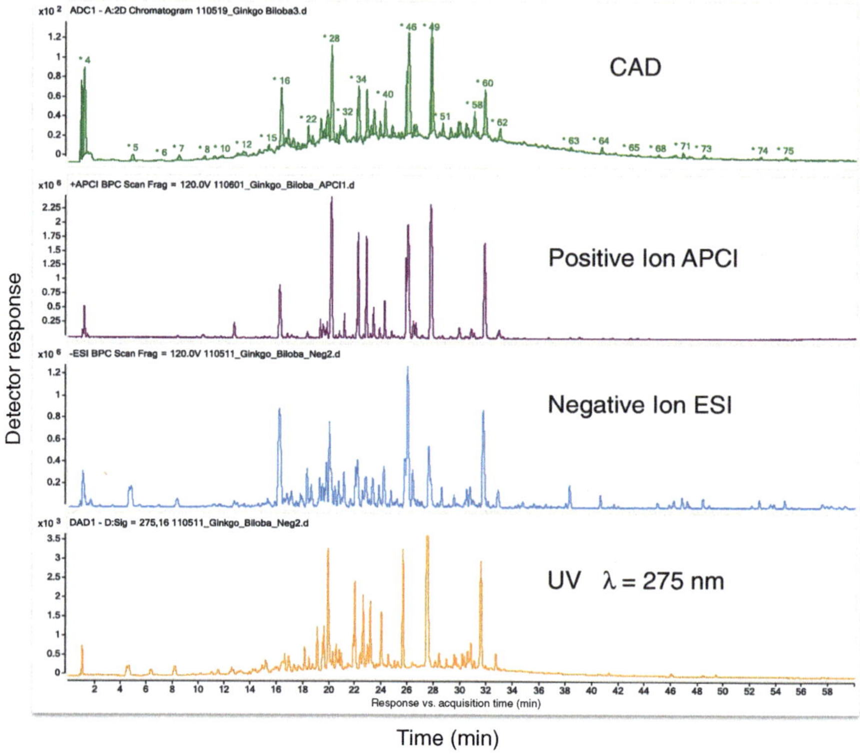

Figure 13.4 Partial chromatograms from the UHPLC/UV/CAD/QTOF-MS analysis of the *Ginkgo* extract showing the majority of analytes from the 125 minutes analysis. Note peak numbers on the CAD chromatogram. *Source:* [20] / Springer Nature / CC BY 4.0.

or *in vivo* safety assessments prior to marketing the botanical medicine. To enable this work, the injected sample amounts had to be adjusted to ensure that CAD limits of quantitation (LOQs) reached the threshold of toxicological concern [45].

The practical use of the methodology was shown in three case studies on red clover herb, which is used for menopausal symptoms among other things, on *Pelargonium sidoides* root, which is used to relieve cold symptoms, and on *Scutellaria baicalensis* root, which is used for arthritis, fevers and other ailments [45]. A flow chart was developed to standardize the use of UV/CAD/MS data from each constituent for hazard analysis and eventual clearance by regulatory agencies.

Rutz et al. [46] used a similar multidetector approach without IGC aimed to generate metabolite profiling data from natural extracts, categorizing them based on semiquantitative information that distinguishes between major and minor metabolome. The emphasis was on accurate, efficient annotation of each peak and characterization of peak purity. Even with the semiquantitative data acquired without IGC, CAD was deemed useful for isolation of new natural products that do not have chromophores.

The approach from the 2018 study with a strong emphasis on quantification accuracy has been implemented for the analysis of grape seed extracts [47]. The LC-UV-CAD-MS setup with IGC again allowed for quantitation of nonvolatiles with a surrogate standard using CAD. In contrast, MS would have required a standard for every component and UV would have required a standard for every component class. CAD enabled two crucial quantitative assessments. For one, CAD was critical for determining which substances were present at levels above a threshold of toxicological concern. Substance amounts were expressed relative to the entire amount injected onto the column and each single component was assessed for toxicity. With this focus on single components, each peak would ideally contain one substance, but the multicomponent samples were not completely baseline separated on the 150 mm × 2.1 mm, 1.7 μm particle size C18 column. MS was used to discern coeluting peaks because of its faster data acquisition rate and specificity.

Some substances could not be identified in the MS. For these unknowns, the most prevalent of which was 1.2% of the total sample peak area by CAD, no toxicological information was available. The safety assessment had to be made conservatively for these unknowns, which was an acceptable solution that allowed the safety assessment to be completed. If questions had remained about the safety of the botanical product, more MS/MS experiments could have been run to identify more of the unknowns. Samples commonly had to be run on the multidetector system multiple times to accommodate all the MS experiments, such as full scan ESI and APCI in positive and negative modes, as well as MS/MS experiments.

The second crucial quantitative assessment enabled by CAD was the calculation in samples from multiple vendors of the percentages of total polyphenols, tannins, non-tannin polyphenols such as (epi)catechin and oligomeric proanthocyanidins, gallic acid, polar compounds, and nonpolar compounds (Figure 13.5). CAD was critical for determination of the true composition of the samples. Levels of the active ingredient, proanthocyanidins, were more accurate by CAD than by UV because UV underestimated the sugars and fatty acids in the sample. Similarly, the determination was more accurate by CAD than MS because CAD revealed tannins that were missed by MS.

The inclusion of GC-FID and GC-MS, as well as LC-UV-CAD-MS allowed analysis of volatile components (GC-FID and GC-MS), as well as nonvolatile and semi volatile components (LC-UV-CAD-MS). The mass balance was 91% by CAD and an additional 2.6% by GC-FID. Eight substances were identified by

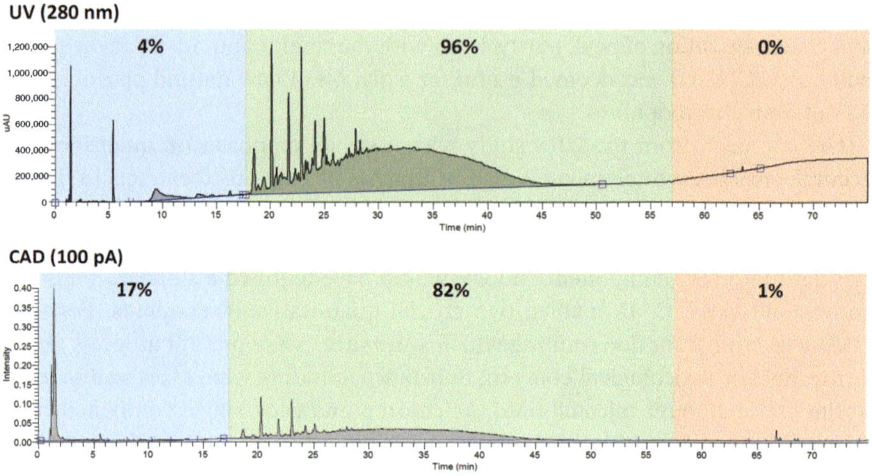

Figure 13.5 Comparison of chromatograms showing the improvement in estimation of polar compounds (blue), polyphenols (green), and nonpolar compounds (red) when analyzed by CAD (bottom) as opposed to UV detection at 280 nm (top). *Source:* [47] / Frontiers Media SA / CC BY 4.0.

GC-FID that were too volatile for CAD, including glycerol, dihydroxyacetone, and substituted phenols.

Still yet to be investigated is the quantitation of substances in the botanical sample using calibration curves with authentic standards at the CAD limit of detection. This strategy would involve absolute quantitation rather than quantitation relative to the total peak area. Some of the botanical substances may be lost as volatiles with CAD and some may be retained too well on the column. An absolute quantitation will indicate if 100% of the injected substance appears in the chromatogram.

Another example of the use of this methodology followed up on initial UV-CAD-MS data collection from a *Poria cocos* mushroom sample with an in silico assessment of the seven peaks for genotoxicity and two *in vitro* tests for genotoxicity; the Ames test and the human lymphocyte micronucleus assay. This analysis confirmed the absence of genotoxic potential in the sample [43].

13.4.2 Comprehensive Analysis by CAD: Extractables and Leachables

Studies designed to ensure that no potentially harmful substances are leached into the pharmaceuticals we use, the food we eat, and the water we drink involve complex samples. Detection of unknown or unexpected peaks in such samples requires multiple orthogonal detectors. Beyond mere detection, laboratories also frequently require identity confirmation and quantitation of these compounds to determine their nature and whether they are below acceptable concentration limits. The focus of this section is an informative example showing CAD as part of a standardized safety assessment scheme for intravaginal devices, such as menstrual cups [48]. Although the multidetector setup used to generate this dataset is now commonplace, this is the first known use of CAD to deliver toxicity data for a medical device to a regulatory agency. The CAD data is generated in the first, "biocompatibility and chemical safety assessment" step, of four steps.

The rigorous safety assessment, which has subsequently also been applied to tampons [49], included a review of the biocompatibility data for the silicone raw material from the supplier, extractability profiles of the menstrual cup after boiling multiple times and wiping with the recommended wipe product, an exposure-based risk assessment.

The same multidetector method was run on samples from the silicone raw material, Silicone QP 1-40 LSR, which had the best possible medical-grade classification (Class VI) according to the U.S. Pharmacopeia, and on samples from the final product. Because there were no "unique" or "common but significantly higher" substances in the final product relative to the raw material (Figure 13.6), the manufacturer could use an FDA Guidance on the use of ISO 10993 [50] to successfully argue for the safety of the finished product. The comprehensive data showed that the product was safe and well tolerated upon intended use, a conclusion that was confirmed during post-market safety surveillance.

The previously published method [47], originally developed for botanical medicines as discussed above, combines UV, MS, CAD, and two pumps. The two

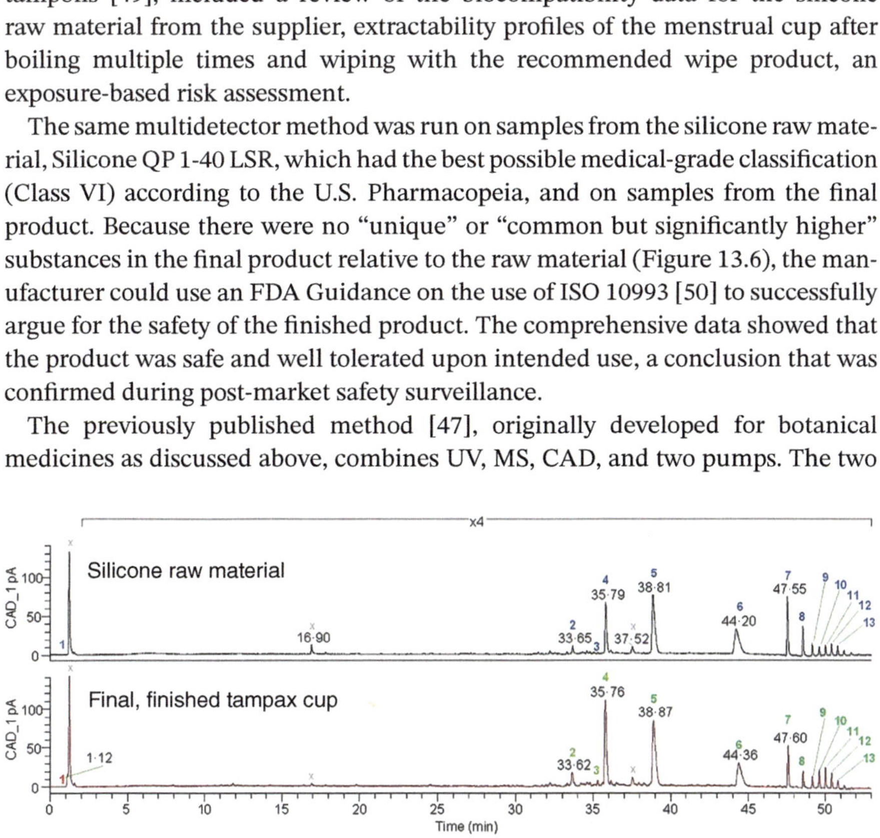

Figure 13.6 Chromatograms of the silicone raw material extract versus the final menstrual cup product upon extraction at 50 °C for 2 hours. Peaks denoted with an X are either below the threshold of toxicological concern (TTC) in both the silicone raw material and the final product or in the solvent and/or blank control. Green numbers indicate peaks of the silicone raw material > peaks of the final product; red numbers indicate peaks of the silicone raw material < peaks of the final, finished Tampax Cup. ">" and "<" were determined overall with the standard deviation of multiple replicates. There are no peaks in the chromatogram of the cup that are not also present in the chromatogram of the silicone [48]. *Source:* [48] / with permission of Elsevier.

pumps provide the analytical gradient and inverse gradient/makeup flow for CAD and MS to ensure a uniform response that is unaffected by the gradient. The near uniform response of CAD allows straightforward quantification without individual reference standards. The uniform response of CAD was critical for analysis of the polymeric, sometimes cyclic, siloxane extractables from the menstrual cup. The response factor used for all peaks in the chromatogram was generated using the averaged response of each of five standards, tryptophan, a Val-Tyr-Val tripeptide, kaempferol, reserpine, and terfenadine, at each of five concentrations between 500 and 7500 ng/mL. The standards passed the criteria for percent relative error (<25%) and $R^2 \geq 0.95$. After quantification by CAD, any peak showing an unknown present at a level greater than 0.63 μg/cup was identified by high-resolution MS so that the toxicological impact could be assessed in detail. This analytical screening level was calculated from a conservative threshold of toxicological concern of 0.025 μg/kg body weight per day for potentially genotoxic substances, which corresponds to a theoretical 10^{-5} excess lifetime risk of cancer.

The 0.63 μg/cup threshold translated to 1260 ng/mL for extracts that had been redissolved in 0.5 mL of solvent after evaporation, which was between the highest and lowest measured calibration curve points of 500 and 7500 ng/mL. Extrapolation beyond the measured calibration curve is not recommended for CAD because of the slightly nonlinear nature of the response. The limit of detection is generally better than extrapolated because response factors for nonvolatiles increase as the analyte concentration decreases. This topic is expanded upon in Chapter 1.

Thirteen CAD peaks were larger than 0.63 μg/cup. Only one peak was larger for the final product than for the raw material. This peak was identified as a salt cluster peak containing sodium, magnesium, and calcium ions that was not of toxicological relevance.

Beyond the polysiloxanes (silicone) in this example, CAD is generally well-suited for analysis of complicated, chromophore-deficient, potentially toxic oligomers that can be extracted from plastic materials [25]. Plastic oligomers can be found and quantified by CAD in substances such as polyphenylsulfone baby bottles [51], polyester-phenol-coated infant food jar closures [52], polyurethane-containing wound dressings [26], polyethylene iv bag parts [17], polypropylene syringes and filters, complex elastomer bottle stoppers [53], polyvinyl chloride-based iv tubing, siloxanes in menstrual cups, polystyrene tracheal tubes, polyamides [25], and polyarylsulfones [25] to name a few. Careful databases of CAD-MS-UV data for pharmaceutical and food contact polymers exist [21, 25]. The potential oligomer forms are so numerous and structurally diverse that standards do not exist for most of them, so CAD is an ideal detection technology. CAD has been well described for its ability to accurately quantify analytes using surrogate standards [54]. A test for nonvolatility of unknowns as described above in Section 13.3.2 and Eckardt et al. [25] improves the reliability of the quantification with a surrogate standard by flagging semi volatile unknowns.

Other extractables and leachable substances that are routinely analyzed by CAD include surfactants, antioxidants, plasticizers, slip agents, acid scavengers, lubricants, and crosslinking agents [24].

The great improvement in uncertainty thresholds due to the quantitation capabilities of CAD has made CAD-UV-MS multidetector setups common for extractables and leachables analysis. The systematic and broadly applicable safety assessment that includes CAD to produce medical device toxicity data for regulatory purposes has already been applied to tampons [49] and has further cemented the acceptability of CAD data in regulatory applications.

13.5 Conclusions

LC does benefit from a universal detector capable of detecting all analytes with uniform response factors, regardless of their physicochemical properties. Charged aerosol detectors are best in class to fulfil this demand and hence provide a useful contribution to setups for comprehensive analysis, specifically for quantification of unknowns or analytes that lack respective calibration standards. However, universal detection technologies like CAD cannot provide additional structural information on compounds that generate a detector signal, nor do they provide information on the purity of peaks. For these reasons, they can be very beneficial when used in configurations that combine orthogonal separation or detection. Comprehensive LC × LC and specific heart-cut 2D-LC solutions deliver such orthogonal separation capabilities and are successfully applied in combination with CAD. Such setups can be advantageous for metabolomic studies, complex natural products and botanical samples analysis, or general unknown impurity quantification in multicomponent samples, to name only a few examples. The most important approach to comprehensive analysis is the combination of orthogonal detectors. A powerful instrumental setup for this is the combination of DAD as the common standard detector, CAD as the device for comprehensive quantification even in the absence of calibration standards, as well as MS for compound identity confirmation of known analytes. If a HRMS or advanced hybrid MS technique is applied, this also offers the potential of unknown compound identification in non-targeted analysis. Relevant application fields are general characterization for toxic, genotoxic, or other harmful ingredients. This ranges from extractable & leachable analysis in pharmaceutical and food products to potential contaminants in implants, e.g., for plastic surgery. The highly beneficial uniform response requirement of CAD for quantitation of unknowns, even for methods with gradient elution is generally fulfilled through the addition of a complementary gradient profile downstream of the separation column to account for a constant mobile phase composition in the detection device. Also, the MS can partially benefit from this approach. As CAD and MS are both destructive techniques, such a setup requires a split of the column effluent to connect to both detectors simultaneously. If this is implemented in conjunction with IGC, a perfect radial homogenization upstream to the splitting device is critical to account for a constant split ratio as well as a constant mobile phase composition in both detectors. These application examples and innovative technical solutions confirm how charged aerosol detectors are a significant addition to the analyst's arsenal for detecting and quantifying a wide range of compounds separated by LC, even for highly complex samples, in particular when a dependable analysis of hazardous contaminants in the absence of calibration standards is required.

References

1 Jiang, Y., Hascall, D., Li, D., and Pease, J.H. (2015). A charged aerosol detector/chemiluminescent nitrogen detector/liquid chromatography/mass spectrometry system for regular and fragment compound analysis in drug discovery. *Journal of Chromatography. A* 1411: 63–68.

2 Kanbayashi, Y., Tsutsumi, Y., Mizukoshi, T., and Yamaguchi, H. (2020). Kanbayashi 2020 NMR. *Bunseki Kagaku* 69 (3): 151–155.

3 Li, Y., Hewitt, D., Lentz, Y.K. et al. (2014). Characterization and stability study of polysorbate 20 in therapeutic monoclonal antibody formulation by multi-dimensional ultrahigh-performance liquid chromatography-charged aerosol detection-mass spectrometry. *Analytical Chemistry* 86 (10): 5150–5157.

4 Song, H.H., Ryu, H.W., Kim, H.S. et al. (2016). A metabolomics approach to identify factors influencing their activity relative to oleanolic acid contents in Korean mistletoe types. *Journal of Functional Foods* 22: 64–72.

5 Schimek, D., Francesconi, K.A., Mautner, A. et al. (2016). Matrix removal in state of the art sample preparation methods for serum by charged aerosol detection and metabolomics-based LC-MS. *Analytica Chimica Acta* 915: 56–63.

6 Stoll, D.R., Li, X., Wang, X. et al. (2007). Fast, comprehensive two-dimensional liquid chromatography. *Journal of Chromatography. A* 1168 (1–2): 3–43.

7 Grübner, M., Dunkel, A., Steiner, F., and Hofmann, T. (2021). Systematic evaluation of liquid chromatography (LC) column combinations for application in two-dimensional LC metabolomic studies. *Analytical Chemistry* 93 (37): 12565–12573.

8 Suk Yoon, C., Taik Lee, K., and Author, C. (2010). The use of the online two-dimensional liquid chromatography coupled with a universal detector for

the screening of non-volatile potential migrants in food packaging materials. *Korean Journal of Packaging Science & Technology* 16 (1): 9–18.
9 van Beek, F.T., Edam, R., Pirok, B.W.J. et al. (2018). Comprehensive two-dimensional liquid chromatography of heavy oil. *Journal of Chromatography. A* 1564: 110–119.
10 Takahashi, R., Nakaya, M., Kotaniguchi, M. et al. (2018). Analysis of phosphatidylethanolamine, phosphatidylcholine, and plasmalogen molecular species in food lipids using an improved 2D high-performance liquid chromatography system. *Journal of Chromatography. B, Analytical Technologies in the Biomedical and Life Sciences* 1077–1078: 35–43.
11 Greiderer, A., Steeneken, L., Aalbers, T. et al. (2011). Characterization of hydroxypropylmethylcellulose (HPMC) using comprehensive two-dimensional liquid chromatography. *Journal of Chromatography. A* 1218 (34): 5787–5793.
12 Zhang, K., Li, Y., Tsang, M., and Chetwyn, N.P. (2013). Analysis of pharmaceutical impurities using multi-heartcutting 2D LC coupled with UV-charged aerosol MS detection. *Journal of Separation Science* 36 (18): 2986–2992.
13 Long, Z., Guo, Z., Liu, X. et al. (2016). A sensitive non-derivatization method for apramycin and impurities analysis using hydrophilic interaction liquid chromatography and charged aerosol detection. *Talanta* 146: 423–429.
14 Pistorino, M. and Pfeifer, B.A. (2008). Polyketide analysis using mass spectrometry, evaporative light scattering, and charged aerosol detector systems. *Analytical and Bioanalytical Chemistry* 390 (4): 1189–1193.
15 Sperber, B., Gutmann, M., Kehrein, J. et al. (2025). Impurity profiling of PEGylated myristoyl diglyceride, DMG-PEG 2000, a functional excipient used in mRNA lipid nanoparticle formulations. *European Journal of Pharmaceutics and Biopharmaceutics* 214: 114762.
16 De Pra, M., Ispan, D.A., Meding, S. et al. (2023). Degradation of polysorbate investigated by a high-performance liquid chromatography multi-detector system with charged aerosol and mass detection. *Journal of Chromatography. A* 1710: 464405.
17 Pinedo-Gonzalez, P., Liu, C., Hower, D. et al. (2023) Confident extractable testing of medical device components using a new automated parallel extraction and evaporation sample preparation system and the multidetector approach. Thermo Fisher Scientific Application Note 001950.
18 Ilko, D., Holzgrabe, U., Bio, P. et al. (2014). Validation and application of an HPLC-CAD-TOF/MS method for identification and quantification of pharmaceutical counterions. *Pharmeuropa Bio & Scientific Notes* 81–91.
19 Pecio, Ł., Otify, A.M., Saber, F.R. et al. (2022). *Iphiona mucronata* (Forssk.) Asch. & Schweinf. A comprehensive phytochemical study via UPLC-Q-TOF-MS in the context of the embryo- and cytotoxicity profiles. *Molecules* 27: 7529.

20 Baker, T.R. and Regg, B.T. (2018). A multi-detector chromatographic approach for characterization and quantitation of botanical constituents to enable in silico safety assessments. *Analytical and Bioanalytical Chemistry* 410 (21): 5143–5154.

21 Jordi, M.A., Khera, S., Roland, K. et al. (2018). Qualitative assessment of extractables from single-use components and the impact of reference standard selection. *Journal of Pharmaceutical and Biomedical Analysis* 150: 368–376.

22 Martin, M., Heidorn, M., Steiner, F. et al. (2011) Ensuring High Sensitivity and Consistent Response in UHPLC-MS Analyses. Dionex Corporation Publication LPN 2773-01.

23 Kostiainen, R. and Kauppila, T.J. (2009). Effect of eluent on the ionization process in liquid chromatography-mass spectrometry. *Journal of Chromatography. A* 1216 (4): 685–699.

24 Pinedo-Gonzalez, P., Liu, C., Hower, D. et al. (2022) Multi-detector platform for comprehensive identification and quantitation of extractables and leachables. Thermo Fisher Scientific Application Note 001401. *Thermo Fisher Application Note.*

25 Eckardt, M., Kubicova, M., and Simat, T.J. (2018). Universal response quantification approach using a corona charged aerosol detector (CAD) – application on linear and cyclic oligomers extractable from polycondensate plastics polyesters, polyamides and polyarylsulfones. *Journal of Chromatography. A* 1572: 187–202.

26 Lovejoy, K., Yu, X., Kraus, C. et al. (2023) A reliable method using a UHPLC/UV/CAD/MS multidetector platform for routine quantification and library matching of extractable and leachable compounds including medical-grade PVC tubing and polyurethane-containing wound dressings. Thermo Fisher Scientific Application Note 001630.

27 Meding, S., Lovejoy, K., Swart, R. et al. (2018) A multi-detector platform comprising UV/Vis detection, charged aerosol detection, and single quadrupole mass spectrometric detection for comprehensive sample analysis. Thermo Fisher Scientific Application Note 72869.

28 Green, H.S., Li, X., De Pra, M. et al. (2020). A rapid method for the detection of extra virgin olive oil adulteration using UHPLC-CAD profiling of triacylglycerols and PCA. *Food Control* 107: 106773.

29 De La Mata-Espinosa, P., Bosque-Sendra, J.M., Bro, R., and Cuadros-Rodríguez, L. (2011). Discriminating olive and non-olive oils using HPLC-CAD and chemometrics. *Analytical and Bioanalytical Chemistry* 399 (6): 2083–2092.

30 López-Ruiz, R., Jimenez-Carvelo, A.M., and Cuadros-Rodríguez, L. (2024) Synergistic use of three complementary detectors coupled to reverse phase liquid 2 chromatography for olive oil authentication based on the phospholipid fraction. SSRN, (preprint), 1–31.

31 Pei, X.C., Zeng, X.B., Li, D.Y. et al. (2025). The change rule of lipid oxidation and hydrolysis driven via water in Antarctic krill oil: based on association colloid formation. *Food Chemistry* 463: 141448.

32 Ismaiel, L., Fanesi, B., Kuhalskaya, A. et al. (2023). The determination of triacylglycerols and tocopherols using UHPLC–CAD/FLD methods for assessing the authenticity of coffee beans. *Food* 12: 4197.

33 Singh, J., Tareq, F.S., and Luthria, D.L. (2024). Comparative investigation of untargeted and targeted metabolomics in turmeric dietary supplements and rhizomes. *Food* 14 (7): 1–16.

34 Waidyanatha, S., Cristy, T., Pierfelice, J. et al. (2022). Working with the natural complexity: Selection and characterization of black cohosh root extract for use in toxicology testing. *Food and Chemical Toxicology* 160: 112769.

35 Plante, M., Bailey, B., and Acworth, I.N. (2014) Determination of olive oil adulteration by principal component analysis with HPLC-charged aerosol detector data. Thermo Fisher Scientific Poster Note 70689.

36 Lv, G.P., Hu, D.J., Cheong, K.L. et al. (2015). Decoding glycome of Astragalus membranaceus based on pressurized liquid extraction, microwave-assisted hydrolysis and chromatographic analysis. *Journal of Chromatography. A* 1409: 19–29.

37 Thermo Scientific (2026) Charged Aerosol Detection Bibliography. https://www.thermofisher.com/CADpublications.

38 Chen, X., Luo, Z., Liu, X. et al. (2022). Marsdenia tenacissima (Roxb.) Moon injection exerts a potential anti-tumor effect in prostate cancer through inhibiting ErbB2-GSK3β-HIF1α signaling axis. *Journal of Ethnopharmacology* 295: 115381.

39 Gaudin, K., Valls-Fonayet, J., Cordazzo, R. et al. (2024). Separation of polyphenols by HILIC methods with diode array detection, charged aerosol detection and mass spectrometry: application to grapevine extracts rich in stilbenoids. *Journal of Chromatography. A* 1736: 465422.

40 Waidyanatha, S., Collins, B.J., Cristy, T. et al. (2024). Advancing botanical safety: a strategy for selecting, sourcing, and characterizing botanicals for developing toxicological tools. *Food and Chemical Toxicology* 186: 114537.

41 U.S. Food and Drug Administration (2017) FDA's Predictive Toxicology Roadmap.

42 Mahony, C., Bowtell, P., Huber, M. et al. (2020). Threshold of toxicological concern (TTC) for botanicals - concentration data analysis of potentially genotoxic constituents to substantiate and extend the TTC approach to botanicals. *Food and Chemical Toxicology* 138: 111182.

43 Sun, B., Huang, B., Sica, V.P. et al. (2021). A genotoxicity assessment approach for botanical materials demonstrated with Poria cocos. *Food and Chemical Toxicology* 156: 112521.

44 Eom, H.Y., Park, S.Y., Kim, M.K. et al. (2010). Comparison between evaporative light scattering detection and charged aerosol detection for the analysis of saikosaponins. *Journal of Chromatography. A* 1217 (26): 4347–4354.

45 Little, J.G., Marsman, D.S., Baker, T.R., and Mahony, C. (2017). In silico approach to safety of botanical dietary supplement ingredients utilizing constituent-level characterization. *Food and Chemical Toxicology* 107: 418–429.

46 Rutz, A. and Wolfender, J.L. (2023). Automated composition assessment of natural extracts: untargeted mass spectrometry-based metabolite profiling integrating semiquantitative detection. *Journal of Agricultural and Food Chemistry* 71 (46): 18010–18023.

47 Sica, V.P., Mahony, C., and Baker, T.R. (2018). Multi-detector characterization of grape seed extract to enable in silico safety assessment. *Frontiers in Chemistry* 6: 334.

48 Sica, V.P., Friberg, M.A., Teufel, A.G. et al. (2022). Safety assessment scheme for menstrual cups and application for the evaluation of a menstrual cup comprised of medical grade silicone. *eBioMedicine* 86: 104339.

49 Hochwalt, A.E., Abbinante-Nissen, J.M., Bohman, L.C. et al. (2023). The safety assessment of tampons: illustration of a comprehensive approach for four different products. *Frontiers in Reproductive Health* 5: 1–23.

50 U.S. Food and Drug Administration (2023) Use of International Standard ISO 10993-1, “Biological evaluation of medical devices-Part 1: Evaluation and testing within a risk management process” Guidance for Industry and Food and Drug Administration Staff Preface Public Comment.

51 Eckardt, M., Greb, A., and Simat, T.J. (2018). Polyphenylsulfone (PPSU) for baby bottles: a comprehensive assessment on polymer-related non-intentionally added substances (NIAS). *Food Additives & Contaminants: Part A* 35 (7): 1421–1437.

52 Eckardt, M., Hetzel, L., Brenz, F., and Simat, T.J. (2020). Release and migration of cyclic polyester oligomers from bisphenol A non-intent polyester–phenol-coatings into food simulants and infant food–a comprehensive study. *Food Additives & Contaminants: Part A* 37 (4): 681–703.

53 Yu, X., Zdravkovic, S., Wood, D. et al. (2009). A new approach to threshold evaluation & quantitation of unknown extractables & leachables using HPLC/CAD. *Drug Delivery Technology* 9 (3): 50–55.

54 Schilling, K. and Holzgrabe, U. (2020). Recent applications of the charged aerosol detector for liquid chromatography in drug quality control. *Journal of Chromatography. A* 1619: 460911.

14

CAD in Drug Discovery

Wesley W. Barnhart[1], *Troy T. Handlovic*[1], *Brian A. Lanman*[1], *David Bauer*[2], *and Imad A. Haidar Ahmad*[1]

[1] *One Amgen Center Drive, AMGEN Inc., Thousand Oaks, CA, USA*
[2] *Amgen Research, Cambridge, MA, USA*

CHAPTER MENU

14.1 Summary

- Drug discovery is a time- and labor-intensive endeavor that typically entails the purification and testing of thousands of compounds in the search for safe and efficacious drug candidates.
- A variety of analytical workflows are incorporated into the drug discovery process.
- The charged aerosol detector (CAD) is a powerful enabling tool in discovery analytical workflows, allowing for the detection of poorly ionizing compounds as well as compounds with weak chromophores.
- CAD can also enable the quantitation of materials lacking authentic standards and is particularly useful in the microgram-scale high-throughput purification of libraries.

14.2 Introduction

The discovery and development of new drugs is a complex, multi-step process, which can take more than a decade and cost over a billion dollars [1]. The discovery of new drugs frequently begins with molecular screening efforts to identify compounds capable of bringing about a desired pharmacological effect. Such efforts aim to explore structurally diverse chemical space by screening large collections of compounds ("libraries") to identify compounds capable of eliciting specific effects in given biochemical or cellular assays.

Analytical and purification workflows play a critical role in facilitating the drug discovery process. Analytical testing is crucial in confirming the identity, purity, and physicochemical properties of potential drug candidates. Chromatography-based assays often play a key role in such analyses, supporting reaction monitoring, purity determination, and physicochemical property assessment (e.g., permeability) [2–6]. Purification workflows additionally play a key role in supporting drug discovery efforts, enabling the isolation of test compounds from compound libraries (or reaction mixtures) to support testing workflows. Chromatography-based purification processes extend beyond the simple removal of undesired impurities (e.g., reaction byproducts, catalysts, and starting material, etc.) from reaction mixtures (i.e., "achiral purification"), and also includes the isolation of specific stereoisomers of target compounds (i.e., "chiral purification") [7].

Speed of analytical and purification processes—both in terms of turnaround times for individual compounds and development times to identify conditions for the analysis and purification of new compounds—is critical to the success of discovery research given frequently limited resources to support multiple discovery projects and often intense competition around promising/well-validated therapeutic targets. Accordingly, automated high-throughput (HT) analytical and purification workflows play an outsized role in discovery research workflows [3, 7, 8]. In this chapter, we discuss the flexibility, wide applicability, and speed-of-operation that chromatography-based workflows leveraging the charged aerosol detector (CAD) bring to discovery-stage research. In particular, we illustrate how the uniform response of CAD across a wide range of compounds enables rapid and reliable quantitation of compounds lacking UV chromophores or authentic standards [9], thereby addressing frequent challenges faced in the purification and analysis of screening hits frequently emerging in the early stages of drug discovery.

14.3 Streamlining Drug Discovery

Arriving at optimal drug candidates from compounds emerging from initial library screening hit discovery efforts is a lengthy, time-consuming, and complicated scientific endeavor [10]. From the thousands of compounds evaluated during the hit optimization process, only a handful (5–20) typically progress to detailed preclinical studies, and only 1–2 of these are ultimately selected for clinical development [11]. Given the large number of compounds to be evaluated in such efforts and the high attrition rate, efficiently prioritizing and assessing compounds at each stage of the discovery research workflow is essential in maintaining competitiveness, both in terms of cost and time.

As illustrated in Figure 14.1, many technologies and workflows have been developed to expedite this drug discovery process, including computer-aided approaches to drug design [12] and recent applications of artificial intelligence (AI) and machine learning (ML) algorithms to discovery workflows [10]. In this chapter, we focus on the critical role played by analytical and purification workflows in supporting the timely evaluation and progression of compounds through discovery screening and optimization workflows. Although beyond the scope of this chapter, it should be noted that such analytical and purification workflows have themselves been facilitated through the application of computational tools to speed physicochemical properties characterization (e.g., solubility [13] and lipophilicity [14] predictions) as well as in the rapid identification/optimization of chromatographic separation conditions [4, 15, 16]. Beyond just supporting timely compound progression through discovery workflows, data emerging from the analytical and purification workflows described in this chapter will additionally play a key role in supplying the inputs needed to train and validate the AI/ML models of the future.

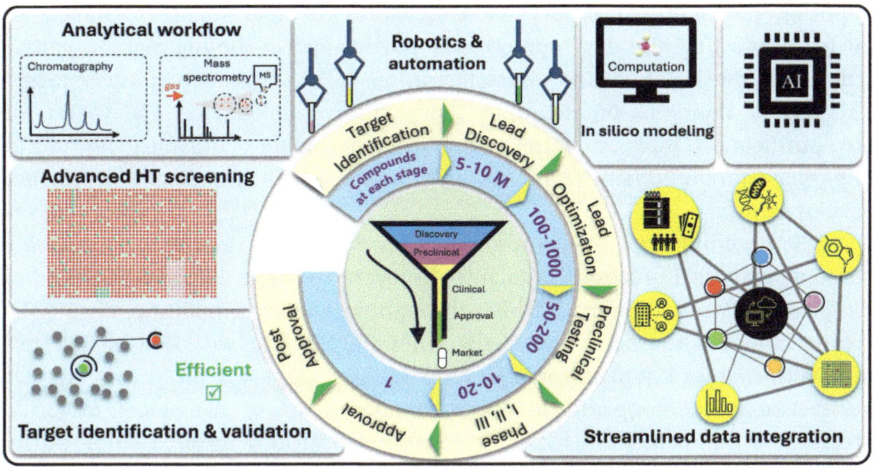

Figure 14.1 Parameters driving speed in drug discovery.

14.4 Chromatography-Based Workflows in Drug Discovery

Drug discovery and development relies upon a wide range of chromatographic workflows at both preparative and analytical scales throughout the journey of a molecule from screening hit to marketed therapeutic [17, 18]. Underpinning such workflows are a range of chromatographic and analyte detection technologies. Liquid chromatography (LC) represents one of the most commonly used chromatographic technologies. It is typically coupled to both ultraviolet (UV) and mass spectrometric (MS) detectors to allow for more descriptive analyses (i.e., mass annotation) of compounds. LC-MS has been utilized to analyze a wide range of therapeutic modalities including antibodies, proteins, peptides, nucleic acids, proteolysis targeting chimeras (PROTACs), and traditional small molecules [19]. An extensive array of techniques [i.e., reversed-phase liquid chromatography (RPLC), hydrophilic interaction liquid chromatography (HILIC), size exclusion chromatography (SEC), etc.] are available to address specific measurement and purification needs [19]. Other types of chromatography, such as sub/super-critical fluid chromatography (SFC), are also routinely used in discovery, with SFC seeing particularly wide use as a chromatographic technology due to its speed and sustainability (i.e., reduced organic solvent consumption). SFC is compatible with MS and is a powerful tool for chiral separations due to the physicochemical properties of pressurized CO_2.

Chromatography-based analyses can be used to determine purity, profile impurities in a sample, measure residual solvents, or determine physicochemical attributes. Additionally, chromatography-based purification plays an equally important role in drug discovery where it enables the rapid isolation of molecules for further testing and development. HT purification at the microgram-scale is crucial for the isolation and characterization of hits from screening libraries. The use of liquid chromatography-charged aerosol detection (LC-CAD) for post-purification analysis in these environments enables universal and accurate quantitation which facilitates the selection of candidates for further development.

As captured in Figure 14.2, chromatography-based workflows underpin a wide range of analytical processes critical to discovery research. Purity assessment of samples can be in the form of high-throughput analysis, quality control (QC) testing, and enantiomeric excess determination. These workflows often use high-performance liquid chromatography (HPLC) coupled to an MS to allow for purity assessment along with mass confirmation. Because many small molecule drugs are chiral compounds [20], it is often critical to determine chiral purity (frequently reported as enantiomeric excess (e.e.)). Chirality affects the binding of a molecule to its target and can impact both on- and off-target effects [20, 21].

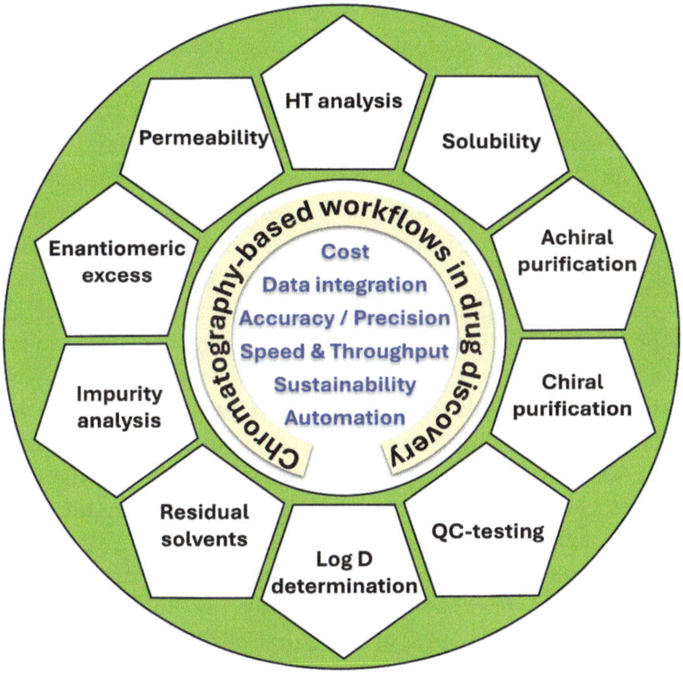

Figure 14.2 A variety of workflows that rely upon chromatography are listed. These processes range from analytical to purification. Key considerations in the implementation of such workflows are listed at the center of this figure.

Enantiomeric excess can be measured using HPLC or SFC by comparing the peak areas of the two enantiomers after they are separated on a chiral stationary phase [21]. Permeability, solubility, and Log*D* values are physicochemical attributes that can be approximated via high-throughput chromatography. Permeability is indicative of the ability of compounds to diffuse through biological membranes to reach intracellular targets [22]. Solubility is a critical attribute that is used to guide compound selection during drug development and impacts compound performance in *in vitro* assays as well as in *in vivo* systems. Log*D*, a measurement of lipophilicity at a specific pH, can also be estimated using chromatographic methods (ChromLog*D*), potentially offering testing throughput advantages over shake-flask-based methods [22, 23]. Here, Log*D* values along with the retention factors of reference standards determined via RPLC are used to establish calibration curves that can be used to determine ChromLog*D* values for unknown compounds. It has been shown that chromatographic analyses allow for the ability to handle complex mixtures better than the original shake-flask measurements [22]. Residual solvent analysis (typically done via gas chromatography (GC) [24]) is another crucial chromatographic workflow used to characterize compounds selected for in vivo profiling [25]. Residual solvents from compound synthesis and handling that are retained in a sample can impact the *in vivo* behavior and safety profile of a compound; threshold levels for residual solvents in a sample are specified by the International Council for Harmonization (ICH) [24].

Aside from its role in supporting diverse analytical workflows, chromatography additionally plays indispensable roles in supporting the purification of compounds prepared during the drug discovery process. Purification workflows encompass both chiral and achiral separations, typically performed using either HPLC (reversed- or normal-phase) or SFC. Many compounds investigated during the drug discovery process will be newly synthesized and may contain residual starting material, side products, reactants, and other unwanted compounds. Developing chromatographic methods to fully resolve all the components of the crude reaction mixture, including chiral and achiral compounds, is challenging. Mass-directed achiral purifications are usually applied for isolating a compound of interest from all other components in the sample. Chiral separations are used to purify stereoisomers, and method development relies heavily on empirical results. Although work is ongoing to predict suitable chiral stationary phases (CSP) for efficient compound purification computationally, this technology currently remains in its infancy [26]. As a result, chiral purification workflows typically require significantly more empirical screening than is typically required for achiral purifications.

As mentioned in Figure 14.2, several factors underpin all of the analytical and purification workflows described above. One is a focus on the implementation of automated workflows wherever possible, as such automation improves the efficiency and increases throughput of such workflows, enabling faster and cheaper generation of data/insights to drive the discovery process. Data integration is another overarching consideration, as automated curation of data from diverse workflows both enhances the accuracy with which compounds can be annotated and facilitates the holistic evaluation of compound properties. Cost and sustainability additionally factor into the choice of analytical and purification techniques, with reduced solvent consumption and reduced analytical times contributing to both shorter turnaround times and reduced costs (both fiscal and environmental). Such considerations inform efforts to enable analytical and purification workflows on ever smaller scales, such as described in the section on microgram-scale purification, below.

14.5 Use and Benefits of CAD in Discovery

While most chromatographic-based methods rely on the use of UV and MS for detection, these techniques have limitations when quantitating compounds which lack a chromophore, present inadequate ionization efficiency, or have unknown response factors (i.e., detector response/unit concentration of analyte). CAD is an excellent alternative in these circumstances, as CAD provides a nearly uniform response (i.e., minimum analyte-to-analyte variability) across a wide range of compounds and concentrations [27]. It has been demonstrated that CAD can be utilized to quantitate nonvolatile compounds in the absence of authentic standards [28]. This makes CAD an excellent choice for high-throughput quantitation of purified compound libraries during drug discovery. In such settings, the goal is to understand the overall compound/library purity to minimize false positive (or negative) assay results resulting from inaccurate estimates of tested compound concentrations [29]. Using UV or MS detection for compound quantitation in such settings is typically not feasible as there are too many compounds (hundreds to thousands) and too little of each compound (<1 mg total) to build the compound-specific calibration curves required for traditional quantitation. When using CAD for quantitating purified materials from these libraries, it is important to regularly check the variability of standard injections and track the results. Looking at data obtained each time the historical variability is checked can give a better understanding of performance and reproducibility, as well as to ensure confidence in the quantitative data [28]. Figure 14.3 provides properties of drug discovery compounds and the benefits of using CAD to analyze such compounds.

Analysis of the purified materials from libraries is just one area of drug discovery that utilizes CAD. Charged aerosol detectors have also been used successfully in

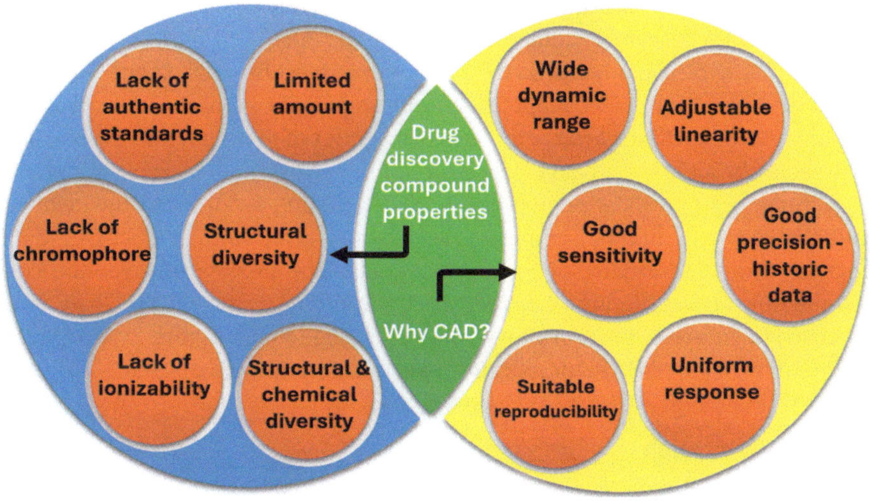

Figure 14.3 Properties of drug discovery compounds and the suitability of the CAD as an analytical tool.

the analysis of diverse analytes including triglycerides [30], natural products [31], phospholipid formulations [32], and carbohydrate polymers [33]. This diversity of analytes further illustrates the broad applicability of CAD to analytical processes across the drug discovery process.

14.6 Microgram High-Throughput Purification Platform Workflow

As previously noted, accurate quantification of compound purity is critical in accurately assessing the biological activity of a given compound [34, 35]. Library screening efforts frequently conducted at the early stages of drug discovery programs present considerable challenges to such purification workflows given the large number of compounds to be purified, the diverse structures of these compounds (which complicates not only purification but also compound quantification workflows), and the small scales (often microgram or less) on which such compounds are prepared [34, 35]. Focusing on the compound quantification challenge, library compounds frequently have unknown absorptivity values, may or may not have strong chromophores, and are in quantities that are insufficient for physical weighing or calibration curve generation for quantitation. As previously noted, such conditions seem well-suited to the application of LC-CAD (given its demonstrated utility in quantitating nonvolatile compounds without authentic standards), and indeed, CAD has previously proven an effective solution in such applications [36, 37]. Processes for the purification, analysis, and quantitation of chemical libraries have been reported and are typically designed for a range of half a milligram to several milligrams [34, 38–41].

In the following pages, we describe a customized workflow developed to rapidly purify and quantify the concentrations of compounds prepared in microscale libraries. Because these libraries involve micrograms of materials and are purified in a high-throughput process, we refer to this workflow as the microgram high-throughput purification platform (μgHTPP). The overall workflow involves multiple steps, including synthesis, purification, reformatting, post-QC, data analysis, and the assaying of purified products. The process involves the collaboration and coordination between multiple groups, each within their own specialty, to complete the workflow from beginning to end. Figure 14.4 provides an overview of the μgHTPP process from purification through post-QC.

14.6.1 Purification

Microscale library samples are synthesized in 384-well plates, in a highly automated fashion. A single 384-well plate is handled at a time for a purification

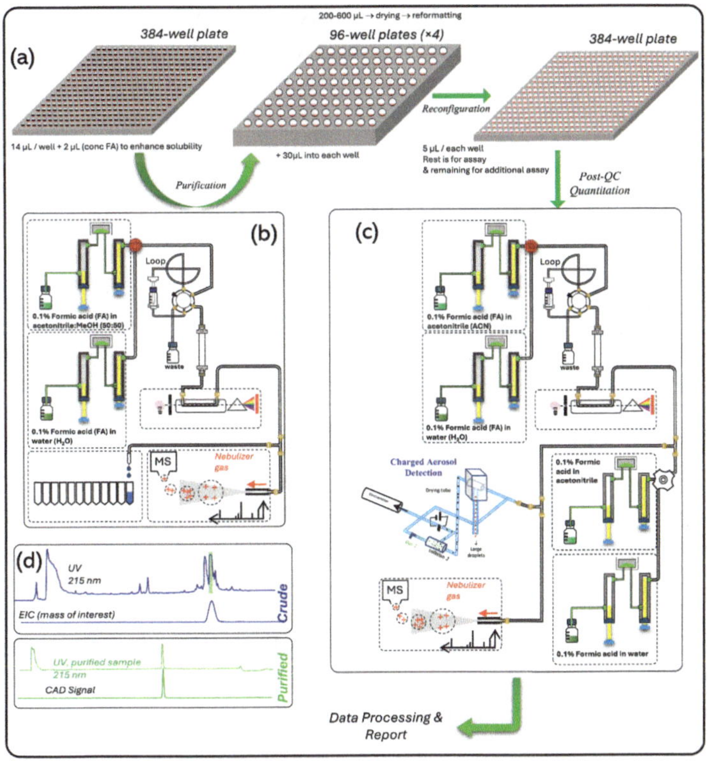

Figure 14.4 (a) Graphical overview of the µgHTPP workflow from sample submission (384-well plate), purification (4 × 96-well plates), reformatting (384-well plates), and post-QC analyses. (b) Diagram of the purification system (Waters ACQUITY Arc) containing a sample manager with flow-through needle (SM-FTN-R), quaternary solvent manager (QSM-R)), column oven, PDA detector (2998), isocratic solvent manager (ISM), fraction manager (WFM), and mass spectrometer (QDA). One of the following stationary phases is used: (i) XSelect CSH C18 2.5 µm, (ii) XSelect CSH PFP 2.5 µm, (iii) YMC-Triart PFP 3.0 µm, or (iv) XBridge BEH C18 XP 2.5 µm all in a 4.6 × 100 mm column format. The mobile phase is 0.1% FA in (A) H_2O and (B) 50 : 50 ACN:MeOH, gradient conditions are 95 : 5 (A:B) at 0 minutes, 5 : 95 at 6 minutes, 5 : 95 at 6.5 minutes, and 95 : 5 at 6.6 minutes. The total run time is 8.5 minutes, column temperature is 40 °C, and flow rate is 1.5 mL/min. (c) Diagram of the LC-MS-CAD system (Agilent Technologies 1290 UHPLC, Thermo Scientific Corona Veo RS CAD) for post-QC analysis including a high-speed pump (G7120A), two valve drive units (G1170A), diode array detector (DAD, G4212A), a multicolumn thermostat (G7116B), a single-quad mass spectrometer (G6130B), and a 1200 series binary inverse gradient pump (G1312A). The ACQUITY UPLC BEH C18 stationary phase in a 2.1 × 50 mm column with 1.7 µm particles is used. The mobile phase is 0.1% FA in (A) H_2O and (B) ACN. Gradient conditions are 0 minutes 95 : 5 (A/B), 0.50 minutes 95 : 5, 3.50 minutes 5 : 95, 4.00 minutes 5 : 95, and 4.01 minutes 95 : 5. After the DAD, the effluent is mixed with an inverse gradient flow of the opposite gradient composition. The total run time is 5 minutes, column temperature is 40 °C, and the flow rate for both the pumps is set to 0.6 mL/min. The DAD is set to 80 Hz with a 0.063 seconds response time, and the CAD evaporation temperature is set to 35 °C with a 1 second filter and a sampling rate of 50 Hz. (d) Profiles of both the crude mixture (UV and EIC signals) and the purified fraction (UV and CAD signals).

request (Figure 14.4a), but a library often consists of multiple plates. Initial processing of sample plates for the μgHTPP process consists of plate centrifugation (500 rpm) for ≥1 minute, followed by the addition of 2 μL of formic acid to each well to enhance compound solubilization. The plate is then resealed, shaken (800 rpm) for ≥20 minutes to ensure complete solubilization of all samples, and finally re-centrifuged (500 rpm) for ≥1 minute to place the solution at the bottom of each well. The entire content of each well is injected into the purification system. Analytical-scale equipment is employed for these purifications, due to the small quantities and volumes (μL) of material per well. This also helps to reduce solvent, chemical costs, and liquid waste generation. A Waters ACQUITY Arc LC-MS is used for this step of the process. The overall diagram of the system components is depicted in Figure 14.4b with full experimental details in the figure caption.

Mobile phase (RPLC with FA:H_2O:MeOH:ACN) and gradient conditions are standardized for the purification while the stationary phase choice typically varies between charge surface hybrid (CSH) C18 or pentafluoro-phenyl (PFP). For situations where an embedded positive charge would be detrimental (i.e., negative charge present on the molecule in the mobile phase) the non-charge surface hybrid versions of PFP and C18 columns were utilized. UV data is generated and monitored (Figure 14.4d); however, collection is mass-directed, ensuring that only the mass of the product ($z = 1$ or 2) triggers the collection (Figure 14.4b). For cases where the +2 charge-state of the product is within 3 Da of a starting material, only the $z = 1$ charge-state is utilized to trigger collection. This avoids potentially collecting starting material instead of product. Each well is collected in a pre-determined location in a Costar 3960 plate (96 well, 2 mL total well capacity) as a single fraction, thus making the process one-for-one (i.e., one purification plate well is collected for one synthesis plate well, as a single fraction). Up to four 96-well plates can be utilized for each purified 384-well plate (Figure 14.4a).

14.6.2 Drying and Reformatting

Solvent removal is required after purification and prior to any additional steps. The mobile phase present in the collected fractions is removed using an HT-12 Genevac. This is typically performed overnight with a run time of 8 hours. After drying, plates are subsequently reformatted for post-QC analysis. Reformatting involves the addition of a fixed, pre-determined volume of DMSO to each well to solubilize the contents, and then the transfer of the resulting solutions to a pre-set location within a 384-well plate (Figure 14.4a), with all steps being performed using an automated liquid handler. Typically, three plates in total are generated in the reformatting process: post-QC, assay, and "remainder" plates. Plates are subsequently stored at 4 °C if work will not be performed within a few days.

14.6.3 Post-QC Analysis

The post-QC plate generated in the reformatting step is analyzed via LC/MS-CAD. A sequence is generated one of two ways: (i) AutoSynth (proprietary Amgen software) is used to generate the plate map information as a CSV file, or (ii) Katalyst (ACD/Labs) is used to generate the CSV file for upload to OpenLab ChemStation (Agilent). Prior to injection, the post-QC plate is centrifuged at 500 rpm for ≥1 minute, shaken at 800 rpm for ≥20 minutes, and then centrifuged again at 500 rpm for ≥1 minute. Instrumentation utilized for post-QC analysis is provided in Figure 14.4c with experimental details in the caption. An ethylene-bridged hybrid (BEH) C18 column is the standard column used for post-QC analysis with a standardized FA:H_2O:ACN gradient.

Blanks are injected at the beginning of the sequence, along with column- and calibration-check standards. The column-check standard is made up of five compounds (caffeine, chrysin, di-*N*-pentyl phthalate, glafenine•HCl, and labetalol•HCl) covering a wide range of retention times to verify adequate performance of the column and the chromatography system. Another calibration check includes separate injections of amcinonide, bendroflumethiazide, ketoprofen, and chloramphenicol dissolved in DMSO to verify the CAD's performance [28]. After confirming that the system, column, and CAD are working properly, the first several injections are monitored to ensure proper correlation between the plate map and the physical plate. Data collected includes UV, CAD, and MS signals. UV data is used to determine purity, CAD peak area is used to calculate amount (mass) of product, and MS allows for the correlation of the results with the expected exact mass. Example UV and CAD chromatograms are shown in Figure 14.4d.

It should be noted that the present purity assessment and quantitation workflow represents only one of several possible approaches for post-purification QC. Other previously reported workflows include (i) quantitating compound fractions via CAD during collection and assessing purity by NMR following drying and reformatting steps [40] (this process was limited to 48 samples per plate); (ii) using 2D-LC to purify in the first dimension while conducting the purity analysis and quantitation (via CAD) in the second dimension [38] (this process was able to accommodate both 96- and 384-well reaction plates); and (iii) conducting purity analysis and quantitation by LC-CAD analysis without prior sample drying [41] (this analysis was limited to 96-compounds at a time). As suggested by these examples, the exact post-purification QC process can be tailored to provide the needed information for a given application.

14.6.4 Data Analysis

Following compound QC, data analysis can be performed either manually or through a combination of automation and select manual processing. For manual processing, chromatograms are analyzed using the software supplied by the chromatography vendor. For automated processing, several software packages (e.g., Katalyst by ACD/Labs, ASPro by Virscidian, etc.) are available. The overall aim of the data analysis step is to confirm mass (i.e., correlate MS data with the expected mass of a given product), calculate purity, determine the amount of product present in a particular sample well (e.g., in μg) and sample concentration (e.g., in mg/mL or mM). All of this information is then reported in an output CSV file. Whether the data is processed manually or automatically, each well is generally reviewed during data processing as a precautionary measure. The product and starting materials are tracked as part of the data review portion of the workflow, so it is important that all components are tracked during the analysis. For certain applications, it may be necessary to confirm the absence of starting material from purified wells (e.g., where the presence of starting material will adversely impact the assay readout). The percentage of remaining starting material is determined from UV analysis; however, the amount is not quantitated. Quantitation is performed for target compounds (products) in addition to purity assessment.

The limited quantities of product present in sample wells limit the options available for mass quantification. For non-microscale library processing, the final amounts of collected products would be determined by gravimetric analysis (e.g., weighing using an analytical balance). For μgHTPP, however, physically weighing product samples is not possible due to the small amounts of material generated and the fact that collection is done in a plate instead of individual tubes/vials. Therefore, LC-CAD is used to determine the amount of product collected in each well following mass-directed purification. Assuming uniform response, the CAD peak area is used to calculate the amount of material present based on a calibration curve generated using multiple standards.

14.6.5 CAD Optimization, Calibration, and Results

For proper quantitation via LC-CAD, multiple prerequisites must be met. Figure 14.5 gives an overview of the multiple factors involved in the optimization of a CAD, along with the analytical benefits derived from using a properly calibrated CAD (e.g., better reproducibility, less noise, more uniform response, etc.). High-purity mobile phases and solvents are required to minimize signal variability and presence of unwanted peaks. At the most fundamental level, ensuring the use of volatile and high-purity solvents are critical, since CAD detects nonvolatile [42]

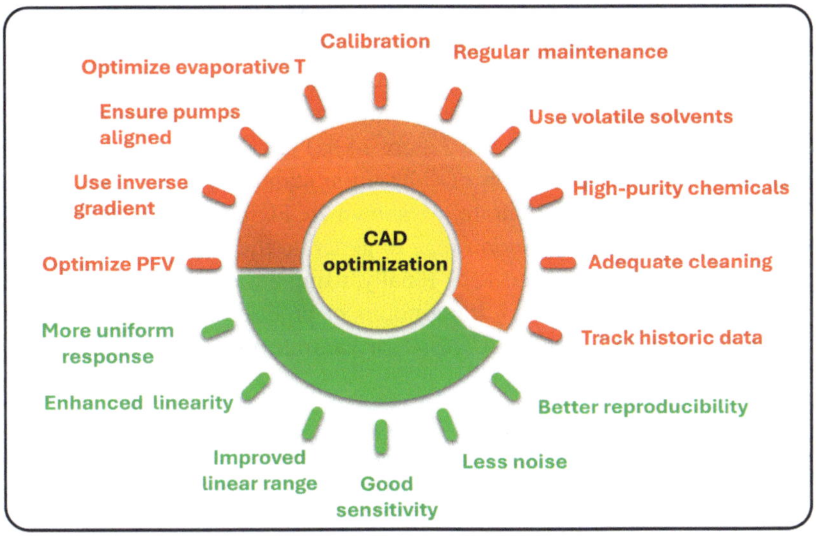

Figure 14.5 Multiple factors to consider in CAD optimization (red) and benefits obtained from compound quantification using a well-calibrated CAD (green).

and semi-volatile [43] compounds, in addition to impurities that may be present in them. Regular maintenance (i.e., cleaning/flushing the system) is critical to prevent detector contamination and to track historic performance of the system [28]. When using gradient elution, it is essential to use an inverse-gradient pump due to the dependence of CAD response to the mobile phase composition [36, 44, 45]. Confirmation that the pumps are aligned to produce an isocratic environment where they meet prior to entering the detector is also essential, and there are ways to visualize this after initial volume calculations and adjustments are completed [28]. Multiple calibration standards must also be employed to confirm the generated response factor and variability are within accepted limits. Subsequently, linearity and reproducibility are assessed. By ensuring such precautions are taken, the LC-CAD system can be relied upon to provide robust and reproducible compound quantitation across diverse compound chemotypes.

A further important consideration in CAD calibration involves optimization of the power function value (PFV) before establishment of the final calibration curve. Since CAD response is non-linear due to the dependence of signal strength on (i) the charge deposited on the surface of dried particles formed during nebulization and (ii) the changing size of nebulized particles as a function of analyte concentration [28, 46], CAD firmware includes a built-in power function to mathematically transform the detector signal and provide increased linearity over a wide dynamic range (see Chapter 3 for more details). A classic way of optimizing linearity involves evaluating the variability in the response factor (RF) across the concentration range at various PFVs, as shown in Figure 14.6a. The "optimal" PFV should result in a nearly horizontal RF line across test concentrations, indicating that detector response is ~invariant across test concentrations (e.g., as is the case for a PFV = 1.3 in Figure 14.6a) [28, 47]. Once the PFV has been appropriately set, a collection of calibration standards (e.g., amcinonide, bendroflumethidazide, and tolnaftate, as shown in Figure 14.6b) are injected over a predetermined concentration range to generate the data to construct the calibration curve at the optimized PFV. Finally, using a this calibration curve at the optimized PFV (Figure 14.6c), the amount of product in each well can be determined based on CAD peak area. (Note: for accurate quantitation, it is critical

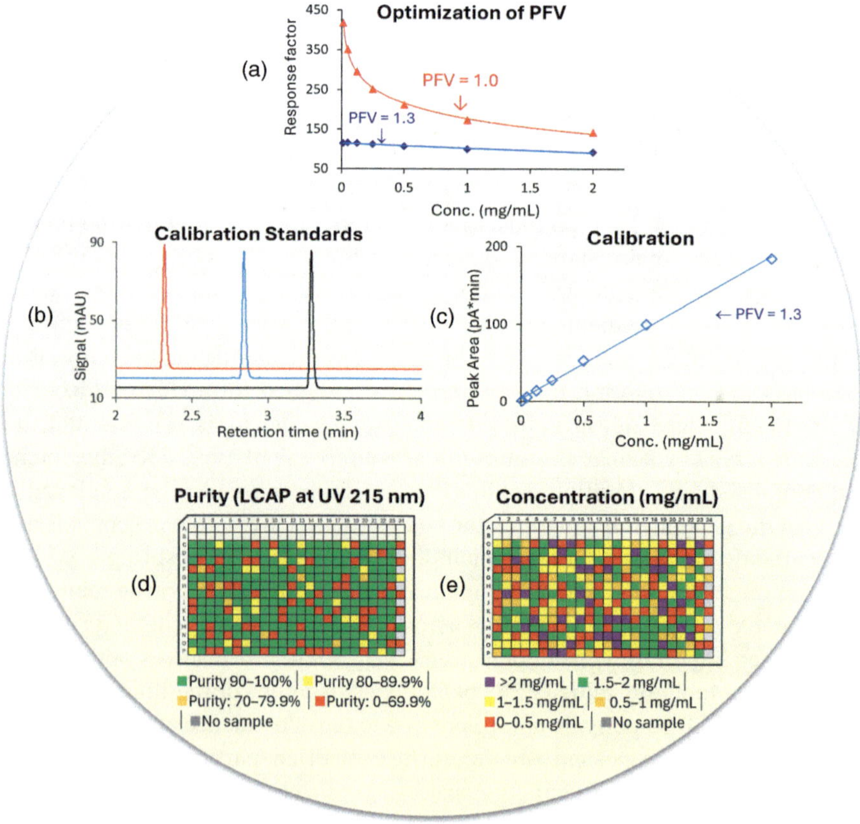

Figure 14.6 CAD PFV adjustment (a) and response calibration (b,c) to enable compound quantification in a microtiter plate (d,e).

to ensure that the calibration range covers the range of sample concentrations being analyzed [47]. For example, in Figure 14.6c, the calibration range for the post-QC process is from 0.0125 to 2 mg/mL, using 1 μL injections.)

Figure 14.6d and e show examples of a typical post-QC plate with a color-coded representation of compound purities and final concentrations (mg/mL). Purity is determined from the relative peak areas of the UV detector, while concentration (or mass, since the injection and diluent volume are known) is determined from CAD analysis. Purity and concentration values are reported in a CSV results file, which can be easily converted into heat maps to help visualize the outcome of the analyses.

14.7 Conclusion

As it has been demonstrated, drug discovery is a complex process that incorporates the use of many chromatographic-based workflows and utilizes various detectors to provide the necessary data. The use of CAD as a detector, however, extends the capabilities of quantitation into processes that cannot be as adequately solved by other means. Assessment of purity and quantitation of product concentrations can be performed in a high-throughput manner for libraries. Quantitation of these samples is possible due to the benefit of utilizing CAD as an integral part of the post-QC portion of the μgHTPP workflow.

References

1 Hughes, J.P., Rees, S., Kalindjian, S.B., and Philpott, K.L. (2011). Principles of early drug discovery. *British Journal of Pharmacology* 162 (6): 1239–1249. https://doi.org/10.1111/j.1476-5381.2010.01127.x.

2 Pidgeon, C., Ong, S., Liu, H. et al. (1995). IAM chromatography: an in vitro screen for predicting drug membrane permeability. *Journal of Medicinal Chemistry* 38 (4): 590–594.

3 D'Atri, V., Fekete, S., Clarke, A. et al. (2019). Recent advances in chromatography for pharmaceutical analysis. *Analytical Chemistry* 91 (1): 210–239.

4 Makey, D.M., Shchurik, V., Wang, H. et al. (2021). Mapping the separation landscape in two-dimensional liquid chromatography: blueprints for efficient analysis and purification of pharmaceuticals enabled by computer-assisted modeling. *Analytical Chemistry* 93 (2): 964–972. https://doi.org/10.1021/acs.analchem.0c03680.

5 Haidar Ahmad, I.A. (2022). Automated column screening and computer-assisted modeling for analysis of complex drug samples in pharmaceutical

laboratories. *Chromatographia* 85 (10): 977–984. https://doi.org/10.1007/s10337-022-04192-6.

6 Görög, S. (2018). Critical review of reports on impurity and degradation product profiling in the last decade. *TrAC Trends in Analytical Chemistry* 101: 2–16. https://doi.org/10.1016/j.trac.2017.09.012.

7 Losacco, G.L., DaSilva, J.O., Ahmad, I.A.H. et al. (2022). Parallel chiral sub/supercritical fluid chromatography screening as a framework for accelerated purification of pharmaceutical targets. *Journal of Chromatography A* 1674: 463094. https://doi.org/10.1016/j.chroma.2022.463094.

8 Vervoort, N., Goossens, K., Baeten, M., and Chen, Q. (2021). Recent advances in analytical techniques for high throughput experimentation. *Analytical Science Advances* 2 (3–4): 109–127.

9 Shi, L., Shen, G., Chai, R. et al. (2025). A review of polysorbate quantification and its degradation analysis by liquid chromatography coupled with charged aerosol detection. *Journal of Chromatography A* 1742: 465651. https://doi.org/10.1016/j.chroma.2024.465651.

10 Hasselgren, C. and Oprea, T.I. (2024). Artificial intelligence for drug discovery: are we there yet? *Annual Review of Pharmacology and Toxicology* 64 (1): 527–550.

11 D. Sun, W. Gao, H. Hu, S. Zhou, Why 90% of clinical drug development fails and how to improve it?, *Acta Pharmaceutica Sinica B* 12(7) (2022) 3049–3062. https://doi.org/10.1016/j.apsb.2022.02.002.

12 Sadybekov, A.V. and Katritch, V. (2023). Computational approaches streamlining drug discovery. *Nature* 616 (7958): 673–685. https://doi.org/10.1038/s41586-023-05905-z.

13 Qi, X., Zhao, Y., Qi, Z. et al. (2024). Machine learning empowering drug discovery: applications, opportunities and challenges. *Molecules* 29 (4): 903.

14 Peteani, G., Huynh, M.T.D., Gerebtzoff, G., and Rodríguez-Pérez, R. (2024). Application of machine learning models for property prediction to targeted protein degraders. *Nature Communications* 15 (1): 5764. https://doi.org/10.1038/s41467-024-49979-3.

15 Tirapelle, M., Chia, D.N., Duanmu, F. et al. (2024). In-silico method development and optimization of on-line comprehensive two-dimensional liquid chromatography via a shortcut model. *Journal of Chromatography A* 1721: 464818. https://doi.org/10.1016/j.chroma.2024.464818.

16 Handlovic, T.T., Roy, D., Farooq, M.Q. et al. (2025). In silico modeling enables greener analytical and preparative chromatographic methods. *Green Chemistry* 27 (1): 109–119.

17 Ahmad Dar, A., Sangwan, P.L., and Kumar, A. (2020). Chromatography: an important tool for drug discovery. *Journal of Separation Science* 43 (1): 105–119. https://doi.org/10.1002/jssc.201900656.

18 D'Atri, V., Barrientos, R.C., Losacco, G.L. et al. (2025). Trends in pharmaceutical analysis: The evolving role of liquid chromatography. *Analytical Chemistry* 97 (9): 4706–4727. https://doi.org/10.1021/acs.analchem.4c06662.

19 Malarvannan, M., Ravichandiran, V., and Paul, D. (2024). Advances in analytical technologies for emerging drug modalities and their separation challenges in LC-MS systems. *Journal of Chromatography A* 1732: 465226. https://doi.org/10.1016/j.chroma.2024.465226.

20 N. Senkuttuvan, Krishnamoorthy, R., Sarkar, S. et al. (2024). The significance of chirality in contemporary drug discovery-a mini review, RSC. *Advances* 14 (45): 33429–33448. https://doi.org/10.1039/d4ra05694a.

21 Mane, S. (2016). Racemic drug resolution: a comprehensive guide. *Analytical Methods* 8: 7567–7586. https://doi.org/10.1039/c6ay02015a.

22 Koch, G., Engstrom, A., Taechalertpaisarn, J. et al. (2024). Chromatographic determination of permeability-relevant lipophilicity facilitates rapid analysis of macrocyclic peptide scaffolds. *Journal of Medicinal Chemistry* 67 (21): 19612–19622. https://doi.org/10.1021/acs.jmedchem.4c01956.

23 Young, R.J., Green, D.V.S., Luscombe, C.N., and Hill, A.P. (2011). Getting physical in drug discovery II: the impact of chromatographic hydrophobicity measurements and aromaticity. *Drug Discovery Today* 16 (17–18): 822–830. https://doi.org/10.1016/j.drudis.2011.06.001.

24 Wang, L. (2024). Determination of four residual solvents in cefepime by headspace gas chromatography. *Results in Chemistry* 7: 101269. https://doi.org/10.1016/j.rechem.2023.101269.

25 Golembiovska, O., Voskoboinik, O., Berest, G. et al. (2021). Method development and validation for the determination of residual solvents in quinabut API by using gas chromatography. Message 2. *Pharmacia* 68: 53–59.

26 Hong, Y., Welch, C.J., Piras, P., and Tang, H. (2024). Enhanced structure-based prediction of chiral stationary phases for chromatographic enantioseparation from 3D molecular conformations. *Analytical Chemistry* 96 (6): 2351–2359. https://doi.org/10.1021/acs.analchem.3c04028.

27 Sun, P., Wang, X., Alquier, L., and Maryanoff, C.A. (2008). Determination of relative response factors of impurities in paclitaxel with high performance liquid chromatography equipped with ultraviolet and charged aerosol detectors. *Journal of Chromatography A* 1177 (1): 87–91. https://doi.org/10.1016/j.chroma.2007.11.035.

28 Barnhart, W.W., Farooq, M.Q., and Ahmad, I.A.H. (2024). A simplified tutorial on charged aerosol detection: understanding the basics, optimization, and troubleshooting. *Journal of Chromatography Open* 6: 100181. https://doi.org/10.1016/j.jcoa.2024.100181.

29 Quinn, A., Ivosev, G., Chin, J. et al. (2024). High-throughput compound quality assessment with high-mass-resolution acoustic ejection mass spectrometry:

an automatic data processing toolkit. *Analytical Chemistry* 96 (21): 8381–8389. https://doi.org/10.1021/acs.analchem.3c05435.

30 Infantes-Garcia, M.R., Verkempinck, S.H.E., Guevara-Zambrano, J.M. et al. (2021). Development and validation of a rapid method to quantify neutral lipids by NP-HPLC-charged aerosol detector. *Journal of Food Composition and Analysis* 102: 104022. https://doi.org/10.1016/j.jfca.2021.104022.

31 Wang, C., Chao, I., Qin, Y. et al. (2022). Comparison for quantification of eight components in Alpinia officinarum Hance by using high-performance liquid chromatography coupled with diode array detector and charged aerosol detector with individual and substitute reference compound. *Journal of Pharmaceutical and Biomedical Analysis* 210: 114545. https://doi.org/10.1016/j.jpba.2021.114545.

32 Tam, J., Ahmad, I.A.H., and Blasko, A. (2018). A four parameter optimization and troubleshooting of a RPLC – charged aerosol detection stability indicating method for determination of S-lysophosphatidylcholines in a phospholipid formulation. *Journal of Pharmaceutical and Biomedical Analysis* 155: 288–297. https://doi.org/10.1016/j.jpba.2018.03.067.

33 R. Ghosh, P. Kline, HPLC with charged aerosol detector (CAD) as a quality control platform for analysis of carbohydrate polymers, *BMC Research Notes* 12 (2019) 268. https://doi.org/10.1186/s13104-019-4296-y.

34 Hettiarachchi, K., Hayes, M., Desai, A.J. et al. (2019). Subminute micro-isolation of pharmaceuticals with ultra-high pressure liquid chromatography. *Journal of Pharmaceutical and Biomedical Analysis* 176: 112794. https://doi.org/10.1016/j.jpba.2019.112794.

35 Barhate, C.L., Donnell, A.F., Davies, M. et al. (2021). Microscale purification in support of high-throughput medicinal chemistry. *Chemical Communications* 57 (84): 11037–11040. https://doi.org/10.1039/D1CC03791A.

36 Górecki, T., Lynen, F., Szucs, R., and Sandra, P. (2006). Universal response in liquid chromatography using charged aerosol detection. *Analytical Chemistry* 78 (9): 3186–3192. https://doi.org/10.1021/ac060078j.

37 Hendrick, C.E., Jorgensen, J.R., Chaudhry, C. et al. (2022). Direct-to-biology accelerates PROTAC synthesis and the evaluation of linker effects on permeability and degradation. *ACS Medicinal Chemistry Letters* 13 (7): 1182–1190. https://doi.org/10.1021/acsmedchemlett.2c00124.

38 Hettiarachchi, K., Streckfuss, E., Sanzone, J.R. et al. (2022). Microscale purification with direct charged aerosol detector quantitation using selective online one- or two-dimensional liquid chromatography. *Analytical Chemistry* 94 (23): 8309–8316. https://doi.org/10.1021/acs.analchem.2c00750.

39 Dykstra, K.D., Streckfuss, E., Liu, M. et al. (2021). Synthesis of HDAC inhibitor libraries via microscale workflow. *ACS Medicinal Chemistry Letters* 12 (3): 337–342. https://doi.org/10.1021/acsmedchemlett.0c00596.

40 Baranczak, A., Tu, N.P., Marjanovic, J. et al. (2017). Integrated platform for expedited synthesis–purification–testing of small molecule libraries. *ACS Medicinal Chemistry Letters* 8 (4): 461–465. https://doi.org/10.1021/acsmedchemlett.7b00054.

41 Ginsburg-Moraff, C., Grob, J., Chin, K. et al. (2022). Integrated and automated high-throughput purification of libraries on microscale. *SLAS Technology* 27 (6): 350–360. https://doi.org/10.1016/j.slast.2022.08.002.

42 Schilling, K. and Holzgrabe, U. (2020). Recent applications of the charged aerosol detector for liquid chromatography in drug quality control. *Journal of Chromatography A* 1619: 460911. https://doi.org/10.1016/j.chroma.2020.460911.

43 Russell, J.J., Heaton, J.C., Underwood, T. et al. (2015). Performance of charged aerosol detection with hydrophilic interaction chromatography. *Journal of Chromatography A* 1405: 72–84. https://doi.org/10.1016/j.chroma.2015.05.050.

44 Behrens, B., Baune, M., Jungkeit, J. et al. (2016). High performance liquid chromatography-charged aerosol detection applying an inverse gradient for quantification of rhamnolipid biosurfactants. *Journal of Chromatography A* 1455: 125–132. https://doi.org/10.1016/j.chroma.2016.05.079.

45 Vehovec, T. and Obreza, A. (2010). Review of operating principle and applications of the charged aerosol detector. *Journal of Chromatography A* 1217 (10): 1549–1556. https://doi.org/10.1016/j.chroma.2010.01.007.

46 Handlovic, T.T., Roy, D., Barnhart, W.W., and H. Ahmad, I.A. (2024). Role of the power function value in linearity and universality for charged aerosol detectors: theoretical elucidations from a validated model. *Analytical Chemistry* 96 (40): 16045–16052. https://doi.org/10.1021/acs.analchem.4c03714.

47 H. Ahmad, I.A., Blasko, A., Wang, H. et al. (2021). Charged aerosol detection in early and late-stage pharmaceutical development: selection of regression models at optimum power function value. *Journal of Chromatography A* 1641: 461997. https://doi.org/10.1016/j.chroma.2021.461997.

15

Applications of Charged Aerosol Detection for Characterization of Industrial Polymers

Paul Cools[1,2] *and Ton Brooijmans*[1]

[1] *DSM Coating Resins, Waalwijk, The Netherlands*
[2] *The Dow Chemical Company, Freeport, TX, USA*

CHAPTER MENU

15.1 Introduction

Molecular analysis of macromolecules and polymers is an intriguing field within the analytical world. In general, a synthetic polymer does not only consist of the intended macromolecules but is also often a complex mixture of the macromolecules combined with additives, polymerization aids, and residual amounts of raw materials and their corresponding impurities. Especially in industrial polymer grades, the analysis of the additives, residuals, and impurities is a complicated task. In addition, the analysis of the polymer molecules is not

☆ *This chapter is presented here as published in the first edition of this book. The described experiments all used earlier generation CAD instruments, which did not have the ability to select a power function value (PFV). An optimal PFV allows response linearization, which is important to obtain more accurate polymer distributions. See Chapters 3 and 4 for additional details.*

straightforward. Various types of sample preparation and separation techniques are necessary to obtain information on the general composition of polymers.

In this chapter the focus will be put on the application of charged aerosol detection (CAD) for the characterization of industrial synthetic oligomers (molecules consisting of only a small number of repeating units) and larger polymer molecules and their corresponding distributions. The analysis of the additives and others, including sample preparation, will not be addressed in this discussion.

For the characterization of polymer molecules, liquid chromatography (LC) is used to separate the molecules based on molar mass and chemical composition. Size exclusion chromatography (SEC) [1] is used to obtain information on molar size distributions or molar mass distributions (MMDs), and gradient polymer elution chromatography (GPEC) [2–8], or gradient LC [9], is used to obtain information on chemical composition. Another LC technique applied especially for oligomer/polymer distributions is liquid chromatography at critical conditions (LCCC) [10–12]. With LCCC polymer molecules can be separated according to end-group functionality fairly independent of the molar mass. For block copolymers, LCCC can be used to obtain information on block length and block length distribution [13, 14].

Detection in general and in particular for polymer molecules is a key concern. Especially in combination with gradient techniques, detection of polymers is not straightforward. The use of solvent displacement detection techniques, such as evaporative light scattering detection (ELSD) [3, 15] and CAD [16–18], is essential for the application of GPEC. For isocratic techniques, such as SEC, refractive index (RI) is a generally applied solution; nevertheless, evaporative detectors can also be used.

This chapter will describe the use of CAD with SEC and GPEC. Qualitative and quantitative aspects will be discussed. Polymers with various composition (various styrene–methyl methacrylate ratios) synthesized by emulsion polymerization will be used to discuss qualification and quantification with CAD in comparison to the ultraviolet (UV) detector and differential refractive index (dRI) detector. Besides, emulsion polymers, polymer standards with low polydispersity were applied. Finally, the importance and applications of CAD in combination with mass spectrometry (MS) will be discussed in particular for the characterization and interpretation of polymer and oligomer distributions.

15.2 Liquid Chromatography of Polymers

In general, LC of polymers is divided into two types, gradient LC and isocratic LC. Under isocratic conditions, the eluent conditions do not change over time. SEC is the traditional isocratic LC method used to characterize MMDs [1, 14].

Another isocratic LC technique for polymers is LCCC [10–14]. Solvent-gradient LC or GPEC [2–9] is mainly used for the separation of polymer molecules according to polarity. The polarity of a polymer molecule depends on the type (or types) of repeating unit, and the number of repeating units or molar mass [8]. Besides solvent gradients, temperature gradients can be used [19].

Under isocratic conditions, the retention of polymers can be divided into three modes: the mode in which separation is governed by adsorption (adsorption mode), a mode where the separation is governed by exclusion (exclusion mode), and an intermediate mode called the critical condition mode where the exclusion contribution is exactly suppressed by the adsorption contribution [10, 13]. The adsorption contribution and the critical conditions depend on the eluent strength of the mobile phase, the chemistry of the repeating unit of the polymer molecule, the chemical type of column material, and temperature. Adsorption of a molecule will result in an additional retention. In comparison with regular organic molecules, polymer molecules are significantly larger in size (in solution), and the number of functional groups (repeating units) available for adsorption is considerably larger. Hence, the adsorption of a polymer molecule is stronger. In addition, a polymer molecule in solution can have dimensions larger than the biggest pore diameter present in the column, especially when standard analytical column particles are used. Consequently, a polymer molecule in solution has considerably less column volume and surface available than those available for the mobile phase molecules. The mobile phase molecules can enter all the pores, and therefore with constant mobile phase flow, the polymer molecule will elute much sooner, depending on the pore size distribution of the used column. Whether adsorption occurs or not, exclusion of polymer molecules from the pores of the chromatographic column will always take place. In other words, if the contribution of adsorption in isocratic LC is secondary to exclusion, the polymer molecules will elute before the injected solvent molecules, and separation takes place according to size. The larger molecules will elute first, followed by smaller molecules. When the adsorption contribution is significant, for example, at weak eluent conditions, the adsorption of the repeating units will retain the polymer molecules to such an extent that the polymer molecules will elute later than the simultaneously injected solvent molecules. In this mode, the adsorption mode, the polymer molecules will be separated according to the number of repeating units; with the increasing number of repeating units, stronger adsorption will occur and therefore the retention time will increase (as can be explained by Martin's rule [20]). The retention sequence in adsorption mode is opposite to the retention sequence in exclusion mode; small molecules will elute before the larger molecules. By controlling the mobile phase composition in combination with temperature, the adsorption can be controlled in such a way that the adsorption effect completely compensates for the exclusion effect, resulting in a repeating

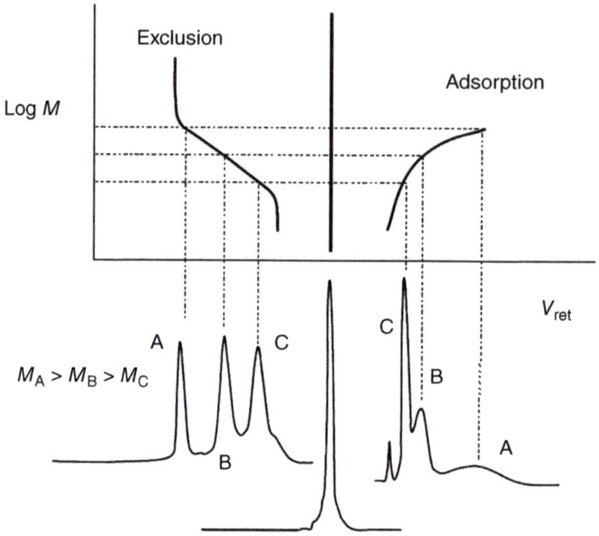

Figure 15.1 Isocratic retention modes. The retention behavior of three different polystyrene SEC standards ($M_A > M_B > M_C$) at three different conditions on analytical silica C18 column is shown: (left) eluent tetrahydrofuran, (middle) tetrahydrofuran/water 87 : 13 (at 30 °C), and (right) eluent tetrahydrofuran/water 75 : 25 [11]. *Source:* Cools et al. [11]. Reproduced with permission of Taylor & Francis.

unit independent of retention, or in other words retention independent of molar mass. This mode is also called LCCC [11].

The different retention modes are visualized in Figure 15.1. Three polystyrene standards normally used for conventional SEC calibration with different molar mass ($M_A > M_B > M_C$) are injected on a silica C18 analytical column at three different conditions: exclusion mode, 100% tetrahydrofuran THF; adsorption mode, 75/25% THF/water; and critical mode, 87/13% THF/water at 30 °C [11].

The extent of adsorption depends on the eluent strength of the mobile phase in combination with the type of column and repeating unit. By changing the eluent strength from weak to strong, adsorption can be controlled. The conditions in which the adsorption is considered to be absent are the conditions ideal for SEC. With SEC, polymer molecules are separated by molecular size and not necessarily molecular weight. The size depends on molar mass via the Mark–Houwink and Newton hard sphere theory [1]. When a homopolymer (consisting of only one type of repeating unit) is considered, the size can be directly calculated into molar mass. Nevertheless, for copolymers (consisting of at least two types of repeating unit), the size depends not only on the number of repeating units but also on the composition (ratio of the different units in combination with the total number of repeating units) [1].

A small adsorption contribution of repeating units can result in (isocratic) irreversible adsorption of the total polymer molecule. Irreversible adsorption is a common phenomenon in the LC of polymer molecules if weak eluent conditions are applied. Consequently, by controlled change of the eluent strength to stronger conditions (e.g., going from MeCN to THF on a silica C18 column), the polymer molecules can be desorbed, and separation can occur according to the type of repeating unit or polymer. This is actually the separation mechanism of gradient LC or GPEC [5–7].

15.3 Solvents

Since polymer molecules are difficult to dissolve, various organic solvents have to be used as eluent. The solubility of polymers is extensively described by the Flory–Huggins Theory [21, 22]. The organic solvents used for polymer analysis are not commonly applied LC eluents like water, MeCN, or methanol (MeOH). Mobile phase or solvents necessary to analyze polymers with LC are shown in Table 15.1.

In order to separate polymer molecules, the molecules need to be dissolved on a molecular basis. The molecules need to be uncoiled and this can only be achieved under extremely dilute conditions. The solvent needs to be sufficiently strong not only for solvation but also for uncoiling of the molecules. Using a good solvent will result in a solution of separate swollen molecules (e.g., THF and polystyrene). Weak solvents enable solvation of the polymer molecules but cannot bring the molecule in a complete uncoiled condition (e.g., MeCN and polystyrene). This does not result in single molecules but a cluster of molecules. For weak solvents even solvation does not occur (e.g., water and polystyrene). The solubility of a polymer in a solvent or solvent combination depends on the molar mass and the chemical type of the polymer molecules. High molar mass needs stronger solvents or solvent compositions. Hence, depending on the chemical type of the polymer molecules, different solvents have to be used. Polystyrene and most poly(meth)acrylate polymers dissolve in the most commonly used solvents like THF. Polar polymers such as polyurethanes and (crystalline) polyesters need more polar solvents such as DMAc and HFIP. Polysaccharides need DMSO to dissolve. Polyolefins need nonpolar organic solvents to dissolve and, in all cases, elevated temperatures higher than 150 °C. The applicability of organic solvents suitable for polymer analysis with CAD needs to be studied in more detail. It is unknown whether CAD is resistant to solvents like HFIP, DMAc, and DMSO. For THF, CAD has already been optimized/modified and filtration of the solvents with a 0.1 μm filter is necessary. This will be discussed further in the succeeding sections.

Table 15.1 Overview of used solvents/eluents for LC polymers applications.

Solvent	LC application	Polymer application	Remarks
Water	SEC [1]	Peptides, cellulose, polyacrylic acid	Polar and/or ionic polymers
Tetrahydrofuran (THF)	SEC [1]/GPEC [3]	Polystyrene, polyacrylates	Non-polar to medium polar polymer
Dimethyl acetamide (DMAc)	SEC [1, 23]	Polyurethanes, polysaccharides	Polar polymers
Hexafluoro isopropanol (HFIP)	SEC [7]/GPEC [7]	Polyurethanes, polyesters, polyamides, polycarbonates	Polar, hydrogen bonding, crystalline polymers
Dimethyl sulfoxide (DMSO)	SEC [1, 23]	Dextrans, starches (polysaccharides)	
Chloroform/dichloromethane	SEC [1]/GPEC [6]	Polystyrene	Non-polar, non-crystalline
Trichloro benzene (TCB)	SEC [1]/GPEC [24, 25]	Polyolefins	High temperature applications (>100 °C) Non-polar crystalline polymers

15.4 Quantitative Detection of Polymer Molecules

The most preferred concentration detector is universal in selectivity and sensitivity and therefore useful for quantifying a broad range of components. With a universal detector, the qualitative and quantitative comparison of complex samples can be performed. In LC, a universal detector is missing in general. The mobile phase of LC regularly contributes to the detector signal and can block the signal entirely.

15.4.1 Ultraviolet Detection

The most commonly used LC detector is the UV detector. UV detection can detect molecules that contain chromophore groups, and in many cases, this includes the LC mobile phase. Consequently, the UV extinction of the analyte present in low concentrations will be surpassed by the extinction of eluent. The baseline signal of the UV detector is always auto-zeroed in order to detect the difference between the dissolved analyte and the used mobile phase. The background extinction of the mobile phase should not be too high. UV detection can be used in isocratic methods as well as in gradient elution. In gradient elution the mobile phase gradient is noticeable over the whole range of wavelengths in the baseline. Not all types of polymers contain chromophore groups that can be detected in solvents like THF. Most polymers absorb UV light below 230 nm, and the UV extinction of THF will hamper the detection. This makes the UV detector not suitable to detect, for instance, acrylic polymers in THF. HFIP is more UV transparent and could be used to analyze polyacrylates. Unfortunately, HFIP is not a good solvent for less-polar polymers like polystyrene. Consequently, for polymers, UV detection is generally not as suitable as universal detection.

15.4.2 Differential Refractive Index Detection

Another commonly used LC detector is the dRI detector. By monitoring the change of the RI of the solution in the sample flow cell compared with the RI of the pure eluent in the reference cell, small differences in RI can be used to detect polymer molecules in solution. The dRI can be used as universal detector when the solvent is chosen wisely (big RI difference between analyte in solution and eluent) and isocratic conditions are applied [16]. Consequently, the dRI is most suitable for SEC separations. Unfortunately, the dRI cannot be used in gradient applications; the change in RI caused by the eluent gradient will hamper the detection of polymer molecules in solution. In the past, efforts were made to modify the dRI optics to use the RI detector in gradients [26], unfortunately without commercial success.

15.4.3 Evaporative Detection

Since the LC eluent is causing most of the problems during the detection, eluent evaporation detectors seem to be the solution. In the past, the evaporation techniques mainly focused on the evaporation of the eluent by spraying and applying heat. The moving-belt flame ionization detector (FID) for LC has been studied, but without commercial success [27]. Another technique is evaporative light scattering (ELS) detection, introduced successfully in the late 1980s. With evaporative detection, the eluent is evaporated and the analytes preferably form particles that can either be detected by light scattering or by corona charging [16]. The corona-charging technique is still undergoing optimization and development, and the introduction of the Nano Quantity Analyte Detection (NQAD) [28] also shows that the development on laser detection in combination with evaporation is not mature. The application of NQAD will not be part of the discussion.

As expected, evaporation of the eluent also has drawbacks. By evaporating the eluent, low-molar mass components will also evaporate, and this hinders the detection as particles. Consequently, low-boiling components cannot be detected by these techniques. By applying lower temperature and better nebulizing techniques, the application temperatures of evaporative detectors have decreased significantly (from boiling point of eluents to about 30 °C).

Polymer molecules are extremely suitable in combination with these detectors. Oligomers (molecules consisting of only a few number of repeating units) might still be affected by the evaporation steps, but the main part of a polymer distribution should be forming particles and therefore be detected. The ELS detector turned out to be a useful detector and definitely contributed to the success of GPEC [3, 7]. Nevertheless, the ELS detector was not perfect with respect to sensitivity and universal applicability. Especially for quantification of distributions, the early ELS detectors were limited. The new ELS detectors with optimized evaporation principles at lower temperatures are significantly more sensitive; however, the new generation of ELS detectors still results in nonlinear concentration dependency [15].

15.4.4 Charged Aerosol Detection

By using CAD, combining evaporation at low temperature and corona charging of the formed particles, analytes can be detected [17, 18]. CAD is considered to be universal [16] and can be used under isocratic and gradient elution conditions.

CAD has three major drawbacks. Similar to other evaporative detectors, volatile components will not be detected due to the evaporative conditions, and the sensitivity significantly depends on the eluent solvent composition. Another drawback is the high sensitivity toward particles present in the eluents. Accordingly, the quality of the eluents should be monitored wisely. These drawbacks are similar to the modern ELS detectors. Fortunately, with the introduction of μLC equipment, the need for high-purity eluents becomes more important. As experienced with μLC, the presence of particles in the mobile phase causes problems; the presence of particles can induce clogging of the system due to the small internal diameters of the capillary, low dead volume, and small column particle sizes. For the generally applied eluents, like water and MeCN, μLC quality grades are available. In most cases, the eluent is filtered with a 0.1 μm filter instead of a 0.2 μm filter. THF at first was not applicable with μLC equipment due to incompatibility issues. With the introduction of CAD, the quality of the THF was not suitable for μLC and incurred problems with CAD. This issue has been solved by an extra filtration step over 0.1 μm filters [29]. However, impurities, erosion of the equipment, and aging of the THF can have a significant effect on CAD signal. Besides the eluent, column bleeding can also give rise to baseline effects. Therefore, the use of good quality columns with low column bleed, for example, Symmetry® (Waters) or Chromolith® (Merck Millipore), is essential. In practice, cleaning of the CAD detector needs to be performed on a regular basis to avoid malfunction of the flow regulation.

15.4.5 Molar Mass-Dependent Detection

Besides concentration detection, molar mass-dependent detectors can be used. The most generally applied detectors are viscosity and light scattering. Both detection techniques have to be used under isocratic conditions and, from the signal, molar mass information can be extracted. Nevertheless, in order to calculate molar masses, a concentration detector, like the dRI, is necessary.

15.4.6 Mass Spectrometry

Another technique that has been introduced for the detection of polymer molecules is MS [30]. Of course, MS is not only a detector as such but is also considered as a separate analytical technology. Nevertheless, MS can be used to quantify components, especially in combination with LC. The issues with MS on selectivity and sensitivity are significant but can be overcome by hyphenation with LC. The detection of polymers or polymer molecules has one main difference in comparison with the classic organic compound detection: polymers emerge as a distribution. As a result, the discussion on having a universal detector becomes even more important. With LC the molecules are separated according to polarity or molecular size; hence, in all cases a distribution is obtained.

In the next paragraphs, the application of CAD in SEC and GPEC will be discussed with special attention to measuring distributions. CAD will be compared with UV and dRI detections. Some quantitative and qualitative aspects will be described with respect to chemical type and molar mass of the polymers. Furthermore, in the last paragraphs, the combination of LC with MS and CAD will be discussed in more detail.

15.5 Size Exclusion Chromatography and Charged Aerosol Detection

To compare the different signals (UV, CAD, dRI) in SEC, model copolymers were synthesized with different methyl methacrylate (MMA) and styrene (S) ratios. The polymers are made by emulsion polymerization using sodium lauryl sulfate (surfactant), ammonium persulfate (initiator), and lauryl mercaptane (chain transfer agent). Different copolymers were synthesized: pMMA, pS-co-pMMA 25 : 75, pS-co-pMMA 50 : 50, pS-co-pMMA 75 : 25, and pS. Well-defined pS SEC standards are used for conventional calibration. pS and pMMA standards ($M_w \sim 5$ kg/mol) are also used for illustration purposes.

All polymers were dissolved in THF (unstabilized, Biosolve, μLC-MS quality) at various concentrations: approximately 0.01, 0.1, 1.0, and 10 mg/mL. The SEC experiments were performed with two SDVB Mixed B columns (Agilent) with guard column, on an Alliance 2695 separations module (Waters) with a 2998 Photo Diode Array detector (PDA) (Waters), a 2410 dRI (Waters) and the Corona® *Plus* (Dionex). The UV signal at 254 nm was used. THF (unstabilized, Biosolve, μLC-MS quality) was used as eluent without modifiers and a flow of 1 mL/min, and an injection volume of 100 μL.

In Figure 15.2, overlays of the chromatograms of low-molar mass pS (white) and pMMA (gray) standards with approximately similar concentration (10 mg/mL) obtained with CAD, UV, and dRI detectors are shown. If the signal is independent of the composition, the peak areas should be comparable. In the CAD chromatograms, typical baseline effects are visible. As discussed before, these artifacts can be caused by the quality of the THF or column bleeding. In this case the artifacts are probably caused by the virgin SDVB columns. A clear tailing also occurs in the CAD signal for both pMMA and pS. In comparison, the UV and dRI show less or no tailing.

The tailing in the CAD signal can be explained by broadening effects and dead volume probably related to the hardware of the CAD. The evaporation step and the corona charging might cause extra broadening, making the early model CAD (Corona Plus) less suitable for use in UHPLC applications. All experiments described in this chapter were performed with a Corona Plus version. To solve the broadening effect, the Corona ultra was introduced. However, this needs to be studied further. In addition to the dead volume, the tailing can also be a result of a

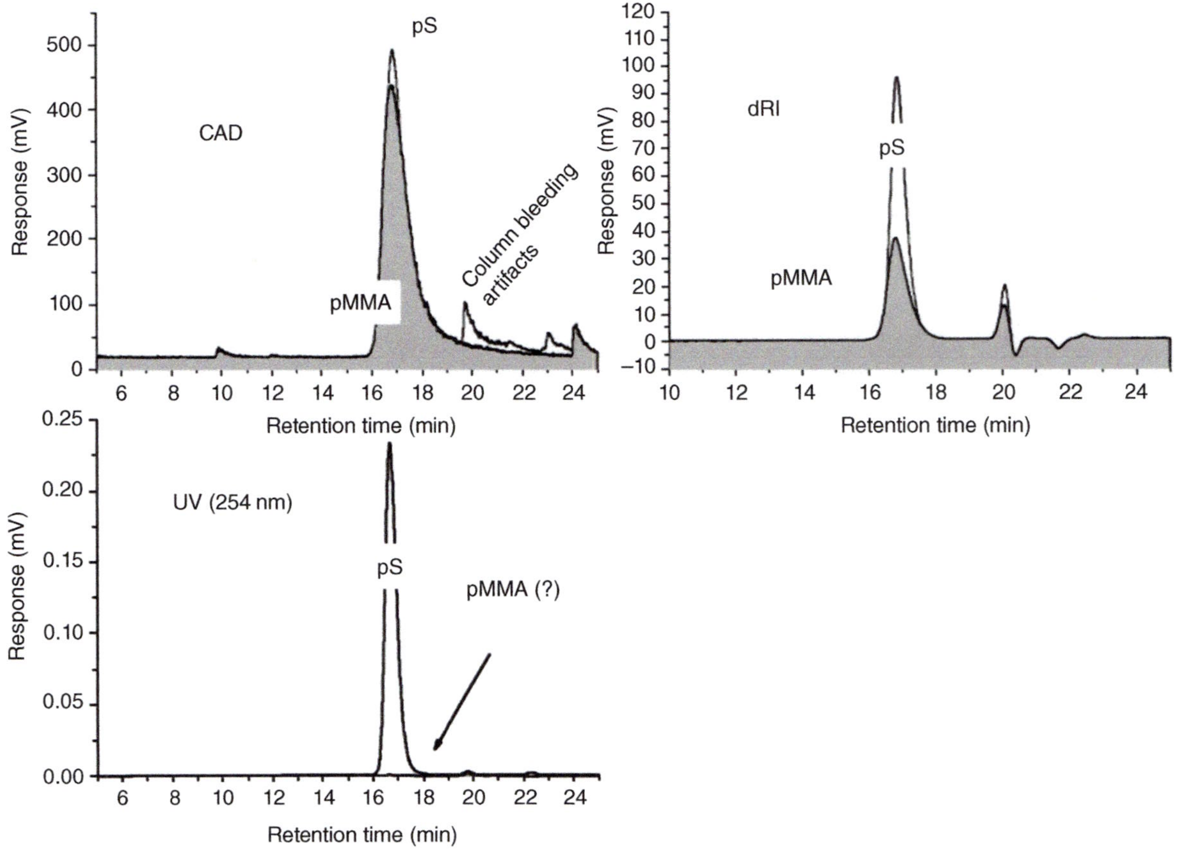

Figure 15.2 **Comparison** of CAD, dRI, and UV signals of homopolymers, a pS (white) standard ($M_w \sim 5$ k/mol) and a pMMA (gray) standard ($M_w \sim 5$ k/mol) at similar concentration.

contaminated detector. Cleaning the detector and the particle filters on a regular basis is advised.

In Figure 15.2, it is clearly shown that the areas obtained with CAD, UV (254 nm), and dRI signals are different in pS and pMMA standards. The UV detector gives no response for the pMMA standard, while the peak area obtained with dRI is significantly larger for the pS standard than for the pMMA standard. CAD shows the most universal behavior; the areas are comparable for the copolymers and the homopolymers pMMA and pS, despite the tailing. The peak tailing is less pronounced due to the broader distribution compared to both standards. Similar findings are obtained when the copolymers are analyzed and the results are shown in Figure 15.3.

The UV response (measured at 254 nm) is highly dependent on styrene content, since MMA has no UV contribution. The dRI signal is more universal but still shows a significant dependency based on the composition. This is attributed to the difference in dn/dc of S and MMA units; the dn/dc of pS is approximately twice as large compared with that of pMMA (0.17 versus 0.08). In general, the dRI is used as a universal detector. However, the dRI signal is dependent on composition, which could result in problems with respect to the quantitative analysis of heterogeneous copolymers, especially when distributions are concerned. CAD response appears to be independent of the composition. The dependency of concentration on the response of the different detectors is studied by observing total peak areas. The peak areas for the different copolymers and detectors are determined for various concentrations (0.01, 0.1, 1.0, and 10 mg/mL) and are depicted in Table 15.2.

First of all, it should be noticed that the concentrations are of the total polymer distributions and not of a single (polymer) molecule. In other words, due to the broad distribution, the concentration of a single molecule is decreased significantly and might be a factor of 100 less concentrated. Therefore, the detection levels appear to be limited. At the lowest concentration, none of the detectors showed a response due to the previously discussed distribution-dilution effect. In Figure 15.4, the data depicted in Table 15.2 are plotted (peak area versus the concentration) for the different detectors and copolymers.

Both UV and dRI results show a linear dependency across the measured concentration range; CAD shows a skewed curve (probably logarithmic dependency). Due to the limited data points, an exact fit was not performed. For the discussion this is not relevant; the fact is that the relation is not linear, which poses as a major drawback of CAD. A possible correction on these types of data could be the use of log–log scales, but this is not present in commercial SEC software. For single components, using a log–log scale might be a solution, but for distribution this is less suitable. The UV and dRI detectors show a significant dependence on the composition, but for CAD this dependency is apparently absent.

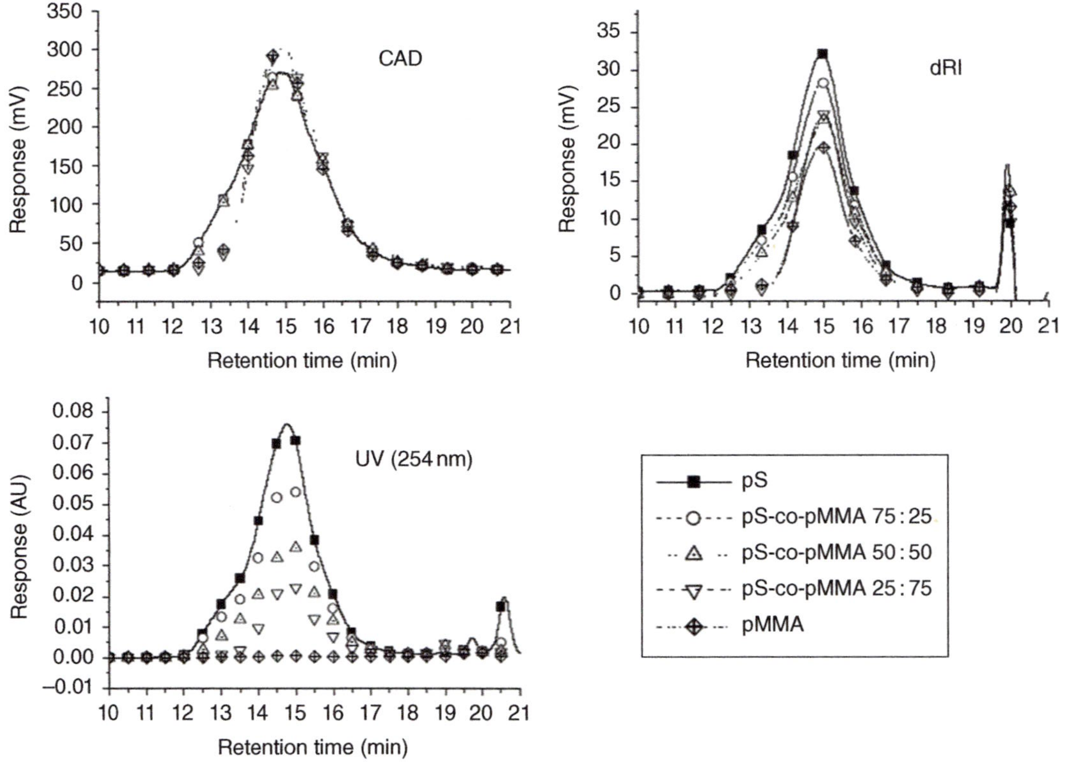

Figure 15.3 **Chromatograms** of copolymers using CAD, UV, and dRI. The UV and dRI clearly show dependency on sample composition.

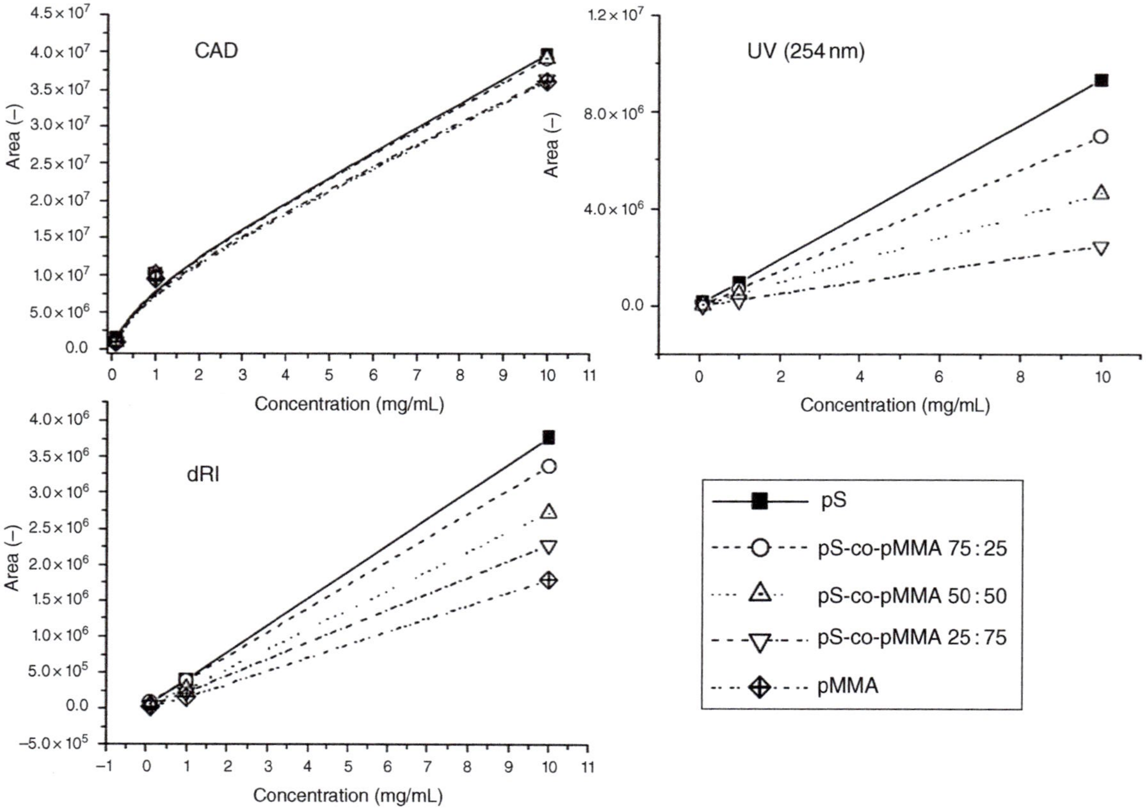

Figure 15.4 **Concentration** dependency of different copolymers. Different concentrations are used 0.01, 0.1, 1.0, and 10 mg/mL. Samples are analyzed using CAD, UV, and dRI detectors.

Table 15.2 Peak areas of different polymer samples obtained with UV, CAD, and dRI detection.

	Load (mg)	Area UV254	Area CAD	Area dRI
pS	1	9.3E6	4.0E7	3.8E6
	0.1	9.2E5	1.0E7	4.0E5
	0.01	1.3E5	1.5E6	6.1E4
	0.001	n.d.	n.d.	n.d.
pS-co-pMMA 75 : 25	1	7.0E6	3.9E7	3.4E6
	0.1	6.7E5	1.0E7	3.8E5
	0.01	5.6E4	9.8E5	9.2E4
	0.001	n.d.	n.d.	n.d.
pS-co-pMMA 50 : 50	1	4.6E6	3.9E7	2.7E6
	0.1	4.6E5	1.0E7	2.7E5
	0.01	3.9E4	9.8E5	6.1E4
	0.001	n.d.	n.d.	n.d.
pS-co-pMMA 25 : 75	1	2.4E6	3.6E7	2.3E6
	0.1	2.4E5	9.9E6	2.2E5
	0.01	1.9E4	1.0E6	4.1E4
	0.001	n.d.	n.d.	n.d.
pMMA	1	n.d.	3.6E7	1.8E6
	0.1	n.d.	9.5E6	1.5E5
	0.01	n.d.	1.0E6	2.7E4
	0.001	n.d.	n.d.	n.d.

n.d., not detected.

A major concern for CAD results is the nonlinear response. This means that molecules present in a sample at low concentrations get overestimated compared with the major constituents. This can have a big influence on the MMD calculation since the peak tails are overestimated. In addition, the tailing effect as can be seen in Figure 15.2 also has a negative contribution. This can be observed in Figure 15.5 and Table 15.3. The MMD calculated by the dRI and CAD are different, resulting in deviating average molecular weight numbers. The calculated molar masses are shown in Table 15.3.

The MMDs are calculated for the dRI and the CAD signals via conventional calibration with pS standards. The MMD for the UV signal was not included for practical reasons. The number average molecular weight (M_n) is significantly lower for

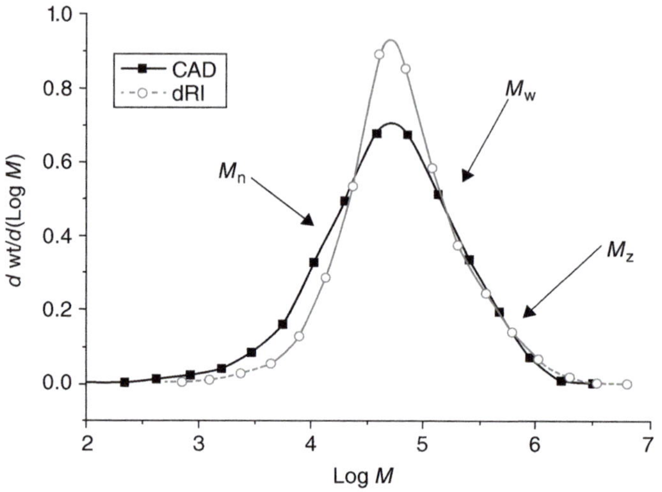

Figure 15.5 **Comparison** of molar mass distributions (MMD) of pS-co-pMMA 50 : 50 copolymer as determined by CAD and dRI.

Table 15.3 Average molecular weights of copolymers determined by CAD and dRI.

	CAD (kg/mol)			dRI (kg/mol)		
Samples	M_n	M_w	M_z	M_n	M_w	M_z
pS	15	127	446	29	139	541
pS-co-pMMA 75 : 25	17	129	457	30	141	556
pS-co-pMMA 50 : 50	14	115	394	28	128	509
pS-co-pMMA 25 : 75	15	68	168	27	71	175
pMMA	16	79	225	32	85	257

CAD results for all polymer samples, due to the tailing of the CAD signal. Tailing in the chromatogram will result in a lower M_n. The effect of the tailing is also shown in the M_w and M_z; however the effect is less pronounced.

In conclusion, the SEC results show that all detectors have their drawbacks; UV is the most selective (the least universal) detector, and dRI shows some compositional dependency. This is especially an issue when chemically heterogeneous distributed copolymers are analyzed. In many industrial polymer grades, the molecules are built up with several monomers. CAD appears more universal with respect to composition but overestimates the tail of the distributions. Eventually, in all cases, the calculated MMDs are not the correct representation of the reality, and the MMDs should be interpreted with caution. This problem cannot be overcome by using viscosity and/or light scattering detection, since the application of these molar mass-sensitive detectors has to make use of a concentration signal in order to calculate "absolute" molar masses [1]. Ultimately, the application is and should be based on which detector is preferable.

15.6 Gradient Polymer Elution Chromatography and CAD

As described in Section 15.2, gradient LC (GPEC) is commonly used to analyze polymer compositions. GPEC can be used to separate a polymer sample into monomers, additives, oligomers, and polymer molecules, and therefore specific information on the polymer composition can be obtained. GPEC can be used to analyze specific target components, such as additives and residuals, as well as the main polymer molecules [2]. The separation mechanism is different from SEC, since polymer–column interactions, undesired in SEC, are used to separate polymer molecules based on monomer type or functional groups.

For the experiments described in the next paragraphs, GPEC is used in reversed-phase mode (polar eluents in combination with a nonpolar column). A sample is injected into the column in weak eluent/solvent conditions, for example, aqueous conditions on Si C18 column. The sample will precipitate on top of the column/frit. Successively, a controlled change of eluent composition/strength is applied, resulting in the dissolution and desorption of the compounds and polymer molecules. Commonly used gradients are water/THF or water/MeCN/THF gradients. Other solvents such as methanol, isopropanol, and HFIP can also be used. Under normal phase conditions, solvents such as chloroform, heptane, and TCB are used.

As already discussed, under gradient conditions dRI detection is not an option. In general, UV detection and/or solvent evaporative detectors, for example, ELS detector and CAD, are used. The polymers described in Section 15.5 have also been analyzed by GPEC. The used system is a 2695 separation module (Waters), with a Symmetry C18 150 × 4.6 mm 3.5 μ column (Waters). Injection volume was 5 μL. A binary gradient (see Table 15.4) was applied (A = water + 0.1 v% trifluoroacetic acid [TFA], B = THF unstabilized) at a flow rate of 0.5 mL/min.

In Figures 15.6 and 15.7, the GPEC results of the various copolymers are shown. With GPEC the copolymers are separated according to composition and, to lesser extent, molar mass. The retention time increases with increasing amount of S units (higher S : MMA ratio). The UV signal is only dependent on the styrene content, resulting in decreasing area with decreasing styrene content. Consequently, UV 254 nm cannot be used for qualification and quantification of

Table 15.4 Gradient conditions in GPEC mode.

Time	%A	%B
0	100	0
40	0	100
45	0	100
50	100	0
60	100	0

A, water + 0.1 v% TFA; B, THF.

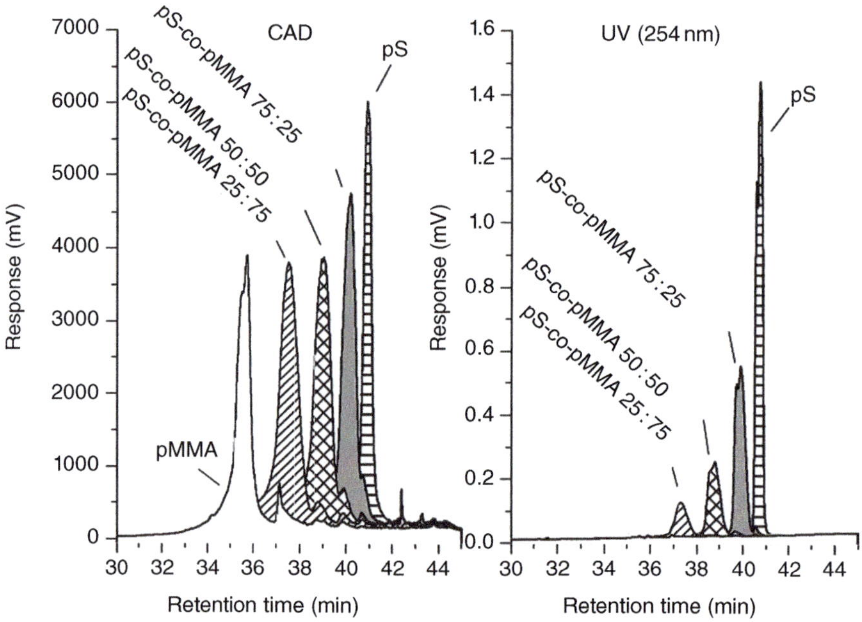

Figure 15.6 **Overlay** of chromatograms of various copolymers and homopolymers of styrene and MMA obtained under gradient conditions shown in Table 15.4.

these copolymers. Even when other UV wavelengths are used, the composition dependency on the response remains significant. Theoretically, at lower wavelengths, MMA units should be detectable, but the different extinction coefficients of styrene and MMA and the UV absorbance of THF still result in inadequate quantification of the distribution. Obviously, the response of the UV can be corrected for the amount for the S : MMA ratio, but this is less favorable. First observation of CAD chromatograms shows that the responses of the various

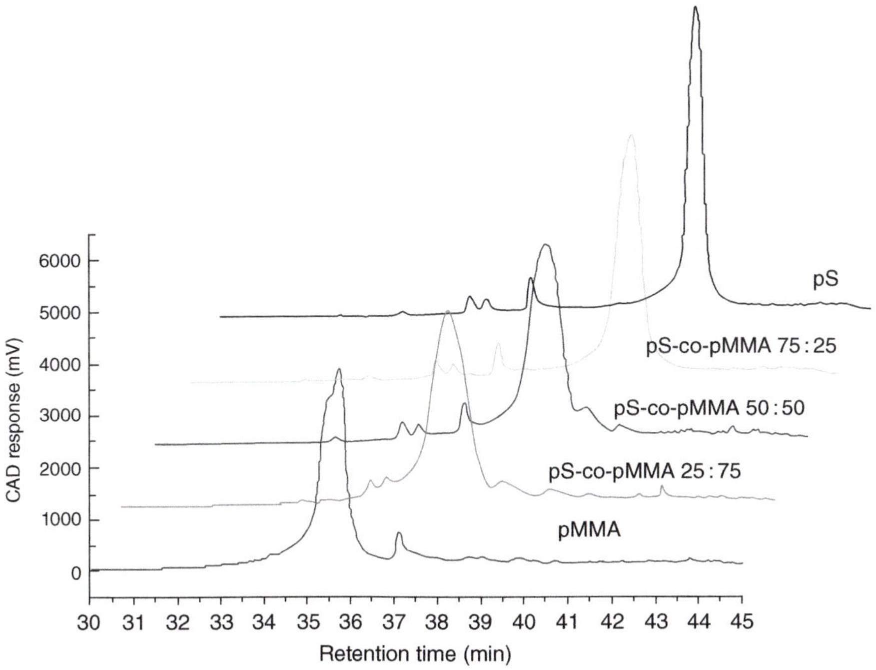

Figure 15.7 **Overlay** of CAD-GPEC chromatograms of the copolymers and corresponding homopolymers. The copolymers clearly show a broader distribution, resulting in a larger peak area due to the nonlinear concentration dependency of CAD.

copolymers are comparable. The response of evaporative detectors is known to be dependent on the eluent composition [15]. For the used gradient and the copolymer compositions, this dependency is not evident.

The peaks of the homopolymers pS and pMMA appear to be narrower than the peaks of the copolymers. The homopolymers pS and pMMA do not have a chemical composition distribution, which is the case with the copolymers. For clarity reasons, the chromatograms at highest concentration of the various copolymers are shown in a different perspective in Figure 15.7.

The broadening of the peaks is mainly caused by system broadening and separation based on molar mass. The fronting of the peaks can be explained by a broad MMD. The tailing effect as detected with SEC cannot be observed in the GPEC-CAD chromatograms. The extra peaks on the tail can be explained by the monomer feeding conditions during the polymerization. Since all polymers are dissolved at similar concentrations, the peak areas should be comparable. In Figure 15.8, the peak areas of CAD and UV chromatograms for the copolymers and homopolymers are plotted versus concentration of the distribution.

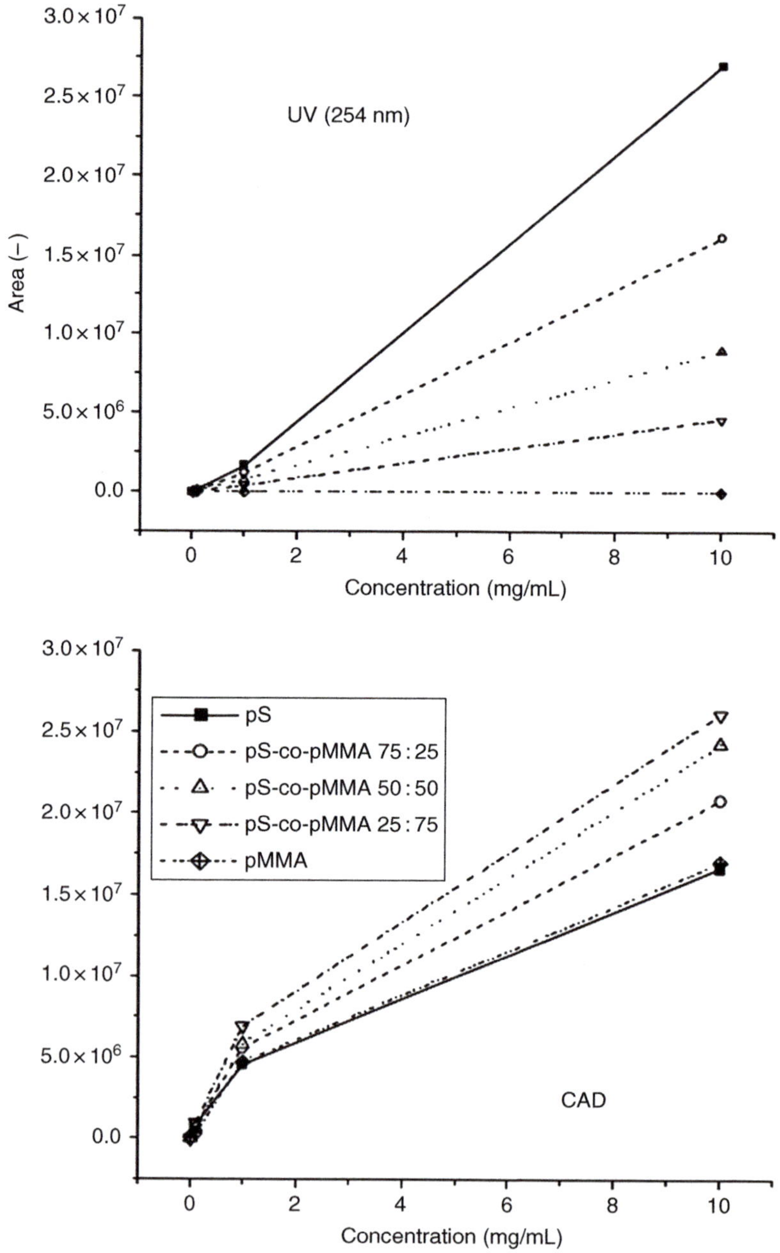

Figure 15.8 **Concentration** dependency on peak area of homopolymers and copolymers of the UV (254 nm) and CAD.

With respect to the comparison of the detectors, similar results are obtained compared to SEC. The UV signal obviously showed a strong compositional dependency, and CAD detector shows a nonlinear dependency on the concentration. In the case of GPEC, the CAD signal is less universal. For the homopolymers pS and pMMA, the peak areas are comparable, indicating universal behavior. The peak areas obtained for the copolymers show a nonuniversal behavior, perhaps due to the nonlinear concentration behavior. In addition, the gradient is always visible in the baseline. The system peaks are present due to eluent and column contaminants. With UV the background extinction increases with increasing THF composition, and the response of CAD changes with the eluent composition.

The drawbacks mentioned previously hamper the universal application of UV and CAD in combination with gradient elution. However, CAD (or other evaporative detectors) is an essential detector to be used with gradient elution. Especially for the analysis of copolymers, the UV cannot be used, therefore evaporative detectors are crucial, although not ideal. As mentioned in the SEC section, a log–log scale could resolve the nonlinear behavior of CAD, but again this is more appropriate for single component analysis, and this is out of the scope of this discussion.

15.7 Liquid Chromatography Combined with UV, CAD, and MS Detection

Besides using UV, dRI, and evaporative detection, MS can also be used to detect the separated polymer molecules. LC-MS can rapidly provide information on several types of issues, such as raw material comparison, contaminations, or compositional analysis. In combination with LC, electrospray ionization (ESI) [30] is the most used system. Matrix-assisted laser desorption ionization (MALDI) [31] is applied for polymer molecules, but due to the challenging sample preparation, MALDI cannot be used in line with LC [31]. Due to the limitations of MS, polymer analysis is primarily focused on the compounds in the oligomer range ($M < 10$ kDa). In general, MS can be divided into three steps: the ionization step, the analyzing/separation step, and the detection step. The time-of-flight (TOF) analyzer is the most common for polymer analysis due to the broad application range (=1,000,000 Da) [30, 31]. Higher molar masses are applicable, but above 10 kDa, only broad peaks are observed and no information can be obtained anymore on the repeating unit or end-groups. Additionally broad distributions are difficult to analyze directly with MS.

A large range of LC techniques is coupled with ESI-TOF-MS, the main technique being reversed-phase gradient elution LC. But SEC can be coupled with MS. The separation power of SEC is limited compared with the polarity-based separation such as reversed-phase LC, but it does provide direct information on the distribution since the molecules are separated according to molar size, making it significantly more straightforward.

15.7.1 LC-ESI-TOF-MS System at DSM Coating Resins

The LC-ESI-TOF MS system is a Dionex U3000 system equipped with two gradient elution pumps, two isocratic pumps, one photo-diode UV detector, one CAD detector, and one ESI-TOF mass spectrometer (MicroTOF, Bruker). ESI is used for the introduction from the LC to MS and the ionization. The analyzer is a TOF analyzer, and the detector is an analog microchannel plate detector. With ESI many polar and slightly polar molecules can be ionized. With TOF the molar mass range is broad, and in reflectron mode, detailed information of separate molar masses and isotope patterns can be obtained up to 10 kDa. The detector enables quantitative analysis of the formed adducts with the constraint that a reference sample of the single component/adduct must be analyzed to establish a concentration calibration curve. A schematic representation of the combined LC-MS and SEC-MS systems is shown in Figure 15.9.

In practice, the MS and LC acquisition is controlled by Bruker software, but the LC modules are controlled by Chromeleon software in order to obtain full control of the LC possibilities. In this way, SEC-MS and gradient LC-MS can be performed on one single system. Besides the efficient use of the LC system, the software-controlled performance of the TOF calibration with an HCOONa solution at any chosen moment (including at the start of the gradient LC or SEC run) is unique. The use of alternate adduct salts delivered by a gradient

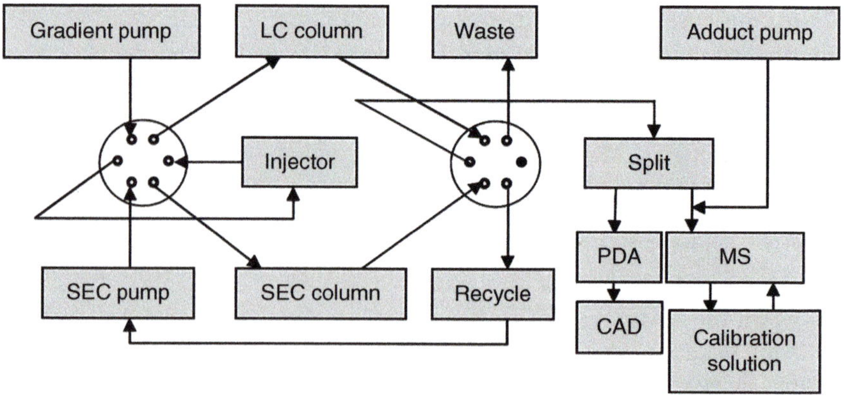

Figure 15.9 **Blueprint** of LC-MS system. The system consists of an ESI-TOF-MS combined with two gradient pumps, two isocratic pumps, one single column oven, a PDA, and a CAD.

pump, post-column added after the splitting, enhances the possibilities of the MS identification.

In general, LC-MS systems are dominated by the MS part with low flexibility in the LC configuration. In this case the LC configuration is leading. In other words, all feasible detection techniques are used in combination with MS. The use of CAD in combination with MS appeared to be crucial. In general, SEC is performed in combination with only a dRI detector, and gradient elution is applied in combination with UV and CAD. The dRI is highly sensitive to fluctuations of the eluent composition, the temperature, and the flow. Consequently, the dRI is not suitable for use in this type of system where isocratic and gradient applications are combined. The LC-MS system had to be versatile in changing solvents/eluents and switching modes. One of the alternatives for dRI detection is CAD, combining the concentration dependency with fast adaptation to solvent changes. As shown in the previous paragraphs, CAD can be used for quantification in SEC systems as well as gradient LC systems.

The selected LC-MS system is in many ways comparable with the system used by Nielen et al. [30] but is considerably more advanced due to the continuing developments in MS: more accurate molar mass determination, higher sensitivity, and better quantitative analysis over several concentration decades. However, the limitations of ESI-TOF-MS still arise: molar mass limitation of 10 kDa, multiple charges resulting in complex spectra, and ion selectivity and suppression. Nevertheless, the information found in the low molar mass region can be used to obtain information on the total MMD. For instance, the presence of low molar mass cyclic oligomeric species can have a large influence on the application properties of polyesters [6, 7]. In addition, hydrolysis and side reactions will occur over the complete range of the MMD. Consequently, characterization of the oligomeric region is worthwhile and, in most cases, the only option to obtain this information. The combination of detectors, UV, CAD, and MS, enables the system to get optimal information from each separation. Together with Fourier transform infrared spectroscopy, gas chromatography mass spectrometry, and nuclear magnetic resonance, LC-MS is used for the elucidation of microstructures. In many ways, the system is flexible without becoming too difficult to use. However, during maintenance the system can be demanding to service.

15.8 Typical Examples of Industrial Applications Using LC-MS-CAD

Within DSM Coating Resins, the LC-MS-CAD system is used in general as primary LC method development for all types of components. Besides the polymer molecules, standard components, such as solvents, raw materials, and impurities, can be analyzed using this system. The combination of the detectors makes it possible to identify and quantify molecules. The combination of separation by LC and detection by UV, CAD, and MS demonstrates to be a versatile tool to unravel complex materials and obtain essential information on batch-to-batch variations, kinetic studies, and impurity profiling. Within resins applications, raw materials, intermediate products, and end products usually consist of oligomer or polymer distributions. Commercial raw materials are often technical grade; the material seldom consists of one single molecule or even the target molecule. Exceptions are the common vinyl monomers like acrylate monomers that seldom contain more than 1 wt% of impurities.

In this paragraph, some applications of the LC-MS-CAD system will be addressed. The examples can be categorized into raw material analysis, intermediate analysis, and end product analysis. Depending on the application and the analytic question, SEC or gradient LC will be applied in combination with UV, CAD, and MS detection. For all systems the chromatograms of the different detectors will be compared. For UV, different wavelengths can be applied and for MS, the total ion current (TIC) chromatogram as well as mass range or single mass extraction is used. By applying LC-MS-CAD system, strengths of both MS and CAD are used and give complementary information.

15.8.1 Raw Material Analysis

In many polymer production processes, raw materials, which are also polymers, are used. The characterization of these and other raw materials is often a vital step in understanding how raw material structure is related to polymer processing. Information supplied on Safety Data Sheets is often incomplete or inaccurate and often does not explain processing effects as seen in production environments (gelation, exothermic reactions, etc.).

In Figure 15.10, a multifunctional acrylate raw material is shown, measured in SEC-MS-CAD mode. The separation is based on differences in molecular size. The large molecules elute first, followed by small molecules. As can be seen from the chromatograms, the distribution is rather broad. Chemically, these types of materials do not consist of one single molecule but are a mixture of various degrees in functionality and polymerization caused by side reactions. Titrations (e.g., acid number or hydroxyl number) on this type of material would give average functionality data, but average numbers are, in most cases, insufficient information to predict the behavior of the material.

In Figure 15.10, an overlay of signals from the MS, CAD, and two different UV wavelengths is shown. The use of UV is not always an option, since not all molecules contain chromophores. The combined use of all detectors gives most information needed. CAD and UV signals show comparable responses, with an

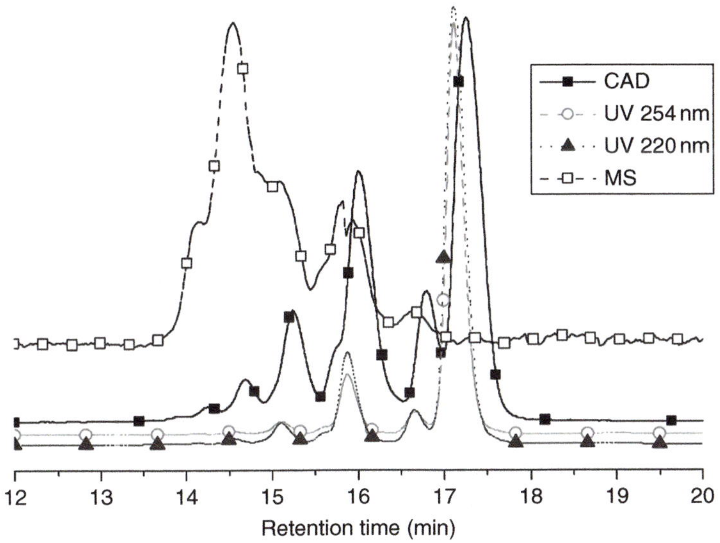

Figure 15.10 **Overlay** of CAD, UV (220 and 254 nm), and MS signals of a multifunctional acrylate obtained with SEC.

underestimation of the UV signal for higher molar mass fraction. The MS signal shows less response in the low-molecular weight region, which can be caused by poor ionization of the molecules, possibly due to ion suppression effects caused by the THF. By comparing the information of the MS (identity of the various constituents) with CAD (quantity of the various constituents), an accurate mass analysis of the material is obtained. It is always preferable to use more than one detector, since assumptions about identity and/or quantity are easily made that may lead to inaccurate conclusions. In Figure 15.10, a distribution is visible, which has a significant fraction of higher molecular weight material. By using the accurate mass data from the MS spectra, the identity of the components can be obtained. The higher molecular weight fraction can cause viscosity to increase resulting in off-spec production. For the analysis of raw materials, the SEC-MS-CAD combination gives detailed insight into the molecular buildup, which is essential to perform successful product and process development.

15.8.2 Intermediates

Besides the analysis of raw materials, the monitoring of polymerizations and analysis of intermediate products are also important.

In Figure 15.11, the results of a gradient LC separation of an intermediate are shown. The intermediate is a typical example of the preparation of a urethane

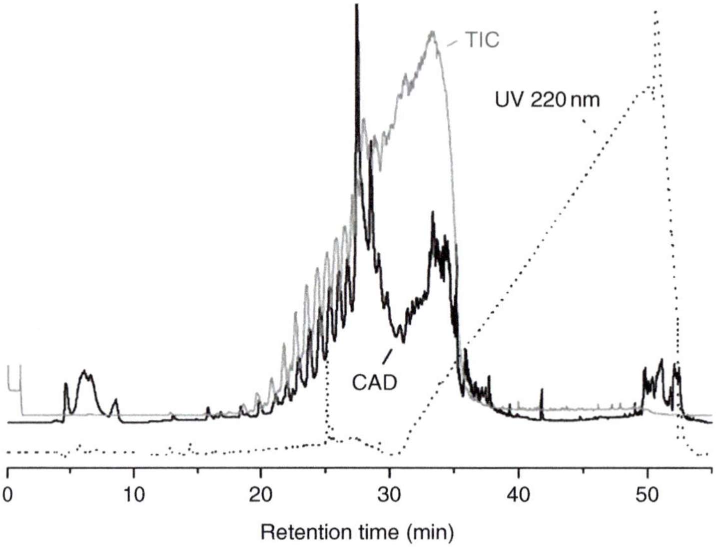

Figure 15.11 **Overlay** of gradient LC chromatograms of an intermediate product of TIC MS and CAD. Gradient as described in Table 15.4 is used.

polymer, synthesized by urethanization of a polyol with a diisocyanate. Since none of the reactants have chromophoric groups, the UV detector only shows the UV signal due to the change in eluents from water to MeCN to THF (dotted line). Both MS and CAD show a distribution of the urethane oligomers and polyol distribution. The MS shows a more intense signal around 30 minutes compared with CAD. This is probably caused by selective ionization of the higher molar mass oligomers (containing the well-ionizable urethane groups), resulting in an overestimation of the higher masses. CAD shows two molar mass fractions, which is in close agreement with theoretical expectation of the composition of the oligomer sample.

15.8.3 End Products

The application of LC-MS-CAD for end products is not so straightforward. As discussed in the SEC and GPEC sections, CAD can be used successfully for poly distributions. However, the MS cannot be used to analyze high-molar mass polymers (>10 kDa). Nevertheless, many resin end products are low-molar mass polymers. In 2007, the European Community introduced the *R*egistration, *E*valuation, *A*uthorization and *R*estriction of *Ch*emical substances (REACh) project [32], which started to regulate and register substances and chemicals. In order to register products in REACh, more detailed information on the chemical composition and the molecular structure needs to be known. The effect of the legislation is to force companies to show that the end products are safe for humans and for the environment. One of the topics within REACh is the polymer definition. If an end product can be registered under the polymer definition, no additional registration is needed. If the end product cannot be registered under the polymer definition, additional tests, especially toxicology tests, need to be performed. Consequently, the end products need to be analyzed in detail to define the product as polymer. A detailed description of the polymer definition can be found on the European Community website [32].

One of the main issues with the polymer definition in REACh is the definition of the repeating units and the quantification of the oligomers with the number of repeating units below $n = 3$. Similar to Toxic Substances Control Act (TSCA) registration (United States substance registration legislation), the preferred method to identify whether a substance falls under the definition of a polymer is SEC, most often combined with a dRI detector. As described in previous paragraphs, a CAD detector can also be used in combination with SEC to quantify distributions. In standard SEC-dRI systems, assumptions have to be made on peak identification, purely based on retention time and reference molar masses. By applying additional detectors such as MS, identification of the different peaks can be performed, resulting in a more accurate definition of the oligomer peaks. This drastically increases the reliability of the REACh analysis, since no more assumptions have to be made on molar mass and number of repeating units. In the following example, an end

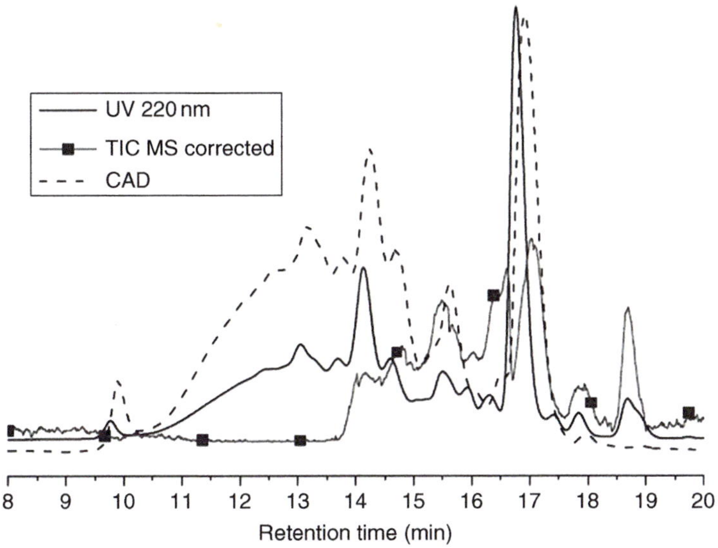

Figure 15.12 **SEC**-MS-UV-CAD chromatograms of oligomeric resin.

product that was subject to REACh registration was analyzed using SEC-MS-CAD. The product contains chromophore groups so the UV detector could also be used.

As seen in Figure 15.12, a wide range of components and a distribution is clearly visible in all signals. The TIC MS signal does not reveal the higher-molar mass species in contrast to the results of the raw material in Figure 15.10. In this case the MS detector is used to identify the different structures present in the product. It was possible to identify the components that have $n \geq 3$ by the MS detector and quantify these as a percentage of the total product by integrating the CAD signal. By using CAD, the end product was successfully classified as polymer under the REACh definition.

15.9 Epilogue

For the analysis of polymeric distributions, CAD has shown to be extremely valuable. CAD is used routinely in the described LC-MS system for troubleshooting and molecular structure elucidation. The LC-MS-CAD system is a valuable addition to the analytical toolbox for the structural elucidation of oligomers and, in combination with nuclear magnetic resonance, is very powerful. CAD turns out to be useful as a concentration detector for distributions in SEC as well as in gradient applications. In combination with an LC-MS system, CAD is better suited than the dRI detector. On the other hand, the quest for a universal detector for LC applications with linear concentration dependency has not ended, since the concentration dependency of CAD is nonlinear. This can be overcome by using log–log scale conversion, but in most commercial software, this is not included, or difficult to achieve. Overall, for the determination of distributions or for comparison purposes, CAD is an important addition to other detectors, such as dRI and UV.

Acknowledgments

The authors would like to thank Harry Philipsen for the useful discussion and scientific input. Tijs Nabuurs en Frank Vaes from DSM Coating Resins (Waalwijk, The Netherlands) are acknowledged for preparing the polymer samples.

References

1 Striegel, A., Yau, W.W., Kirkland, J.J., and Bly, D.D. (2009). *Modern Size-Exclusion Liquid Chromatography, Practice of Gel Permeation and*

Gel Filtration Chromatography, 2e. Hoboken, NJ: John Wiley & Sons, Inc. 978-0-471-20172-4.

2 Staal, W.J., Cools, P., van Herk, A.M., and German, A.L. (1994). Monitoring of originated polymer in pure monomer with gradient polymer elution chromatography (GPEC)®. *Journal of Liquid Chromatography* 17 (14–15): 3191–3199.

3 Cools, P.J.C.H., Maesen, F., Klumperman, B. et al. (1996). Determination of the chemical composition distribution of copolymers of styrene and butadiene by gradient polymer elution chromatography. *Journal of Chromatography A* 736 (1–2): 125–130.

4 Philipsen, H.J.A., Klumperman, B., and German, A.L. (1996). Characterization of low-molar-mass polymers by gradient polymer elution chromatography I. Practical parameters and applications of the analysis of polyester resins under reversed phase conditions. *Journal of Chromatography A* 746 (2): 211–224.

5 Staal, W.J. (1996). Gradient polymer elution chromatography. PhD Thesis. University of Technology Eindhoven, Eindhoven, 90-386-0126-3.

6 Philipsen, H.J.A. (1998). Mechanisms of gradient polymer elution chromatography and its application to (Co)polyesters. PhD Thesis. University of Technology Eindhoven, Eindhoven, 90-386-0578-1.

7 Cools, P.J.C.H. (1999). Characterization of copolymers by gradient polymer elution chromatography. PhD Thesis. University of Technology Eindhoven, Eindhoven, 90-386-0970-1.

8 Schunk, T.C. (1993). Chemical composition separation of synthetic polymers by reversed-phase liquid chromatography. *Journal of Chromatography A* 656 (1–2): 591–615.

9 Glöckner, G. (1991). *Gradient HPLC of Copolymers and Chromatographic Cross-Fractionation*. New York: Springer-Verlag.

10 Gorshkov, A.V., Verenich, S.S., Evreinov, V.V., and Entelis, S.G. (1988). Separation of epoxy resins by functionality types. *Chromatographia* 26: 338–342.

11 Cools, P.J.C.H., van Herk, A.M., German, A.L., and Staal, W.J. (1994). Critical retention behavior of homopolymers. *Journal of Liquid Chromatography & Related Technologies* 17 (14–15): 3133–3143.

12 Pasch, H., Brinkmann, C., Much, H., and Just, U. (1992). Chromatographic investigations of macromolecules in the “critical range” of liquid chromatography: II. Two-dimensional separations of poly (ethylene oxide-block-propylene oxide). *Journal of Chromatography A* 623 (2): 315–322.

13 Entelis, S.G., Evreinov, V.V., and Gorshkov, A.V. (1986). *Advances in Polymer Science* 76: 129.

14 Pasch, H. and Trathnigg, B. (1997). *HPLC of Polymers*. Berlin: Springer.

15 Schultz, R. and Engelhardt, H. (1990). HPLC of synthetic polymers characterization of polystyrenes by high performance precipitation liquid chromatography (HPPLC). *Chromatographia* 29 (5): 205–213.

16 Snyder, L.R., Kirkland, J.J., and Dolan, J.W. (2010). *Introduction to Modern Liquid Chromatography*, 3e. Hoboken, NJ: John Wiley & Sons, Inc.

17 Vehovec, T. and Obreza, A. (2010). Review of operating principle and applications of the charged aerosol detector. *Journal of Chromatography A* 1217 (10): 1549–1556.

18 Hutchinson, J.P., Li, J., Farrell, W. et al. (2010). Universal response model for a corona charged aerosol detector. *Journal of Chromatography A* 1217 (47): 7418–7427.

19 Lee, H.C. and Chang, T. (1996). Polymer molecular weight characterization by temperature gradient high performance liquid chromatography. *Polymer* 37 (25): 5747–5749.

20 Martin, A.J.P. (1950). *Biochemical Society Symposium* 3: 4–20.

21 Flory, P.J. (1953). *Principles of Polymer Chemistry*. Ithaca, NY: Cornell University Press.

22 Huggins, M.L. (1958). *Physical Chemistry of High Polymers*. New York: John Wiley & Sons, Inc.

23 Hoang, N.-L., Landolfi, A., Kravchuk, A. et al. (2008). Toward a full characterization of native starch: Separation and detection by size-exclusion chromatography. *Journal of Chromatography A* 1205 (1–2): 60–70.

24 Macko, T. and Pasch, H. (2009). Separation of linear polyethylene from isotactic, atactic, and syndiotactic polypropylene by high-temperature adsorption liquid chromatography. *Macromolecules* 42 (16): 6063–6067.

25 Roy, A., Miller, M.D., Meunier, D.M. et al. (2010). Development of comprehensive two-dimensional high temperature liquid chromatography × gel permeation chromatography for characterization of polyolefins. *Macromolecules* 43 (8): 3710–3720.

26 Evans, C.E. and McGuffin, V.L. (1990). Evaluation of refractive index artifacts in liquid chromatographic absorbance detection: Extension to non-ideal injection and gradient elution. *Journal of Chromatography* 503 (1): 127–154.

27 Van Doremaele, G.H.J., Kurja, J., Claessens, H.A., and German, A.L. (1991). Flame ionization detection for the determination of the chemical composition distribution of non-UV absorbing copolymers. *Chromatographia* 31 (9–10): 493–499.

28 Allen, L.B., Koropchak, J.A., and Szostek, B. (1995). Condensation nucleation light scattering detection for conventional reversed-phase liquid chromatography. *Analytical Chemistry* 67 (3): 659–666.

29 Biosolve B.V. http://www.biosolve-chemicals.com/page.php?content=product_line_lcms (accessed 30 January 2015).

30 Nielen, M.W.F. and Buijtenhuijs, F.A. (1999). Polymer analysis by liquid chromatography/electrospray ionization time-of-flight mass spectrometry. *Analytical Chemistry* 71 (9): 1809–1814.

31 Staal, B.B.P. (2005). Characterization of (Co)polymers by MALDI-TOF-MS. PhD Thesis. University of Technology Eindhoven, Eindhoven, 90-386-2826-9.

32 European Community (2007). Registration, evaluation, authorisation and restriction of chemical substances (REACH). http://ec.europa.eu/environment/chemicals/reach/reach_intro.htm (accessed 6 March 2017).

Index

d

s

t

u

Printed in the USA/Agawam, MA
September 30, 2026

911251.001